George Patterson

A History of the County of Pictou, Nova Scotia

George Patterson

A History of the County of Pictou, Nova Scotia

ISBN/EAN: 9783337321895

Printed in Europe, USA, Canada, Australia, Japan

Cover: Foto ©ninafisch / pixelio.de

More available books at **www.hansebooks.com**

A HISTORY

OF THE

COUNTY OF PICTOU

NOVA SCOTIA.

By the Rev. GEORGE PATTERSON, D.D.

Author of "Memoir of James M'Gregor, D.D.," "Memoirs of Johnston and Matheson," "The Doctrine of the Trinity, underlying the Revelation of Redemption," &c.

MONTREAL: DAWSON BROTHERS.
PICTOU, N. S.: JAMES M'LEAN & Co.
HALIFAX, N.S.: A. & W. McKINLAY.   St. John, N. B.: J. & A. McMILLAN.
TORONTO: JAMES CAMPBELL & SON.

1877.

PREFACE.

By those for whom this work is intended, no apology for the undertaking will be deemed necessary. As to the manner in which it has been accomplished, however, the author deems some explanations due to himself and his readers. First, he feels it but right to mention the difficulties in the way of obtaining correct information, particularly regarding the early settlement of the county. With the exception of copies of grants and similar papers, preserved in the public records, there is scarcely a document of that period in existence, and these serve very imperfectly to give us an insight into the life of the early settlers. We are thus indebted for all our knowledge of that era, almost entirely to unwritten sources, and the difficulty of obtaining exact information in this way, can only be understood by those, who have made an attempt of the same kind.

The author must, however, here acknowledge valuable aid from a writer, whose name is unknown to him, who under the signature of " Philo-antiquarius," published in the first volume of the *Colonial Patriot*, a series of letters on the early history of Pictou. These letters he has quoted fully, but they do not enter largely into details, and when they do, they are not always accurate.

It is but due to himself to say, that he has spared no effort to gain information. He has ransacked the County and Provincial records, and teased officials with his enquiries; he has plodded his weary way through newspaper files, and works of Colonial history; he has interrogated Micmacs, and, as the Scotch would say, "expiscated" every old man and woman he has met with in the county for years; he has also conducted a large correspondence, and visited various sections of the country in search of facts. To arrive at the exact truth, he has labored as conscientiously, as if he were writing the history of Europe; and though he can scarcely hope, that his work will be found free from all errors, yet he believes that these will not be material.

Farther, as to the execution of the work, he desires to say that the plan adopted has been, to present as full an account of the early settlers of the county and the pioneers in each section of it, with as vivid a picture of their toils, as in his power, even to the exclusion of information that might be desired, regarding more recent events. This course has been followed, partly on the general ground, that these things being longer past, are now more properly the subjects of history, partly because he considers that portion of our annals as most worthy of notice, but especially, because the information regarding it, depends on oral tradition, which in another generation would be lost altogether.

In this course he has also had a special object in view. Of those who have hitherto professed to write the History of Nova Scotia, none have yet attempted fully to delineate the period of British colonization; and yet the author regards this as the most important era in the past of the Province. He has, therefore, attempted to depict the life of the early settlers in Pictou,—to give, as far as he can, the very form and pressure of the age; and as what occurred in one county, was, in a large measure, a repetition of what took place in another, he hopes that his work will thus serve, as a contribution to the illustration of that era in our colonial history.

It is too well known that the history of the county has been disfigured by painful controversies. These could scarcely be ignored in a history like this, but the treatment of them, it will be readily seen, must be a work of difficulty and exceeding delicacy. The course which he has adopted, has been to pass over all contentions of a personal character, but where there seemed questions of importance at issue, to point them out clearly and candidly. And though he could not help, to some extent, viewing these from his own standpoint, yet it has been his aim to look at them from all sides, to endeavour to arrive at the exact truth regarding them, and to judge charitably, where his convictions would lead him to condemn.

There only remains the duty, which the author has great pleasure in discharging, of acknowledging the aid

v.

received from various parties in his undertaking. Of the public officials, who were always ready to facilitate his enquiries, he desires to mention,—in Halifax, Messrs. W. A. Hendry, late of the Crown Land Department, Thomas Robertson, Provincial Secretary's office, Venables, of the Legislative Library; and in Pictou, Messrs. David Matheson, Clerk of the Peace and Prothonotary, and John Ferguson, Registrar of Deeds. He must also record his obligations to Rev. S. T. Rand, for much of the information in the second chapter regarding the Micmacs, to T. B. Akins, Esq., of Halifax, for access to works of the early French voyagers, to Dr. J. W. Dawson, and Ed. Gilpin, jr., Esq., for information as to the geology and mineralogy of the county, to Thomas Millar, Esq., of Truro, for aid in enquiries in Colchester, to the officers of the different coal companies, especially Thomas Blenkinsop, Jesse Hoyt, Roderick McDougall, George Hattie and J. P. Lawson, Esqs., for information regarding the different collieries; and, for various personal reminiscences, to Messrs. Robert Patterson, George Glennie, W. H. Harris and Jas. Hepburn of Pictou, John McKay, Esq., of New Glasgow, and among the departed, Mr. John Douglass of Middle River, James McGregor, Esq., of New Glasgow, and his late father. To the following he is specially indebted for the information regarding the settlements under-mentioned: to Rev. H. B. McKay, for River John, Toney River and Cape John Shore; Rev. William Grant for Earltown and West Branch River John; Rev. J. Watson, for New Annan; Rev. D. B. Blair, for Blue Mountain, Barneys River and Garden of Eden; Rev. Robert Cumming, for St. Marys; and Mr. William Fraser, for Pictou Island. He has largely adopted their words and interwoven them with his own narrative. Many others have rendered him aid, of which he is fully sensible, who, he trusts, will accept this general acknowledgment.

GEORGE PATTERSON.

New Glasgow, N. S., Feby., 1877.

CONTENTS.

CHAPTER I.

INTRODUCTORY.

Boundaries — Divisions — Coast and Harbours — Interior — Geological Structure—Natural History—Name............................ 9

CHAPTER II.

PICTOU IN THE PREHISTORIC PERIOD.

Early visitors—Micmac occupation—Remains—Names—Their wars..... 24

CHAPTER III.

THE FRENCH IN PICTOU.

French Settlements—Remains found—Removal of French—Peace with Indians—Vessel stranded at Carriboo............................ 37

CHAPTER IV.

FIRST ENGLISH SETTLEMENT OF PICTOU—1767-1773.

Schemes for settling Nova Scotia—Philadelphia Company and their grant—Arrival of first settlers—Their labours—Social condition... 46

CHAPTER V.

ARRIVAL OF HECTOR AND SETTLEMENT OF HER PASSENGERS—1773-1776.

Hector's voyage and arrival—Difficulties of settlers—Progress—Settlers from Dumfries, on Prince Edward Island and removal to Pictou.. 79

CHAPTER VI.

PICTOU DURING THE AMERICAN REVOLUTIONARY WAR—1776-1783.

American Revolutionary War—Effects on Trade—Vessels captured—Indian gathering—" Malignant "—Slavery — First settlement at Merigomish.. 98

CHAPTER VII.

FROM THE CLOSE OF THE WAR TILL THE ARRIVAL OF DR. M'GREGOR—1783-1786.

Eighty-second Regiment—Eighty-fourth—Upper settlement, East River—Other arrivals—First settlers of Tatamagouche and River John—Dr. McGregor's arrival... 114

CHAPTER VIII.

DR. M'GREGOR'S EARLY LABOURS—1786-1789.

State of Society—First churches—Pastoral labours—Redemption of slaves —Missionary journeys.................................... 136

CHAPTER IX.

FROM THE COMMENCEMENT OF THE TOWN TILL THE FRENCH WAR—1789—1793.

Governor Patterson's grant—Commencement of town—First ship-building —Immigration—Pictou made a separate district............... 151

CHAPTER X.

FROM THE COMMENCEMENT OF THE FRENCH WAR TILL THE WALLACE ELECTION— 1793-1799.

French Revolutionary War and trade—Rev. Duncan Ross—Population— Social condition—Hunting—Micmacs—First contested election— Discovery of coal.. 168

CHAPTER XI.

COUNTY AND COURT BUSINESS

Building jail—Stocks—Collecting taxes—Sessions regulations—Inferior Court—Supreme Court—First trial for murder 199

CHAPTER XII.

IMMIGRATION AT THE BEGINNING OF THE CENTURY.

Large immigration — Dunoon's passengers — New settlements — Mount Thom, McLennans Mount, New Lairg, &c..................... 222

CHAPTER XIII.

FROM THE BEGINNING OF THE CENTURY TILL THE PEACE—1800-1815.

Timber trade—Drinking—Edward Mortimer and others—War times— Travelling — Dr. McCulloch — First Bible Society — Rev. John Mitchell—Rev. Wm. Patrick—New Glasgow commenced.......... 244

CHAPTER XIV.

IMMIGRATION AND NEW SETTLEMENTS AT THE CLOSE OF THE WAR.

New settlements—Dalhousie Mountain, Earltown, Blue Mountain, St. Marys, &c.. 275

CHAPTER XV.

FROM THE PEACE TILL THE FINANCIAL CRISIS OF 1825-6—1815-1826.

Effects of peace—Year of the mice—Year of frost—Agricultural improvement — Murders — Business men — Financial crisis — Religious societies—Religious divisions...................... 292

CHAPTER XVI.

DR. M'CULLOCH AND THE PICTOU ACADEMY.

Higher seminary projected—Commenced—Opposition—Dr. McCulloch's labours—Native preachers—Institution remodelled—Decay and fall 321

CHAPTER XVII.

FROM THE FINANCIAL CRISIS OF 1825-6 TO THE DIVISION OF THE COUNTY—1826-36.

Trade—Jotham Blanchard and *Pictou Patriot*—First Temperance Society—Improved roads and travelling—Dr. McGregor's death—Election of 1830—First steam navigation—L. and S. Society............. 363

CHAPTER XVIII.

MINES AND MINING INDUSTRIES OF THE COUNTY.

First coal mining—Coal field—General Mining Association—Acadia Company — Intercolonial — Nova Scotia Company — Vale — Iron ores.. 398

CHAPTER XIX.

FROM THE DIVISION OF THE COUNTY TO THE PRESENT TIME—1836-1876.

Pictou as a separate county—Shipbuilding and Capt. McKenzie—Trade, agriculture and manufactures — Religious denominations — Conclusion.. 428

APPENDIX. 449

HISTORY

OF

THE COUNTY OF PICTOU.

CHAPTER I.

INTRODUCTORY.

The County of Pictou lies on the Southern Shore of the Straits of Northumberland, along which, it presents a length of about fifty miles. It extends into the interior to a distance of over twenty miles, being bounded on the South, by the County of Guysborough, on the East by the County of Antigonish, and on the West by the County of Colchester. When originally set off in the year 1792, from Colchester as a separate district, its boundaries were thus described: "Beginning four miles eastward of David Archibald's house, at Salmon River, between Truro and Pictou, as the road now runs, from thence to run north, four degrees west, (by the Magnet,) to the shore of Tatamagouche Harbour, thence from said place of beginning, to run south twenty-seven degrees east, to the southern line of the District of Colchester; thence east by the said line to the western line of the County of Sydney, including all the lands to the eastward and northward of said lines, within the (then) District of Colchester." More exactly its limits are thus described, " commencing at the boundary of the County of Colchester, at the Gulf of St. Lawrence, thence south four degrees east 19 miles, thence south twenty degrees east 26 miles, thence east 25 miles, thence north 26 miles to the shore."

The following are the Latitudes and Longitudes of leading points, determined by a series of observations by officers of H. M. Navy, in the year 1828.

	Lat.	Long.
Pictou Island South Side	45 47 52	62 37 33
Pictou Harbour............	45 41 56	62 42 ....
Pictou Academy............	45 40 20	62 44 28

It is divided into three townships, Pictou, Egerton and Maxwelton. The first of these embraces the western part of the county, from the Colchester County line to the harbour of Pictou. It is separated from the Township of Egerton, by a line commencing at Boat Harbour and running thence south 54° west, till it reaches East River, at what is called the Big Gut, and by another line commencing at Doctor's Island, at the point between the Middle and West Rivers, and thence south 30 degrees west 19 miles, to the Colchester County line. The Township of Egerton embraces all the central portions of the county, and is bounded on the west by the Township of Pictou, as just mentioned, and on the east is separated from the Township of Maxwelton, by a line commencing at the bridge at Sutherlands River, and thence running south to the Guysborough County line. The Township of Maxwelton includes the remaining part of the county.

Its estimated area is as follows :—

Pictou Township.................	215,360	Acres
Egerton "	239,600	"
Maxwelton "	222,400	"

Its coast is indented by several harbours, of which the most important is Pictou, about the centre of its seaboard, and which is the largest, and by far, the best harbour on the Northern shore of Nova Scotia. It has a bar at its mouth with twenty feet of water on it at low tide. The entrance is narrow, but within it expands into a very large and capacious basin, having from five to nine fathoms of water, where a large navy might ride in perfect safety, and with muddy bottom, affording superior holding ground. Its main disadvantage is that it is frozen over from the middle or end of December to the beginning or end of April. Three fine streams, after winding through a fertile district, fall into it, known as the East, Middle and West Rivers, the first navigable for small vessels for five miles from its mouth.

A short distance to the westward is the small harbour of Carriboo,* formed between the main land and two islands, named respectively Big and Little Carriboo. This name is said to have arisen, from some of the first explorers, having seen a herd of caribou on the east point of what is now the Big Island, but which was then a headland connected with the shore, and which they thence called Caribou Point. This harbour has two principal entrances, one between the two islands, the other much narrower but deeper, between the smaller and the mainland. When the first settlers arrived this was the only entrance, what is now the wide entrance being then a sandbeach, over which the sea was beginning to make its way. It has, however, continued its encroachments, till it has entirely separated between the two islands, making a passage half a mile wide with four feet of water on it at low tide. Within the memory of the first settlers the sea has also cut across the beach, which connected what is now the Big Island with the land, and thus formed a third entrance, which, however, is still shallow. † And further changes are

---

* Such is the spelling now commonly adopted, though the name of the animal is generally spelled Caribou.

† The late James Harris used to say that when he first visited the island, he could cross between them by wading to the knee, and he remembered the first storm which cut across the beach connecting the Big Island with the land.

going on. At two if not three places on the Big Island, which were once meadows, cutting considerable quantites of hay, are now only narrow sand beaches, which the sea is wearing away, and which it will soon cut through, and thus convert it into three or four islands. In the days of the Pictou timber trade, vessels of considerable size loaded in this harbour, but it is now but little used. Two small streams dignified as Big and Little Carriboo Rivers, unite their waters about three quarters of a mile from the harbor into which they flow in a deep channel. About 15 miles farther to the westward, is the only other harbour on that side of the county, viz:—River John, being the estuary of the river so named. This harbour is not large and not well sheltered, being exposed to northerly winds, but it has for many years been the seat of a large shipbuilding trade. The River John, (in Micmac Cajje-Boogwek, i. e., flowing through a wilderness, *) flowing into it, drains a large tract of country. Between these two harbours a small stream known as Toney River, with several brooks empty into the Strait.

Proceeding from Pictou Harbour eastward, along the coast, we pass some small harbours, known as Chance, Boat and Little Harbours, and then meet Merigomish, formed by what is called the Big Island of Merigomish, which, however, is connected at its eastern end with the mainland by a sand-beach about three miles long. Here seems originally to have been the entrance to the harbour. The early French explorers in the 17th century speak of this as the entrance, but represent it as becoming choked with sand, so that only small vessels could enter, and that only at high-tide. When the first English settlers arrived, the old Indians could recollect when there was sufficient water to afford passage for their canoes. Now, however, it is a sand beach from an eighth to a quarter of a mile wide, and for some

---

* From *Cajjah*, to be alone.

distance along the centre, judging by the eye, about 30 feet high, covered with a coarse grass and a few plants, such as will grow in that situation. The entrance is now at the west end of the Island, and my opinion is that originally this was connected with the land, but that the sea here cut a new entrance, that in consequence the tidal and river waters flowing in this course, the stream at the east end became too sluggish to keep the passage there clear of the sand accumulating at its mouth, and thus led ultimately to its being closed up. Even yet a very heavy storm will make a passage across it.

The present entrance has a bar with 14 feet at low water, formed by rocky shoals running out from the points on each side. Within, however, it is a large and safe harbour. Once inside, the mariner finds himself well sheltered, but it has this disadvantage, that from the bend in the channel, turning round the end of the Island, the same wind by which sailing vessels can enter, will not bring them up to the upper parts of the harbour. It contains a number of islands, varying in size from a few acres to a square mile in extent. Into it flow as we proceed easterly, Sutherlands, French and Barneys Rivers.

Beside the islands already mentioned and some smaller ones, there lies off the coast at a distance of about eight miles from the entrance of Pictou Harbour, Pictou Island, about five miles long from east to west, with an average breadth of about a mile and three quarters. From each end reefs run out to a considerable distance.

The coast has few dangers for navigation, and these are largely obviated by light houses. Approaching from the east, the mariner first sights Pictou Island light, which is situated on the south-east point of the Island, showing a white fixed light, 52 feet above sea level, and visible 11 miles, from a square white building. Then comes Pictou Harbour light on the south side of the entrance. It is a white fixed light, with a small red light below, 65 feet above sea level, visible 12 miles. The building is octagonal, painted in red and white vertical stripes. Farther to the northward and westward is Big Carriboo Island light, situated on the north-east end of the Island, a white revolving light, showing its greatest brilliancy every minute, 35 feet above high water, and visible 10 miles from a square white building.

The coast is generally low, scarcely in any place forming cliffs, "the Roaring Bull," a point four miles to the eastward of Pictou Harbour, making the nearest approach to one. Both inside and outside the harbours, it is being gradually worn away, the sandstone, which forms the underlying rock, readily yielding to the influence of the waves. At Middle River Point, those who can remember a period of about fifty years, estimate that in that time about 200 feet of the shore has been carried away. The Island there is not now half the size it was within their recollection, and a small island on the Middle River has in the same time been entirely carried away, except a few stones visible at low water. At Abercromby Point, residents calculate that about sixty feet of the bank has been carried away.

This wasting goes on with greatest rapidity under the influence of north-east winds, which cause our highest tides, and drive the water with great force, particularly against the shores on the south side of the Harbour. In this way the banks are undermined, and the frost and rain bring down the superincumbent soil, which is washed away by the waves and tides.

In Merigomish the same thing is observed. Mr. Wm. Dunn, an intelligent resident, estimates that during a period of about fifty years, from sixty to a hundred feet of shore has been carried away on the point formerly occupied by Mr. James Crerar, and in the cove on the front of his own farm. The old ship yard is now almost entirely covered with water, so that where the bow of the vessel rested on the "ways," is now about high water mark.

On the open coast the wasting must be greater. We have already referred to the changes going on at Carriboo Island. But Pictou Island is wearing away with perhaps greater rapidity, though at one [point toward the southeast side, a sand beach is making. The old Indians spoke of a time, when the passage between it and Carriboo Island, now five and three quarter miles wide, was comparatively narrow. At Cole's Point, near the entrance of the harbour, about thirty yards of a bank twenty or thirty feet high has been carried away within a short time, and its foundation is now a shingly beach. Since the light house was built in 1834, about 200 yards of the eastern side of the beach on which it stands, then yielding a coarse hay, is now under water, and had not a breakwater been erected, protecting that building on three sides, it would have been swept away some time ago. At the same time, however, the beach has been making toward the west.

The beach on the north side has also been diminishing, though not with the same rapidity. It is calculated that during the same period it has narrowed to the extent of fifty yards. One great storm carried away about a quarter of an acre in one place. On examination of the ground laid bare, there were found roots of ash trees as they had grown, and with them the skeleton of a bear. The sand, however, after a time again covered the spot.

On the other hand, the land covered by the estuaries of the rivers, and the shallower parts of the harbour are gradually rising. On the West River, by the calculation of those who can remember fifty years, the flats have risen over eighteen inches. At Middle River Point, between the island and the shore, it has filled up to the extent of about eighteen inches, and where it was once too soft to walk on, it is now so hard, that a horse and cart may be driven over it. On the Middle River, similar changes have been noted as in progress. Not only are the marshes rising, but a channel formerly largely used by boats, for which the regular landing place was at Lochead's, is now filled up. In like manner what the residents knew as "the long pool" is now filled with gravel, and generally the creeks on both rivers are estimated as having risen to about the same extent as the surrounding flats.

In the harbour the flats from Middle River point to the channel, are estimated to have risen about a foot in the last fifty years, while residents at Abercromby Point estimate that those off that point have risen at least two feet. The East River, from its greater size carrying down a greater amount of soil, will naturally account for this. Thirty years ago it was considered, that vessels drawing nineteen feet of water might safely load at the loading ground at the mouth of the East River, but since that time a ford has arisen further down, over which, previous to the late dredging operations, it was difficult to take vessels drawing over fifteen. Indeed it is maintained that every part of the harbour, even the channel itself, is becoming shallower. In former years vessels drawing twenty-four feet of water passed over the bar outside the harbour, but this cannot be done now.

In Merigomish the same thing is noticed, particularly in the eastern portion of the harbour, between French and Barneys Rivers. Residents have observed that the flats are widening and the water upon them becoming more shallow. The bottom, too, consists of a rich, soft, fine mud, extending up to the beach itself, evidently brought down by the rivers. On Barneys River from the bridge downward, where people forty or fifty years ago went freely in their canoes, is now in grass.

Along the shore the land is level and not elevated, but in the interior ranges of hills extend in every direction, which, with the various river valleys, by which it is traversed, present scenery of the most beautiful, though not of the grandest description. Some of these hills, such as Frasers Mountain, Green Hill, Mount Thom or Fitzpatricks Mountain, exhibit prospects which in richness and variety, of sea and land, hill and dale, river and shore, field and forest, will compare with any in America. On the western boundary the hills rise to greater elevation, being a continuation of the Cobequid Mountains, while a similar range, not so high, but more rugged in outline, and regarded as a continuation of the South Mountains of Kings and Annapolis Counties, traverses the southern portion of it, and is continued in the Antigonish Mountains.

It has few lakes, compared with some of the other counties of the Province, and these are all small. The principal are Eden, Brora, Sutherlands, and McDonalds Lakes, in the southern portion of Maxwelton township. The others, though small, add in some places a pleasing variety to the landscape.

In the descent of the streams are some pretty cascades, the largest of which and the only one which may be called grand, is on Sutherlands River, about two and a half miles from its mouth. The stream is here about 100 feet wide. In the centre a large rock, on which is a little soil, bearing a few scrubby trees, divides it into two, and each portion descends by three stages to a pool at the bottom. But just below, a perpendicular precipice, which on the left bank rises high over the fall, projects nearly half-way across, so that the parted streams as they reunite are forced through a narrow gorge.

It has but little marsh land, and none to compare in fertility with the dyked marshes of the Bay of Fundy, but along its rivers and brooks is much intervale, and meadow land of excellent quality, while much of the upland, even to the summits of the hills, is fertile. Indeed with the exception of a tract extending from the head of the West River to the County of Guysborough, and some smaller portions elsewhere, the whole is capable of cultivation. Forest fires .have, however, in some instances rendered considerable tracts for a time comparatively barren. Perhaps the largest extent of land of this kind lies between the Albion Mines and Middle River.

Its geological structure may be described in general terms as follows : Across the whole southern side of the county extends a range of hills of Upper Silurian formation, composed principally of beds of quartzite and slates, the latter varying much in colour and texture with masses and dykes of syenite and greenstone. This band which commences on the east at Cape Porcupine and Cape George, is about fifteen miles broad from the east side of the County, till it approaches the East River, when it suddenly bends to the south, allowing the carboniferous strata to extend far up into the valley of the river. Farther west it again widens and so continues beyond the boundaries of the county. Rocks of the same formation are also found further north on its western border, where the Eastern Cobequid hills enter the county at Mount Thom and adjacent hills. In an economical point of view, these rocks derive their chief importance from their valuable iron ores, but a large part of them are covered with a fertile soil.

At the base of these hills are lower carboniferous rocks, chiefly sandstones and conglomerates, over and associated with which, is a series of reddish and grey sandstones and shales, with thick beds of limestones and gypsum, the latter not of the economic importance, of those of Hants County. These can be traced from the upper part of the West River, eastward to the East River, along the valley of which, they enter in the form of narrow bay into the Metamorphic District to the Southward. Eastward of this they continue to skirt the older hills, until they reach the Gulf of St. Lawrence, at Arisaig, beyond the bounds of the county.

To the northward of these older members of the system, there is in some places, especially on the East River, a large development of the productive or middle coal measures, which we shall have occasion to notice more particularly hereafter. The remaining portion of the county, stretching along the straits of Northumberland, consists of newer carboniferous rocks, supposed by Dr. Dawson, to belong to the upper coal measures, or permo-carboniferous series. These formations afford in a great number of places grey freestones, much esteemed for architectural purposes, and also suited for the manufacture of grindstones. Copper ores are found at various localities, the principal being Carriboo river, the West River a little below Durham, the East River a few miles above the Albion Mines, and River John near the village, but none yet in quantities to be of economic value.

Its Natural History need not further be particularly described, as its flora and fauna are the same with the other portions of the Province. The Beaver has become extinct, though the effects of his labours may yet be seen in various places. Most other wild animals have become scarcer, and some, as the Fisher and Marten, are nearly if not quite extinct. We may mention, however, that the Skunk and the Raccoon are recent arrivals. The first appearance of the former was a noted event about fifty years ago. A young man from the Middle River was on his way to the East, to attend a sacramental service; crossing the wilderness land lying between the rivers, he saw an unknown animal, which he attacked vigorously, and with results that may be imagined. He proceeded on his journey, but as he approached the groups surrounding the church, met the averted faces even of friends, and was obliged to return home and bury his clothes. Other scenes of a similar kind took place in other quarters.

The waters along our shores exhibit similar changes in their inhabitants within the historic period. When European voyagers first visited our coast, the Walrus was still found in this latitude; and within the memory of persons still living, the Seal was in such abundance as to be each spring a regular object of pursuit. The first visitors to Pictou describe in glowing terms the size and abundance of the oysters, to be found in our harbour, and the shell heaps on the site of old Indian encampments, corroborate their statements, but now scarcely any are to be found, and these are but small. I am also informed that the clams, which are but little used, are not only becoming fewer in number, but smaller in size.

As to its vegetable productions, occasionally a rare specimen may be found. A short distance from the road up Sutherlands River, toward Antigonish, stands a solitary specimen of a species of spruce, which is not found any where in this part of the Province at least. And I have heard of other instances of trees being found, belonging to species not known to exist anywhere near. Many plants have been introduced by colonists, which have spread and become wild. The introduction of one presents some circumstances of interest. A vessel landing ballast at Mortimer's wharf, a few stalks of a species of ragweed, known in Scotland among the common people as Stinking Willie, were thrown out with it. They had been pulled by some cultivator in the old country and thrown among stones, and with them conveyed to the hold of the vessel, and thus transported across the Atlantic. Some of the seeds took root on the shore, and some fifty years ago the late John Taylor pointed out the plants and warned some bystanders of their character. They laughed at him, and he took no trouble in the matter, but he afterward often expressed his regret that he had not himself set to work and rooted them out, which he might readily have done. About forty years ago, a Highland servant of my father's, pointed out to me a large bunch of it on the shore of the side of the point toward the town. At that time, farmers being interested in getting rid of the oxeyed daisy and other weeds, he remarked that he was mistaken if they did not find ere long that that would be more troublesome than any with which they were contending. Since that time, it has not only occupied all the highways round the town and proved troublesome in the fields of the farmers there, but it has spread to the extremities of the county, and beyond it. I have seen it at Carriboo Island on the north shore, at Tatamagouche to the west, well up the East and Middle Rivers to the southward, and I have pulled a stalk of it on the road by Lochaber Lake to the east.

Its meteorology exhibits little of interest as distinct from the other portions of the Province. Like the rest of the north coast it present a remarkable contrast to the southern, in its almost entire freedom from fog. The mean and extreme temperatures are, however, higher there than here, owing specially to the influence of the Gulf stream. The ice which comes down from the north in spring lingers long off our coast, cooling the air, so as to retard vegetation and impart a rawness to the east winds at that season, which is trying to the health, particularly of persons under any pulmonary weakness. The autumn, however, is much finer than on the south coast, there being much less wet weather, and the southerly gales of that season being felt less severely.

The name Pictou was supposed by many to have been a corruption of Poictou, the name of an old Province of France, and to have originated with the French. In many documents of the early part of this century, even Government plans, it is so spelled, and the old Highland settlers pronounced it in this way. I have heard of educated persons in Pictou, who maintained that this was the proper name, and spelled it so in their correspondence.

But this is a mistake. The name appears spelled as we spell it in the writings of the earliest French voyagers, and there can be little doubt that it is formed from the Indian name, which, according to Mr. Rand, is *Pictook*. The *k* at the end of Micmac names, he says, marks what grammarians call the locative case, expressing at or in. The French generally dropped the *k*. Thus we have Chebooktoo for Chebooktook, and so Pictou for Pictouck, the *ou* being originally sounded as in French.

As most of the Indian names are descriptive, attempts have been made to discover the meaning of this, but as yet we do not think that certainty has been reached. Some have supposed that the word is analogous to *Buctou*, properly *Booktook*, which means a harbour, or more properly a bay or arm of the sea; but this is used only with a prefix, as in *Chebooktou, Richibooktou, Chedabooktou*, &c.

Mr. Rand explains the word differently. He says that the word *Pict* means an explosion of gas, and he supposes the name to have originated from the escape of gas at the East River from the coal lying below. Whenever the noun ends in the sound of *kt*, the regular form of the case locative is the addition of *ook*. Thus *nebookt* means woods, *nebooktook* " in the woods," and thus Pict becomes Pictook, and the *k* being dropped, as just mentioned, we have the name Pictou. It may appear presumptuous to express any doubt regarding a point of Micmac philology, on which Mr. Rand is satisfied. Yet it appears to us a serious objection to his view, that the phenomenon to which he refers was only seen at the East River, to which the Indians gave another name (*Apchechkumooch—waakade*, or duckland) while it seems certain that the name Pictouck was applied specially to the harbour, where no such phenomenon exists.

Others again have supposed, that it is a corruption of the Micmac word, *Bucto*, which signifies fire. That this was the derivation of the name, was a common opinion.

among the early settlers, and I find it asserted by Peter Toney, now about the oldest of the Micmacs in Pictou, and by others of the tribe. Their story or tradition is, that at one time there had been a large encampment up the West River. On one occasion they all left in their canoes on a cruise down the harbour. During their short absence, the whole encampment was burned up, and also the woods for a considerable distance around. No person could tell how the fire originated. They always spoke of the event as the "*Miskeak Bucto*," or big fire, which naturally became associated with the place. When the whites came, hearing the Micmacs speak of it in this way, they corrupted the name and called the whole north side of the harbour, Pictou, because according to this learned Micmac, they could not pronounce it aright. Others adopting the same derivation, have supposed the name to have been given in consequence of a large fire, at what is now the East River mines. When coal was first discovered there, it was covered with from four to six feet of burnt clay and ashes, over which large hemlock trees were growing, and I am informed that the Indians had traditionary accounts of a fire, which continued burning there for some length of time. This view I regard as entirely a supposition, and would consider Mr. Toney's much more probable, on this ground if on no other, that the name was originally given not to the Mines but to the north side of the harbour. But Mr. Rand asserts, that the difference between the words is too decided, to admit of this being the correct derivation.

Another meaning was given by Philo Antiquarius, and also by the late Mr. Howe, as derived from a Micmac. It is that it means anything like a jar or bottle, which has a narrow mouth and widens afterward. We have never received this from Micmacs, but when we have suggested it to them in the form of a leading question, they have assented to it, whether to please us or because it was correct, may not be quite certain. This would well represent the shape of the harbour, and could it be shown to be in accordance with the Micmac language, we would deem it preferable to the others. But when such difference of opinion exists among the learned, we are obliged to leave the matter unsettled.

CHAPTER II.

PICTOU IN THE PREHISTORIC PERIOD.

It is now known that these coasts were visited by the Breton and Basque fishermen during the sixteenth century, and that they traded with the aborigines, supplying them with various implements in exchange for their furs. It is probable that Pictou harbour was then well known to these hardy mariners. The only fact, however, known to us which seem to afford evidence of their presence, was the discovery by Henry Poole, Esq., on the 17th March, 1860, of a piece of wood three and a half feet below the surface of the ground, while the men were engaged in cutting a drain, on what is now the Acadia Company's area at the Albion Mines. This piece of wood, three feet long, showed marks of having been cut by an axe, while the trees growing above the spot were two feet in diameter, and he counted 230 rings of annual growth in the hemlock tree cut down just over it.

The first recorded notices of Pictou, however, are to be found in the voyages of the early French visitors, in the early part of the 17th century. We may here give a description of its shores from an account published in the year 1672, by Monsieur Denys, appointed Governor of the Gulf of St. Lawrence in the year 1654.

"Starting from Cape St. Louis (now Cape George), ten

leagues thence we come to a small river, whose entrance has a bar, which sometimes closes it, when the weather is stormy and the sea piles up the sand at its mouth, but when the river swells it passes over and makes an opening. Only small sloops can enter this river, and it does not run deep into the country, which is tolerably fine and covered with trees." This we take to be the eastern end of the Big Island of Merigomish. " Proceeding westward for about a dozen leagues the coast is nothing but a rugged mass, with the exception of several openings of different dimensions. The land round about is low, it appears fertile, and is covered with fine trees, among which I noticed quantities of oak."

The following is his description of Pictou harbour, or, as he calls it, the river of Pictou:—" Passing these you find a large opening, where there are several cliffs by the side of low headlands or meadows, in which are numerous ponds, where there is so great an abundance of all kinds of game that it is surprising, and if the game there is abundant, the earth is not less beneficent. All the trees there are very fine and large. There are oaks, maples, cedars, pines, firs and every kind of wood. The large river is right at the entrance, and the sloops go from seven to eight leagues within, after which you meet with a small island covered with the same wood, farther than which you cannot proceed without canoes. The country on both sides of the river, for the space of a league toward its source, is covered with pines, large and small, and they are fine trees, as they were down below. There are also along its sides, creeks and " cul de sacs," with meadows, where the chase is capital."

" A league and a half up the river there is a large harbour (we suppose at South Pictou) where you may find large quantities of excellent oysters; some, in one place, are nearly all round, and deeper in the harbour they are monstrous. Among them are some larger than a shoe and nearly the same shape, and they are all very fat and of good taste. And at the entrance of this river, toward the right, half a league from its mouth, there is also a large bay, which runs nearly three leagues into the land, and contains a number of islands, and on both sides you find meadows and game in abundance." For some of these details Mr. Denys seems to have drawn on his imagination.

When first visited by Europeans this, like the rest of the Province, was inhabited by the Micmac (properly Miggumac) tribe of Indians, a branch of the great Algonquin race, which included all the tribes along the Atlantic coast from Virginia to Labrador. Of these the Micmacs were one of the most powerful, occupying not only Nova Scotia and Prince Edward Island, but the whole eastern and northern coasts of New Brunswick and the south side of the St. Lawrence for some distance from its mouth. This extensive territory, known to the aborigines as Miggumahgee,* Micmacland, or country of the Micmacs, was, and indeed is yet, divided into districts, inhabited by tribes, or subdivisions of the race, each under its separate chief, who acknowledged the chief of Oonamahgee, or Cape Breton as their head, his superiority, however, consisting in little more than his being umpire in case of any dispute between the other chiefs, and presiding at any general council. Of these divisions, Pictou was the centre of the district extending along the north shore of Nova Scotia, those belonging to it being known as Pectougawak, or Pictonians. † Merigomish, however, seems to have been their head quarters.

This was a favorable position for them. It was near the fishery in the Gulf; the islands abounded in wild

---

* The classical reader may observe in the termination of this and other names the Greek word *ge*, land or country.

† The others were, besides Cape Breton, Memramcook, and Restigouche to the north, and Eskegawaage, from Canso to Halifax, Sigunikt or Cape Negro, and Kespoogwit or Cape Chignecto, seven in all.

fowl, the rivers swarmed with fish, and the woods in the rear were plentifully stocked with game. Their principal place of encampment was at the foot of Barneys River, on the east side, where they had, when the English settlers arrived, some clearings on which they raised a little Indian corn and a few beans. Other places around, such as the Big Island, some of the smaller islands in the harbour, and some of the points on the shore, were also sites of their encampments, as may yet be seen by the quantities of shells of oysters and other shellfish found on the land, and the stone hatchets and arrow heads still occasionally picked up. Their burying ground, when the English settled, and for how long previous we know not, was near the west end of the Big Island on the south side, a short distance east of Savage Point.* This they used till about forty years ago, and here stood a number of crosses till a recent period. But all the Indians of the county now bury on Chapel Island or Indian Island, an island in the harbour donated to them by Governor Wentworth.

In the map accompanying Charlevoix's work, the mouth of the East River is marked as the site of an Indian village. This must have been situated on the east side, nearly opposite the loading ground, on the farm of the late Jas. McKay, now in possession of J. McGregor and McKenzie. There, close by the river, is a beautiful flat, like a piece of intervale, but higher and very slightly rounded, bounded in the rear by a bank, by which the land rises abruptly to a higher level. Here the land was clear when the English settlers arrived, and for some time after, when it was ploughed, various articles were turned up, such as broken pieces of crockery, a gun barrel, and on one occasion a

---

* This was so called from a Captain Savage, of Truro, who had died while his vessel was lying there, and was buried in the sand on the shore. Either his vessel or another, named the Betty, drifted ashore on the point of the island opposite, which has since been called Point Betty Island.

pewter basin, about eight inches in diameter, with a narrow rim, also five or six table spoons, while around have been found quite a number of stone hatchets, and oyster shells are abundant. These facts show that this place was occupied by them, both before and after the arrival of Europeans.*

The opposite side of the river gives evidence of similar occupancy, in particular a field on the farm of William Dunbar, on being ploughed, has been found covered with oyster shells.

On a point a little lower down the river was another burying place. Here stood at the arrival of the English settlers, and until a recent period, a large iron cross, about ten feet high. Hence the place is still known as Indian. Cross Point, though the locality is known among the Micmacs, as *Soogunagade*, or rotting place.

Here the Indians buried till a few years ago. Many of the graves can still be traced by the rows of flat stones, by which they were originally covered, which have now sunk to the level of the ground or perhaps were always in that position, and are partly overgrown with grass. The water is wasting away the bank, so that human bones may be found exposed on the shore.

Frasers Point, particularly on the farm of Mr. Hugh Fraser, and Middle River Point, especially at McKay's farm, by the shells which the plough turns up, and the stone implements formerly found in abundance, and still occasionally obtained, are shown to have been also places of frequent resort.

The decaying remnant of the Micmac tribe look back

---

* An impression has prevailed that this was a French settlement, and it has even been supposed that some embankments at the Big Gut, a little further up were their work. One hut was found by the English settlers at the latter point, but all the other facts indicate the occupancy of the place by Micmacs, while the slightest examination of the embankments referred to, show that they were not raised by the hand of man, but by the tide assisted probably by ice along the shores of the creek.

on the period referred to as the golden age of their race. Then they held undisputed possession of all these regions, and were a terror to surrounding tribes. They could muster by thousands. They were at peace among themselves, drunkenness was unknown, and the various European diseases, by which they have since been swept away, were unheard of. The land abounded with game and the waters teemed with fish. The forest sheltered them from the storm, and skins of animals afforded the warmest covering by night and by day. " My father," said an old Indian, " have coat outside beaver, inside otter." Thus speaks tradition, and in some respects truly, though it would not be difficult from what we know of savage life, to find another side to the picture.

Though divided into small tribes they could combine to prosecute wars, in which they were frequently engaged with the natives of Maine and New Hampshire, and with the Iroquois and the Mohawks of the St. Lawrence. The wars with the latter occupied a prominent place in the traditions of the Micmacs of Pictou, and they preserve the memory of fierce battles, fought in the neighborhood of Merigomish.

I have lately had evidence that these traditions are not without foundation. Mr. Donald McGregor of the Big Island, in ploughing a spot in his field, where the vegetation was ranker than usual, turned up a human skull. On examination there was found a mass of human bones much decayed, among them a skull, transfixed by a flint arrow head, which yet remained in its place. Along with these remains were a large number of ancient implements, stone axes, flint arrow heads, etc., but none of them giving evidence of intercourse with Europeans. The transfixed skull, and the whole appearance of the place, plainly showed that here the bodies of those who had fallen in some battle, have been heaped together, "in one red burial blent."

I visited the place in 1874. The spot is small, not more than eight or ten feet in diameter, and as soon as the ground is turned, it will at once be distinguished from the surrounding soil, being a loose black mould, containing fragments of bone, so decayed that they can be crushed between the fingers, all, no doubt, once the flesh and blood of brave warriors. This pit, if it can be called such, is very shallow, being not more than fifteen to twenty inches deep. At the bottom I found decayed fragments of the birch bark, in which, according to the custom of the ancient Micmacs, the dead were laid. Below this was a hard subsoil, which plainly had never been disturbed. The shallowness of the pit also indicates that this burial took place previous to the coming of Europeans, when sharpened sticks of wood were their only instruments of digging.

The ground had been so thoroughly dug over before my visit, that it was impossible to ascertain anything as to the arrangements of the bodies, and nearly all the implements had been carried away; but I found a stone axe, which bore the evidence of having been ground to a sharp edge, probably immediately before the encounter in which its owner fell, some fragments of very rude pottery, and a broken tobacco pipe, made of a piece of very finely grained granite rock, the shaping and polishing as well as the drilling of the bowl and stem of which, must have involved much labour.

On examining the ground around, we found that it was the site of an ancient cemetery, in which we found, in addition to such implements as already mentioned, bone spearheads and small copper knives. The burying ground used by the Micmacs till about forty years ago was about half a mile further to the west, but the place we refer to is evidently much older. Indeed, some of the remains seemed to indicate that they belonged to another race, a people of small size, like the Esquimaux. That the Algonquin race came from the south-west is now the received opinion of American Antiquarians, and there are also strong reasons to believe, that the Esquimaux occupied the shores of North America, to a point much farther south than they now do. Charlevoix describes the Micmacs in his day, as maintaining a constant warfare with them, and the probability is that the former on first occupying this region, drove the latter before them, and these remains may be the relics of their conflicts.

One curious fact was manifest in this cemetery, which has not hitherto been noticed in connection with Micmac customs, viz., the use of fire in some way in connection with the dead. Some of the graves give no indication of this, and in one I was able to trace the position in which the body had lain, viz., on its side in a crouching posture. But in other cases the remains were mixed with ashes, small pieces of charcoal and burnt earth, showing the use of fire for some unexplained purpose. In another I found just a quantity of ashes with small fragments of burnt bones, none of them an inch long. The whole had been carefully buried, and were probably the remains of some captive whom they had burned.

We may add that here, as elsewhere, every prominent object, whether hill or river, streamlet or lake, headland or island, had its appropriate designation in their language, which is still in use among them. A few of these names, with the meanings, so far as we have been furnished, we subjoin:—

English Names.	*Micmac Names.*	*Meaning.*
Pictou Island	Cunsunk	
Moody's Point	Poogunipkechk	
Merigomish	Mallegomichk	A hardwood grove.
Carriboo Harbour	Comagun	A decoy place, where they set duck decoys.
Green Hill	Espakumegek	High land.

English Names.	Micmac Names.	Meaning.
Mount Thom	Pamdunook*	A mountain chain.
Middle River	Nemcheboogwek, ...	Straight flowing.
West River	Wakumutkook	Clear water.
East River	Apchechkumooch-waakade	Duckland.
Saw Mill Brook	Nawegunichk	Saw mill brook.
Fisher's Grant	Soogunugade	Rotting place, so called from the old Indian burying ground.
Roger's Hill	Nimnokunaagunikt..	Black birch grove.
Narrow entrance of Cariboo harbour	Tedootkesit	Running into the bushes place, from Tedootkindesink, "he rushes into the bushes."
Toney River	Bucto taagun	Spark of fire.
Shore between Carriboo and Lazaretto	Nemtookawaak	Running straight up.
Little Harbour	Munbegweck	Little Harbour.
Sutherland's Island	Coondawaakadu	A stone quarry.
Morrison's (?) Island	Tumakunawaakade..	Pipe stone place.
Point Betty Island	Mkobeel	Beaver place.

Mr. Rand tells us that the Micmacs regard themselves as the bravest and best of the Indian nations, and boast of success even over the Mohawks. But we know that till very recently the name of a Mohawk, was sufficient to excite the most abject terror in the mind of a Micmac. Tell him that there was a Mohawk at any place, and he would rather than pass it, go miles round even to reach his home. So late as our boyhood, it was an amusement even for children to frighten the Indians by some tale of Mohawks, and they never seemed to get over the feeling of alarm, which their name inspired, and we believe they are not yet free from it. †

One incident of these wars seems well established, viz., the loss by drowning of a large number of warriors of the

* Pamdun is a mountain chain, and Camdun a mountain peak. It is interesting to note here the word *dun*, a hill, which both in Gaelic and Anglo-Saxon and cognate languages denotes a hill, which appears in so many Scottish names, Dunheld, Dunblane, Dunvegan, &c., also in Dumbarton, Dumfries, &c., and another form of which we have in the English Downs.

† Mr. H. B. Lowden who has so long kept the lighthouse, informs me that if a strange canoe is seen passing the entrance of the harbour, the Indians will still come enquiring anxiously about it, and showing fears of an invasion of their old foes.

Mohawk tribe at the little entrance of Carriboo harbour. As we have been able to gather the facts of the story, the Micmacs had concealed themselves in the woods on Little Carriboo Island. Between this and the main land the passage is very narrow, not 200 yards wide. The Mohawks had detected the hiding place of the Micmacs, and supposing that they might readily, by wading or swimming, pass that distance, resolved to cross by night and attack their enemies while they were asleep. But the tide is too powerful for any man to swim across it. The Mohawks, not knowing this, plunged in, and the tide ebbing at the time, they were swept away. In the morning the returning tide brought back their dead bodies, each with tomahawk tied on his head. The Micmacs coming out of their place of concealment, were filled with joy at the sight of their dead foes, and danced in triumph for their deliverance. At the time of the arrival of the English settlers the affair was still fresh in the memory of the Micmacs, and was represented as having taken place only a short time before, during the wars between the English and the French. The late James Harris mentioned that he found two or three iron tomahawks in the sand on the shore of Little Carriboo Island, which at the time were regarded as having belonged to the Mohawks. The place is still named by the Micmacs Tedootkesit, meaning the place of running to the bushes, from the Micmacs taking refuge in the woods.*

* An old resident in the neighborhood informed us that as near as he could guess, about fifty-six years ago, or in the year 1820, an old squaw, one of the most reliable he had known, told him the story, adding that she was the first to discover what had happened. She was at the time a little girl. In the morning, as soon as she had awakened, she had gone to the shore, and there saw the dead bodies. The wind, she said, had been easterly, which would have helped to bring them back to land. She immediately ran back to tell her father, and soon the whole band were at the shore, rejoicing over their fallen foes. Supposing she were seventy years of age when she told the story, and ten when the affair occurred, this would make the date of it 1760, about the time we had supposed.

As illustrative of these times, we shall give a traditionary account of the conclusion of the last war with the Canibas, as the Micmacs call them, the tribe of Indians inhabiting Maine, and extending up to the St. Lawrence, now usually known as the Abenakis. This was related by Peter Toney and taken down by Mr. Rand, and we have reason to believe that the main facts are correct:—

"There had existed for sometime a state of hostility between the Canibas and the Micmacs. Two parties of the former, led by two brothers, had come down to Pictou and had fortified themselves in two block houses, at Little Harbour. These block houses were constructed of logs, raised up around a vault first dug in the ground. The buildings were covered over, had each a heavy door, and were quite a safe fortification in Indian warfare. At the mouth of Barneys River, near the site of the burying ground, the Micmacs were entrenched in a similar fort.*

"There was no fighting for some weeks. The parties kept a careful eye upon each other; there was no friendly intercourse between them, but there was no actual conflict.

"One night a party of the Micmacs went out "torching" (catching fish by torchlight). They were watched by the Canibas, who ascertained that they did not return to their fort after they returned to the shore, but lay down on the bank, about midway between the fortifications of the hostile parties. This was too powerful a temptation to be resisted. Two canoes came upon them, filled with armed men. They were surprised and butchered, except two, who effected their escape.

"These had rushed to the water and swam for life, and

---

* The old Indian fortifications were a sort of palisaded enclosures, formed of trees and stakes driven into the ground between them, with branches of trees interlaced. In times of war the women and children were always kept in such fortifications. After obtaining axes from Europeans they may have constructed one like a block house, as here mentioned. There is a sort of dim tradition of a French fort at Merigomish. We are satisfied that this is a mistake, but probably the idea rose from a Micmac fortification of this kind.

were hotly pursued. But passing a place where a tree had fallen over into the water from the bank, and lay there with a quantity of eelgrass piled and lodged upon it, they took refuge under the eelgrass and under the tree, and their pursuers missed them in the darkness. After the search was abandoned and the canoes had returned, the two men came forth from their hiding place and hastened home to spread the alarm.

"Their dead companions had been scalped and their bodies consumed by fire. This news roused all the warriors, and they resolved immediately to attack the party that had committed the outrage and avenge it. They had a small vessel lying inside the long bar that makes out at Merigomish. This was immediately emptied of its ballast, drawn across the Big Island beach, filled with men, arms and ammunition (for it was since the advent of the French), and immediately moved up to the forts of the Canibas, where it was run ashore. The party was led by a "keenap," a "brave," named Thunder, or *Cakloogow*, or, as this name first rendered into French and then transferred back into Indian, has come down, *Toonale* (Tonnerre). They ran the vessel ashore, and, in his eagerness for the encounter, the chief jumped into the sea, swam ashore and rushed upon the fort without waiting for his men.

"Being a mighty *Powwow*, as well as a warrior, he could render himself invisible and invulnerable, and they fell before him, as we would say, like the Philistines before Samson and his jaw bone of an ass.

"Having despatched them all he piled their bodies into the building and set fire to it, serving them as they had served his friends. When all was accomplished, his wrath was appeased.

"He then, at the head of his men, walked up towards the other fort without any hostile display, and the Abenaki chief directed his men to open the door for them and

admit them in a peaceful manner. This chief had taken
no part in the fray. He had disapproved of the attack
upon the torching party, and had endeavored to dissuade
the other from it. So when *Toonale* entered his fort there
was no display of hostility. After their mutual saluta-
tion, *Toonale* dryly remarked, 'Our boys have been at
play over yonder.'* 'Serve them right,' answers the
chief, ' I told them not to do as they did. I told them it
would be the death of us all.'

"It is now proposed that they shall make peace and
live in amity for the future. A feast is made accordingly
and they celebrate it together. After the eating comes the
games. They toss the *alkestakun*—the Indian dice. They
run, they play ball. A pole is raised at the edge of a
void space, some three hundred yards across. The parties arrange themselves four or five on each side. The
ball is thrown into the air, and all hands dart toward it
to catch it. He who succeeds in catching it before it
strikes the ground darts away to the pole, all on the opposite side pursuing him, and if they can catch him before
he reaches the pole, his party loses, and the one who
seizes him throws up the ball and another plunge is made
after it; it is seized and the fortunate party dashes off
again for the pole, and the excitement is kept up amid
shouts and bursts of laughter, until the game is finished.

"This kind of game at ball is called '*tooadijik.*'
Another kind is called *Wolchamaadijik*, the ball being
knocked along on the ground. 'Did they not wrestle?'
I enquired of my friend Peter. 'Oh, no,' was the reply.
'Wrestling is apt to lead to a quarrel, and they would not,
under the circumstances, run any risk on that score.'

"In all the games the Micmacs get the victory. And,
if they are impartial historians, they usually beat in their
wars with the other tribes and with the whites. Unfor-

---

* Compare 2 Sam. 3, 14.

tunately we have not the records of the opposite parties of Mohawks and Abenakis, but if we may judge from what takes place among other nations, their accounts would present a very different view.

"But, to return to the fort at Little Harbour. After the games were ended, the Caniba chief gives the word *Novgooelnumook*, 'Now pay the stakes.' A large blanket is spread out to receive them, and the Canibas strip themselves of their ornaments and cast them in. The following articles were enumerated by the historian: *Meehoowale*, epauletts, *Pugalak*, breastplates, *Neskumunul*, brooches, *Nasaboodakun*, noserings, *Nasogwadakunul*, finger-rings, *Nasunegunul*, a sort of large collar loaded with ornaments, more like a jacket than a collar; *Epelakunul*, hair binders, *Egatepesoon*, garters, sometimes, as in the present case, made of silver; *Ahgwesunabel*, hat-bands. These articles were piled in and the blanket filled so full that they could scarcely tie it. Then another was put down and filled as full. After this the Canibas returned to their own country. A lasting peace had been concluded, which has never yet been violated, and it is not likely it ever will be."

CHAPTER III.

THE FRENCH IN PICTOU.

The period of Micmac ascendancy in Nova Scotia, was followed by the time of its colonization by the French, and of contention between them and the English for its possession. But at this time Pictou is scarcely ever mentioned. When we consider the resources of the county, and the skill of the French in availing themselves of all the advantages of the country, it seems quite surprising that they had done so little here. But the coal and other mineral resources were in the interior and unknown. Cape Breton was more convenient for the fisheries, and, for agriculture, they had been led by their experience of the richness of the marshes of the Bay of Fundy, to seek that kind of land, of which there is little in Pictou, and that of inferior quality. Besides they had made considerable settlement at Tatamagouche, which, being nearer than Pictou to Truro, was the point of communication by water, between their settlements on the Basin of Minas, and those in Prince Edward Island and Cape Breton. At all events there is not a county in the Province, in which they have left fewer traces of their presence than in Pictou.

Halyburton says, " The French had made a few inconsiderable settlements here previous to the peace of 1763, but upon the reduction of Canada they deserted them, and in a few years they were again covered with wood." All we know of their presence here is by what the first English settlers found on their arrival. We thus learn that their largest settlement was on the big island of Merigomish. A small channel which makes off from the main one there, is still known as the French Channel. It has good water and is well sheltered, and is said to have been used by them for running into with their small shallops, in which they prosecuted the fisheries or traded with the other French settlements. At the head of this were found the remains of several dwellings. Within the memory of persons still living, the foundations of seven or eight could still be traced. There was but little land cleared; but there were gardens or orchards, the bushes in which continued to bear for many years. A variety of articles were picked up here, shovels, knives, spoons, crockery and a few coins. Towards the west of the island the remains were seen of a similar settlement, and among other articles found was the debris of a forge, with axes unfinished and one in the tongs.

They had also a small settlement at the mouth of the French River, from which it derives its name. Here also various articles have been picked up.

A few also had settled at the upper part of Little Harbour, where they also seem to have been employed in fishing. The first English settlers found there the remains of their old dwelling houses. An old man, in 1873, informed me that in his boyhood he had picked up beads and other articles among the ruins, and that some of the first settlers had told him that they had found a brass kettle under almost every chimney. A well was found on what was afterward Lauder's farm, which was long afterward known as the French well. Some traces of them were also found at the harbour of Pictou. A log shanty stood at the mouth of the Middle River, and another on the East River. Some pine had been cut down at the Town Gut and along the stream upward, and the spot where Barrie's (late Dickson's) mill now stands, selected as the site of a mill. The remains of a cellar, which had been well constructed with logs was, for a length of time, to be seen about half way between the Town Gut Bridge and Browns Point.

At what has long been known as the Burying Ground Point, inside the entrance of the harbour, on the north side, now known as Seaview Cemetery, was found a sawpit fallen in, with a log upon it in which the whip saw, much rusted, still remained. It is believed by many, on the assertion of some Micmacs, that this was used as a burying ground by the French, and in the faith of this several Roman Catholics have been buried there, and with their Protestant neighbors sleep their last sleep in peace. The remains of two or three huts were also found near this point toward the entrance of the harbour.

Evidence of their presence was also found at Carriboo.

The remains of three houses were found on the island, and of three or four on the mainland; one at Rod. McLeod's and another at Three Brooks, now Weir's place. Here they had fenced the marsh and used it for pasturing and feeding a few cattle, but they had very little land cleared. They are generally spoken of as having been principally engaged in fishing, but the tradition is that the shores of this harbour then abounded with large oak, which they cut and shipped to Louisburg, where it was largely used in the construction of the city, and probably also in shipbuilding.

Various remains have been found at different places in the county, which tell the tale of the presence of visitors at this period, but which afford us scarcely any further information regarding them. The hilt of a sword, with only a small portion of the blade remaining, and supposed from its appearance, to have been French, was picked up on Carriboo Island, and some soldiers' buttons on the mainland, near the entrance of Carriboo Harbour. Two muskets, with bayonets attached, were dug up at Frasers Point, and the remains of some guns, so decayed that both wood and iron fell to pieces when handled, were turned up by the plough near the Beaches.

The late Mr. Hugh Fraser, some time after he had settled at Middle River Point, turned up with the plough parts of a human skeleton, alongside of which he found a sword, still of such excellent temper that the point could be bent to touch the hilt. Alas for military glory! It was taken to a blacksmith's shop and there made into knives for splitting mackerel. When digging the bank at the east side of the West River for the erection of the bridge at Durham, the workmen came upon the bones of a very large man, covered with a flat stone. In digging a well at Dunbar's, near South Pictou, a skeleton was found about eighteen inches below the surface; the bones were of small size, and were supposed to have belonged to a young person or a female. Other remains of the same kind have been found at other places, all telling of visitors previous to the English settlement. "Only this and nothing more."

Such are all the facts we have been able, after diligent enquiry, to collect regarding the French settlement of Pictou. We had despaired of ever being able to know anything of those of whom these remains speak. Unexpectedly, however, we have become able to give the name of at least one settler. A number of years ago Charles McGee, of Merigomish, coming from the Strait of Canso, as he passed Big Tracadie, lodged at the house of a Mr. Petitpas; during the evening, finding that he was from Merigomish, the conversation turned on the original French settlement, when he learned that Mr. P's father had been one of the settlers there, and his mother, who was then very old and infirm, said, that if able to go to the place, she could yet show them where she had buried a large brass kettle, containing a number of household articles.

Of this era, tradition has preserved some faint reminiscence of a fight between an English and a French man-of-war in the harbor. But the details are given in such different and even contradictory ways, that while I have little doubt of some such affair having taken place, I am unable to give the particulars. The first settlers found in one tree back of the town a piece of chain-shot, and in another a cannon-ball lodged, which they considered as evidence of such an encounter. According to tradition, the French had some guns landed and mounted on the battery hill.

There is also a tradition of the capture, off Pictou island, of a valuable French vessel on the way down from Quebec. Word had been received of the sailing of such a vessel, and accordingly one or two English vessels laid in wait under the island till she made her appearance, when they put out and captured her, but the whole is involved in obscurity.

At what time they left Pictou, cannot be determined exactly. At the time of the expulsion of the Acadians from the district around Truro, then known as Cobequid, Colonel Monckton was ordered to send a detachment to Tatamagouche, to demolish all the houses they found there, together with all the shallops, boats, canoes, or vessels of any kind, etc.; and to give "particular orders for entirely destroying and demolishing the villages of Jediacke (Shediac), Ramsack (now Wallace), etc." How far these injunctions were carried out, we have no information. It is not likely that those employed came as far east as Pictou, but certain it is that all the French settlements along the North Shore of Nova Scotia were abandoned shortly after, and the circumstances in which articles were found leave little doubt that their departure was hurried. It is said that those driven out moved eastward, and formed the settlements of Tracadie and Harbour Bushie, in the County of Antigonish.

As there were no English inhabitants in Pictou during the period referred to, this county was the scene of none of the atrocities inflicted by the Indians on the early English settlements, though there is little doubt that the Micmacs in this quarter had their share with their brethren in the war carried on under the instigation of the French against the English in other parts of the Province. But in the year 1761, on the 15th October, as stated by Mr. Murdoch, a treaty of peace was signed in council with Janneoville Pectougawack (meaning Pictouman), chief of the Indians of Pictouck and Malagoniche (Merigomish), and the way was thus opened for the peaceable occupation of the place by English settlers.

We have not been able to find the record of this treaty, and Mr. Murdoch could not direct us to the source of his information. The name is not Micmac, and we believe it is either a misprint or that the Micmacs have corrupted the French name. At all events, we believe that the party was the same person afterward known as Capt. Toney. He is said to have been a Frenchman, who had adopted the mode of life of the Aborigines, and had acquired such influence over them that he was regarded as a high chief,—that he spoke French well and English tolerably, besides Micmac,—that he has dined at the Governor's table and was able to conduct himself with the politeness of a Parisian. He was the ancestor of the present Toney family among the Micmacs, and they assert that the treaty was made by him in the name of the tribe—that on the part of the English, gun and bayonet, and on the part of the Micmacs, tomahawk, bow and arrow, were solemnly buried in one grave on the Citadel Hill, at Halifax, the latter weapons underneath. Perhaps the name as given by Murdoch may have been a misreading for Toneyville. We may add, that from him Toney River derives its name, but how it came to be connected with him we have not been able to ascertain.

One incident, however, we shall give as connected with this period, which we believe to be well established. Among the first English settlers it was received as a well-known fact, that a French war vessel had escaped from Louisburg during the siege, containing treasure, and that she had been chased into Carriboo Harbor. The entrance being narrow, and the English probably not acquainted with the navigation, did not venture to pursue. As she did not come out, and could not be seen, it was supposed that she had gone ashore in some creek. Accordingly, soon after the arrival of the first English settlers, Dr. Harris and his brother Matthew resolved on a search for her. They set out in a log canoe and paddled down the harbor and round the coast to Carriboo Harbor, thence along the south shore of the harbor till they reached Carriboo River, then up that stream to where it forks. Here they resolved to separate, each following a branch of the river, agreeing that if either should succeed, he should sound a horn to call his brother. Matthew took the Little River, which joins the main stream at a course nearly at right angles with it. On going round the point formed by their juncture, he had proceeded but a short distance when, in a little cove, at what is now George Morrison's place, he suddenly came upon the object of their search, snugly beached. The channel of the river is deep, but somewhat crooked, and those on board must have been thoroughly acquainted with it to have brought her here, and to have selected this spot to run her ashore. So completely concealed was she by the bend in the shore and intervening woods, that Harris was within ten yards of her before she was seen.

He immediately blew his horn, when his brother came, and they gave a cursory examination of their prize. She was a sloop, a neat and trim vessel. She had been armed, but the cannon were out of her, it was supposed, having been thrown overboard. All her rigging still remained on her.

From the position in which she lay, they supposed that there would be no difficulty in getting her off, and they left, intending to return speedily with proper appliances for the purpose. On arrival home, they freely made known their discovery, but before they could return, the Indians had set fire to her. When spoken to about their conduct, they explained that she had been left in their charge by the French owners, with instructions not to touch her unless the English discovered her, but if they did, to burn her at once, which they did.*

---

* About fifty years ago James A. Harris, of Carriboo Island, a son of Matthew, and who frequently told this story as he received it from his father, pointed out to his son James the keel and some of the timbers still standing. One who was present tried them with his axe, and pronounced them to be of American white oak. Probably some remains of her might yet be found in the mud.

It is certain that vessels escaped from Louisburg with treasure during the siege, and there is strong reason to believe that this had contained valuables, which those on board, when they abandoned her, could not carry away and concealed. About the year 1802 a vessel one evening came to anchor off the mouth of the harbour, and a boat with a strong crew put off from her, and was seen going up the river. It was not seen to return, but early the next morning the vessel got under weigh and departed. Shortly after, some of the people going up the river found, at the head of the tide, a place bearing all the marks of their having been at work. There was a hole from four to six feet square, and not very deep, perhaps four feet, with hand-spokes, whose position showed that they had been used in prying something like a chest out of the bottom of the hole.

It is said that on examination the trees around were found to have upon them marks pointing in the direction of where the hole was. This place is at some distance from the place where the vessel was ashore, and on the other branch of the river, but we can easily understand the wisdom of seeking such a place of concealment.

A settler who lived near, is reputed to have found a large sum of money. There have been various stories of the French burying money, which have led parties foolishly to dig in various places. That in the hurry of leaving and in the expectation of returning, they sometimes buried some of their possessions, we have reason to believe, but it was little money they had to bury, and what they had they carried away. We are generally incredulous regarding all stories of money found, but the information we have received, leads us to give some credit to this case. A son of the settler referred to, told a gentleman who reported the case to me, that it was true—that he and his sister, both then children, first found the money under a stump, that it consisted entirely of old coins, strange to him, but whether French or not he did not know ; that they told their father of it, who gathered them, but gave them none of it. The story commonly received is that he took it to his merchant who shipped it to England, both agreeing to say nothing about the matter lest government should claim the amount. The merchant in the meantime supplied the settler abundantly with articles for his family, but afterward failed, so that they received little more for their find. Other facts that we have, give probability to the story.

CHAPTER IV.

FIRST ENGLISH SETTLEMENT OF PICTOU—1765-1773.

During the war on this continent between the English and French, which resulted in the taking of Louisburg and Quebec, and was terminated by the peace of 1763, the settlement of Nova Scotia engaged considerable attention in the old colonies. About the years 1760 and 1761 a considerable number of persons removed from different parts of New England and settled several townships in the western parts of the Province. So little, however, was known of Pictou at this time, that in a description of "the several towns in the Province, with the lands comprehended in and bordering on said towns," drawn up by the Surveyor General in the year 1762, by order of Lieutenant-Governor Belcher, for the information of the Home Government, it is stated that "from Tatamagouche to the Gut of Canso there is no harbour, but a good road under the Isle Poitee (Pictou Island). No inhabitant ever settled in this part of the country, and consequently no kind of improvement."

At the conclusion of the war, a large number of influential persons, not only in the New England States, but in other of the Old Colonies, took up the subject of the colonization of the Province, and it is in this way that Pictou first comes into notice in the early settlement of Nova Scotia. Their views are thus stated in a letter from the Lieutenant-Governor of the Province to the Lords Commissioners of Trade and Plantations, dated 30th April, 1765 :—

" By the late arrival of several persons from Pennsylvania, New Jersey and some of the neighbouring colonies, we have the prospect of having this Province soon peopled by the accession of many settlers from these parts.

" These persons have come on behalf of several associations of commercial people and others in good circumstances, to view the country and examine what advantages the settlement and cultivation of it may produce. By their accounts the considerable numbers of Germans annually imported in the Colonies from whence they come, has so overstocked the good lands, and those situated within any convenient distance of navigation, that not only many of them have lately been obliged to move into Carolina and Virginia, but that there are also now numbers of useless persons among them. And this is not the only motive they have for making settlements in this country for the merchants in those parts are much at a loss to provide an export in return for the British commodities, and, therefore, have turned their thoughts to this Province for fish and hemp, to produce which, of the best kind and greatest abundance, nothing but a sufficiency of labouring people is wanting, and thus those people being employed, they will be sufficiently prevented from any attention to manufactures.

" And indeed, my Lords, what seems to promise the certain acquisition of these great advantages from the present applications, is that these settlements are to be undertaken by people of very sufficient and able circumstances, who propose the establishment of many German families, by which means the annual current of Germans to America will very suddenly be diverted into this Province, from whence it must receive a very considerable degree of strength, for these frugal, laborious and industrious people will not only improve and enrich their property, but pertinaciously defend it.

" Among the several persons who have arrived here with a view to these undertakings is Mr. Alexander McNutt, who has frequently attended at your Lordships' Board. His applications are of a very considerable degree and extent, and he produces many letters from the associations I have before mentioned, soliciting him, in the most pressing manner, to use his utmost endeavours to procure for them the tracts of land for which they apply, and on such conditions as he had obtained at your Lordships' Board the 27th February, 1761, for all such settlers as he would introduce into this Province."

When we remember that at this time the whole of what is now the Western States was still open for settlement, it seems curious to find parties in the Middle States a century ago representing the good soil there as already overstocked, and in consequence seeking land in Nova Scotia. Accompanying this representation was a list of firms or companies, to the number of fifteen, among whom we notice James Lyon, of Trenton, and "Dr. Franklin & Co.," who sought grants of land, some of 100,000 acres and some of 200,000, making altogether 2,000,000 acres. The Dr. Franklin mentioned here is we believe the great Benjamin, who was at that time influential in England and interested in the settlement of this Province. In a petition on their behalf, McNutt says, "that he did engage with several persons in Ireland, Pennsylvania, Virginia, and other parts of His Majesty's dominions, to provide lands in this Province on the terms contained in his proposals, for the settlement of as many families as they would furnish; that the several persons so engaging with him had been at considerable expense and trouble to fulfil their engagement by procuring many families for that purpose, who are now waiting with much anxiety and impatience to transport themselves to this Province."

Among the speculators at that time engaged in taking up land and bringing settlers to the Province, none was more active than Mr. McNutt, who is styled by Halyburton "an enthusiastic adventurer from the North of Ireland," who had already been the means of settling Truro, Onslow, and Londonderry.*

The result of these applications was, that in June of that year it was agreed in Council to reserve 200,000 acres for a company consisting of the Rev. James Lyon,

---

* He was also engaged afterward in settling portions of the County of Shelburne. At the close of his life he resided on an island there, still called McNutts Island, and was drowned crossing to the shore.

McNutt, and thirteen others, principally residing in the city of Philadelphia, of land "between Onslow, Truro, and the lands granted to Colonel DesBarres at Tatamagouche." In July, on their representing that they had at considerable expense and fatigue viewed the 200,000 acres reserved for them, and found that there was not the quantity applied for there, it was resolved that they should have "liberty to choose the aforesaid quantity between Tatamagouche and *Picto*."

At the same time 1,600,000 was reserved for McNutt and his associates at various places, among which is a block of 100,000 acres at Pictou.

At this period land was being granted by order of the British Government to various individuals, principally officers of the army and navy, for services during the war. It was in this way that two years later the whole of Prince Edward Island was granted in one day. Accordingly, on the 15th of October, five grants passed nominally for 20,000 acres each, though in reality containing much more, embracing the whole eastern half of the county.

In accordance with the resolutions above mentioned, there was a lot of 100,000 acres granted on the same day to McNutt and some of his friends, and on the 31st another nominally for 200,000 acres to the Philadelphia Company, commonly known as the Philadelphia Grant, to which we shall have occasion more particularly to refer presently. Thus in one month, and principally in one day, a district not exactly coinciding with the county of Pictou, but embracing the larger portion of it, and also a large part of the county of Colchester, was granted to individuals, the most of it to speculators.

The names of the grantees of the first five mentioned lots are John Major, John Henerker, John Haygens (afterward corrected Godhard Huygens), John Fisher, and John Wentworth. Major's grant fronted on Merigomish Harbor, at its eastern end. To the west of this lay Henerker's lot. In the rear of these and extending still further westward, was Huygens'. Of these parties we know nothing. The only mention of their names we have seen is in a memorial from Sir John Wentworth, in which he petitions against the escheating of these grants, "on behalf of the most noble Duchess Dowager of Chandois and Sir John Henerker, Bart., and member of the British Parliament, heirs and proprietors of certain lands at Pictou and Merigomish, formerly granted to John Henerker and Godhard Huygens."

Fisher, we have heard, was a major in the army. Wentworth, in a memorial at the close of the American war, says, "That your memorialist and said John Fisher were in His Majesty's service in America, and in consequence of their fidelity, and zeal in their duty, were proscribed and exiled from the United States of America, their extensive property in New England confiscated, and their means of improving their estate at Pictou considerably diminished." By letters at that date, it appears that he was then residing in London. His grant is now only interesting as having given its name to that part of the south side of the harbor immediately fronting upon the town; but to show how lands were granted at that time, we may give the description of the grant. It is as follows:

"Beginning at the north-east corner of McNutt's land, at a cove on the east side of Pictou Harbor, and running south 47° east 456 chains on said McNutt's land, thence south 808 chains on the same, thence east 74 chains on lot No. 1 (Huygens grant), thence north 600 chains on lot No. 3 (Wentworth grant), thence east till it meets Merigomish Harbor, thence along the sea-coast and harbor of Pictou to the first mentioned boundary, including the islands in the harbor of Merigomish."

We may mention that the portions of these grants on the shore nearly coincide with what was afterward called the 82nd grant. Westward of Henerker's lot, and fronting on the western part of Merigomish harbor, was what is still known as the Wentworth grant. Of all the grants given in that October, this is the only one which was not escheated. As it is thus the oldest grant in the county, we may give its boundaries as originally described:

"Beginning at a cove in Merigomish harbor, bounded on lot No. 2 (Fisher's grant), and to run west 56 chains, thence south 600 chains (or 7½ miles) on said lot, thence east 352 chains on lot No. 1 (Huygens), thence north 648 chains (over 10 miles) on lot No. 4 (Henerker's lot) to Merigomish harbor, thence to be bounded by said harbor to the first mentioned boundary."

Wentworth, afterward Sir John, was a native of New Hampshire, afterward Governor of that Province, and at a later period of Nova Scotia (1792—1808). He was at that time Surveyor of His Majesty's woods in North America, an office which he continued to hold till the American Revolution, after which he held the same position in the remaining Provinces.

The conditions of these grants were the same, viz.: that the grantees should pay a quit rent of one farthing per acre for the one-half within five years, and the whole to be payable within ten years; and secondly, to settle Protestant settlers upon it in the proportion of one person to every 200 acres within ten years from the date of their grant. These conditions were never fulfilled, and, so far as we know, no effort was made for that purpose by any of the parties except Wentworth, at a period, however, after the time fixed; and, as already mentioned, all the grants except his were escheated.

We may mention here that the only mines reserved on these lands were gold, silver and coal, so that the present owners of the Wentworth grant are proprietors of all other minerals they can find on their land and already portions of it have been found to be rich in iron ore.

On the same day with the date of these grants another passed to Mr. Alexander McNutt, William Caldwell, Arthur Vance and Richard Caldwell, of a tract of land:

"Beginning at a cove on the east side of Pictou Harbour (this must have been near the mouth of the East River) and running south 47°, east 550 chains.

thence south 1,040 chains (13 miles), thence west 872 (nearly 11 miles), thence north till it meets the innermost river of Pictou, thence bounded by said river and harbour of Pictou to first mentioned boundary. Also one other piece beginning at a point bearing north 33 east, from the little island in the harbour of Pictou (this was at Brown's Point), and running north to the sea shore (near Roddicks, Carriboo), thence to be bounded by the seashore and harbour of Pictou to the first mentioned boundary, including Pictou Island."

This was afterward known as the Irish grant. Of the parties to it, except McNutt, all we know is that they are said to belong to Londonderry, in Ireland. It will be seen that it embraced all the southern and western shores of the harbour from Fishers Grant round to the West River, and the land into the interior to the southward to a distance of about 20 miles, embracing both banks of the East and Middle Rivers and the west side of the West, to the distance of about a mile above Durham. It also embraced the block on which the town now stands, commencing at Browns Point and extended round the coast to Carriboo.

It will also be seen that it covered nearly all the most desirable portions of the harbour, and had the first settlement been upon it, the effort might have been more successful, and much of the suffering afterward experienced might have been avoided.

But the grant, which is of special interest, as connected with the early history of Pictou, was the last one mentioned as having passed at that time, usually known as the Philadelphia grant. It is dated the 31st October, and it is to Edmund Crawley, Esq., (for 20,000 acres) the Rev. James Lyon, John Rhea, Richard Stockton, George Bryan, William Symonds, John Wykoff, Isaac Wykoff, Jonathan Smith, Andrew Hodge, John Bayard, Thomas Harris, Robert Harris, and David Rhea for 180,000 acres.

Of these grantees Lyon and the Harrises will come under our notice again. Of the others, we know nothing except that they resided in Philadelphia, from which circumstance the company and the grant derived their name. The following is the description of their land:

"Beginning at the southwest bounds of lands granted to Joseph Frederick Wallet Des Barres, and running thence west 550 chains on ungranted lands, thence south 580 chains on ungranted lands, and on the township of Londonderry, thence east 800 chains on the township of Onslow, thence south 900 chains on said township and on ungranted lands, thence 1,000 chains on ungranted lands, thence north 932 chains, more or less, till it meets the westernmost river of Pictou, thence the course of the said river on the north side, till it meets the westernmost boundary of land granted to Alexander McNutt and associates, thence running north on said lands till it meets the seashore, thence the course of the seashore till it meets the northeast boundary of lands granted to J. F. W. Des Barres aforesaid, thence on his eastern boundary 480 chains, on said lands to the first mentioned boundary; together with the islands adjacent, containing on the whole 200,000 acres, more or less.

"In manner and form following, viz.: one equal undivided tenth part to Edmund Crawley, Esq., and the remaining nine-tenths to and among the others mentioned."

The line of the Des Barres grant referred to, commenced at Point Brule, between two and three miles to the west of the present county line, so that the Philadelphia grant included not only the greater portion of the township of Pictou, but a large portion of the county of Colchester, including part of the River John road settlement to Point Brule, a large part of New Annan, the whole of Earltown and Kemptown, with a considerable portion of Stewiacke. In fact, it would have made a county of itself.

But it will also be seen that on Pictou Harbour it had a very small frontage on the water, including only that part of the shore from Browns Point to the head of the harbour. All the shore from Browns Point eastward round to Carriboo had been obtained by McNutt. This naturally belonged to the Philadelphia Company's grant, and it plainly appears that, while acting for his friends in Philadelphia, he had also been acting for himself and some others, and had managed very unfairly to get this into his own hands. This we know was afterwards the subject of bitter complaint against him, we have no doubt, justly, and, as we shall presently see, it was a great obstacle in the way of the settlement of the place.

The conditions of these two grants are somewhat curious. They were, first, that the grantees should pay a quit rent the same as on the other grants. Secondly, that they plant, cultivate, improve or enclose one-third part within ten years; one other-third part within twenty years, and the other one-third within thirty years ; otherwise, such portions as are not improved to be forfeited. These terms we think simply impossible to be fulfilled. Thirdly, they were to plant, within ten years from the date of the grant, one rood of every 1,000 acres with hemp, and to keep a like quantity of land planted during the successive years. This condition is in a good many grants of the time. It probably originated in the desire of the British Government to be independent of foreign nations, in providing cordage for her marine. But though this is the condition on which so much land in the Province is held, yet, as the late L. Doyle remarked in the Legislature, there is now not sufficient raised in the Province for criminal purposes. The last condition was, that they were to settle one-fourth of the land within one year after the 31st day of November next, in the proportion of one Protestant person to every 200 acres ; one other fourth within two years ; one other fourth within three years, and the remaining fourth within four years, or the land so unsettled should revert to the Crown. The last condition was an excellent one, and had it been carried out to any extent, it would have tended to the rapid settlement of the country. But the time allowed was too short. Altogether, the terms were unreasonably severe. For five months the agents had remained seeking better terms, and then, as the Governor says, " it was with great difficulty they submitted to these terms, which they thought severe, and it was with reluctance they were granted, because they were not strictly conformable to the King's intentions," and they only accepted them on condition that they should be at liberty to avail themselves of such better terms as could be obtained by representations to the Home Government.

It does not appear that McNutt made any attempt to settle his land, and the grant was consequently escheated in the year 1770, but not until the first settlement had been made by the Philadelphia Company, and till his grant had proved an obstacle to their progress. The Philadelphia Company, however, seemed determined honestly to carry out their engagement as far as practicable. In a memorial to the Governor, dated 21st August, 1766, they represent that " they have received many disappointments, in their intentions of settling the lands granted to them between Pictou and the townships of Onslow and Truro, by the misrepresentations of one Mr. Anthill, who had represented the country as rocky, barren and unfit for improvement, and likewise made very injurious representations of the Government of this Province, all which very much prejudiced the persons who had engaged to settle the lands. That they had likewise met with a very great disappointment, on finding that a considerable part of the harbour of Pictou, by some mistake in the survey, was not granted to them, as they expected, all which, with many obstructions from the scarcity of money and the stagnation of trade, occasioned by the Stamp Act taking place at that time, rendered them incapable of making any settlements this year, as intended." In consequence of this, they were allowed to the 1st of June following to settle the first portion of settlers.

On the 5th of May, 1767, seven of the Company, George Bryan, William Symonds, Andrew Hodge, Robert Harris, John Bayard, and John Smith, all of Philadelphia, and Thomas Harris, of Baltimore County, in the Province of Maryland, executed a power of attorney to John Wykoff, of Philadelphia, merchant, and Dr. John Harris, of Baltimore County, empowering them to grant and sell, in the name of the Company, their land, on such terms as they should see fit. The Rev. James Lyon, who was already in the Province, afterward executed a similar paper to Harris.

They also despatched a small brig, called the Hope, Captain Hull, of Rhode Island, with six families of settlers and supplies of provisions for their use. These families consisted of Dr. Harris, the agent and wife, Robert Patterson, who came as a surveyor for the Company, his wife and five children, the eldest nine years old, and the youngest only three months; James McCabe, wife and six children; John Rogers, wife and four children; Henry Cumminger, wife and four or five children, and a sixth family whose name is uncertain. Besides these, Patterson had with him a convict servant. It was customary at that time in the old colonies to sell criminals sentenced to penal servitude, to serve out their sentence in a position similar to that of slaves, to any who might be willing to buy them. Patterson had bought this man for a term of seven years, which, we may here observe, he fully served. Thus the company consisted of twelve heads of families, about twenty children, and one convict servant, and possibly one or two colored slaves.

The Hope sailed from Philadelphia toward the end of May, and called at Halifax to obtain information regarding the coast round to Pictou. Harris's power of attorney is attested there on the 3rd June. Leaving Halifax after a few days' stay, they reached the harbor of Pictou on the 10th of June. The people of Truro had heard of their coming, and five or six young men set out through the woods to meet them and aid in commencing operations. Of these we have heard the names of Samuel Archibald, father of the late S. G. W. Archibald, John Otterson, Thomas Troop, and Ephraim Howard. The two latter we notice from the circumstance that, in passing the mountains on the western border of the county, they named the one Mount Thom and the other

Mount Ephraim, after themselves, names which they have retained to this day.* They reached the harbor the same afternoon that the vessel arrived, and made large fires on the shore about Beck's place to attract her up. Those on board saw the fires and supposed that they were made by savages, of whom they naturally stood in terror. The vessel accordingly stood off and on till next morning, and the company deliberated whether to resist or submit to their mercy. Like true Englishmen, they chose the bolder alternative.

During the night their number was increased by Mrs. Harris giving birth to a son, afterward known as Clerk Tommy, having filled the situation of Clerk of the Peace and Prothonotary for some years. He died in 1809, and was buried in Pictou graveyard, where a monument stood till recently to his memory, on which he was described as " the first descendant of an Englishman born in Pictou."

The next morning they saw the Truro party coming along the shore, and by their spy-glasses, discovered, to their joy, that they were whites, and as the vessel stood in toward the shore, they heard the cheerful hail of friends. That day they landed at the point just above the Town Gut, which had been selected as the site of a town, as the part of the Company's grant nearest to the entrance of the harbour. The prospect was indeed dreary enough. One unbroken forest covered the whole surface of the country to the water's edge. What is now the lower part of the town was then an alder swamp. All around stood the mighty monarchs of the wood in all their primeval grandeur, the evergreens spreading a sombre covering over the plains and up the hills, relieved by the lighter shade of the deciduous trees, with here and there some tall spruce rising like a black minaret or spire above its fellows. But chiefly conspicuous to the eye of the obser-

* Some think that Mt. Thom derived its name from Thomas Archibald, but the tradition we have followed we think well founded.

ver were the tasselled heads of the white pines, for which Pictou was afterwards so long distinguished — their straight stems towering to the height of 150 or 200 feet, " like masts of some huge admiral." * Some of the early grants reserved all pine trees over two feet in diameter, "suitable for His Majesty's navy," but here within sight might be seen probably enough to have masted all the ships, not only of His Britannic Majesty, but of all the navies in christendom. The scene was one on which the eye of the lover of nature might have gazed with delight, but it is needless to say that these settlers looked upon the matter with more practical eyes. The interminable forest only presented itself to them as an insuperable obstacle to their labours, and their hearts sank as they contemplated the idea of wresting a subsistence from the soil so encumbered.

Knowing the hostility which the Indians had maintained to the English almost up to that period, and the cruelties which they practised upon the infant settlements; familiar, too, with the tale of their atrocities in the colonies which they had left, their minds were filled with fear of the savages. Nor was this without reason. The French were not yet without hope of regaining their ancient power over this land by the expulsion of the English, and with this view were still intriguing with the Indians. During the two years previous the latter held meetings in a hostile spirit, and on the last of these occasions had declared their intention not to allow any settlement at Pictou, on the north shore of the Province. †

---

* In the days of the pine timber trade, a tree that would not square a foot to the length of sixty feet would be considered a small tree, not worth taking, while sometimes they stood so close together that the lumberers could not take them all, lest in felling them they would break by falling across one another.

† " The last year they showed how capable the French are of drawing them together whenever they think proper, which they actually did by some means unknown to the Government, for the whole body of Indians were collected

When we add that there was not one English settler on the north shore of the Province, from the Strait of Canso to Bay Verte, or perhaps, even to Miramichi, we may picture the loneliness of the little band, and need not wonder that their hearts sunk within them at the prospect of the toils and dangers before them.

What rendered this disappointment greater was, that highly coloured representations had been made to them of the country to which they were coming ; such as, that they could get sugar off the trees, in fact, they had come with such ideas of the place as are now entertained by the emigrant to California. The advantages of those who, in the years immediately previous, had removed from the old colonies and taken possession of the clearances of the French, and who had exchanged the rocky and barren shores of New England for the rich marshes of the Bay of Fundy, had excited high expectations regarding Nova Scotia. The more bitter, therefore, was their disappointment at the dreary prospect before them. After they landed, Mrs. Patterson used to tell, that she leaned her head against a tree, which stood for many a year after, and thought that if there was a broken-hearted creature on the face of the earth, she was the one. As she looked

---

from every part of the Province, and assembled on an island called Madame, to the north East of Canso, and not far from the head of La Brador, and as they passed through the different townships to their rendezvous they declared they were to meet French forces and threatened to destroy the settlements when they should return. This alarmed the inhabitants to so great a degree that for several weeks together they were kept in continual apprehension, and some part of the time even in arms ; and with difficulty this body of Indians were dispersed, partly by the influence of some gentlemen, and partly upon finding themselves deceived in their expected support from the French.

" This year they have assembled in like manner, although not in so great a body, but with the same disposition, and some of them have, in addition, declared *they will not allow any settlements to be made at Pictou*, and that part of the coast of this continent which lies nearest St. Peters ; but they dispersed upon the Government sending for a Canadian priest who officiates in the Bay of Chaleurs."—*Letter of Lieutenant-Governor, dated 3rd September,* 1766.

upon her little ones left shelterless in the cruel wilderness, among savages deemed still more cruel, she could only cling to her husband with the cry, "Oh, Robert, take me back." So discouraged were the whole band with the state of matters, that the most of them were determined to return in the vessel which brought them; but the captain, after landing his passengers and supplies, slipped out of the harbour in the night and left them to their fate, probably with the concurrence of the agent.

We have no particular account of the subsequent proceedings, but the few facts we have gleaned we shall put together as connectedly as in our power. The first night on shore they spent under the trees, without even a camp to shelter them, but the weather was warm and they did not suffer from the exposure. Their first care, of course, was to provide some shelter, which they did by the erection of rude huts. The agents of the company proceeded to lay out a town where they had landed. A half acre was assigned to each family, No. 1 being McCabe's. At that point the trees were not large. He immediately set to work and cleared his half acre, and his descendants boast that he cut down the first tree in Pictou. He had brought with him what they called a mattock, a heavy instrument, on one side an axe and on the other a grubbing hoe. Instead of chopping down the tree, his practice was to take away the earth from the main roots and cut off all the smaller ones, and then either leave it to fall by the wind or drag it down and out of root. In this way he cleared his lot, and instead of burning the trees, he hauled them out to the tide. He was so prompt that to reward him he was assigned another half acre. He planted his lot with potatoes without ploughing, just placing the seed under the moss, which had not been burned, and which he supposed would serve for manure. The land at this point was inferior, and not having been prepared, only a few weakly sprouts appeared, and in the fall the tubers were not larger than potatoe balls. Whether the others planted more or fared better that season, we have not ascertained, but not likely they did.

Around this point a plot was reserved for a town, and hence the creek close by has ever since been known as the Town Gut. Farm lots were assigned to each settler. Patterson, afterward the Squire, had his where his eldest son, John, afterward resided, and where his great grandson, Henry, and the Fullartons now live, about two miles from town. The remains of his orchard are still standing. McCabe got his where W. Evans now lives, about five miles from town, and another where the late George Murray lived below Durham. But he was a Roman Catholic, and the company's grant bound them to settle their land with Protestants, and hence the deed of his lots is in his wife's name. He had been partly educated for a priest, and managed to gain in this way an influence over the Indians, who pointed out to him the place where he took up his land as rich, which it proved to be.

John Rogers took up his land on Rogers Hill, which derived its name from him, where his grandson recently lived. The situation is a beautiful one, being nearly at the summit of the hill. But it seems singular that, with all the shore unoccupied, he should have gone so far back; but the land there was rich, with a fine lay, and it was on the blazed path to Truro, which he supposed would be the road to Halifax. Some of the apple trees raised from seed which he brought with him from Maryland are still standing.

Among the first efforts of the settlers was the opening of a road, or rather blazing a path, to Truro. It is claimed that this was done by Thomas Archibald, John Otterson, of Truro, and John Rogers, the compass their only guide. The road left the shore at the head of Pictou harbor, above Evans, place, and went over Rogers hill, following nearly the course of the present road through

Rogers settlement, beyond the Six and Eight Mile brooks. I have been informed that the first course was by the North Mountain and down the North River to Truro, which is regarded by many as the shortest line between the two places. But there were difficulties by that route, and a line was opened over Mount Thom, for which Thomas Archibald has always got the credit, which continued to be the regular line of communication between the two places till about the year 1831. After reaching the summit, it descended till it struck the Salmon River, which it crossed at Kemptown, thence proceeding along the north side of the river, along the upland, till within about four miles of Truro, and the remaining part of the distance along the intervale.

It seems even at this early period to have been regularly laid out and duly measured. We find in the deeds given shortly after such descriptions as the following :— " At the south-west corner of land laid out on the Cobyquid Road, between the three and four mile trees." " Another lying on Cobyquid Road, beginning at a stake and stones near the eight mile camp, thence westwardly on said road a mile and a quarter." At what time this road was laid out we are uncertain, but think it was the summer they arrived. It may be mentioned that the various streams on the west side of the West River derived their names from the distance on this road from the point of departure. Thus Forbes Brook was long known as the Half-mile Brook, and so we have the Four, Six and Eight-mile Brooks. It may be added that the farthest-up settler on the Truro side for a long time was Thomas Archibald, generally known as Uncle Tom, whose house was long the home of the traveller.

"In addition," says Philo Antiquarius, "to the difficulties already mentioned, they were constrained to submit to many indignities from the aborigines, who viewed their operations with no friendly eye. These considered the settlers as usurpers of their natural rights, who had encroached on their undoubted property; and it required not a moderate portion of skill on the part of the civilized to gain the good-will of the savage, nor inconsiderable prudence to establish this amicableness when formed." We have heard, for example, of a white man taking a fish from the river, and an Indian taking it from him, saying it was not his. They would enter the houses of the settlers, and help themselves to the cakes that the women might be baking on the hearth, or other provisions, with threatening gestures. The settlers cultivated their friendship by such means as playing draughts, wrestling or by what was perhaps more effective, drinking fire water with them. And though the Indians were fond of working on their fears, when they could do so, they do not seem to have intended to do them any serious injury. In fact, through kindness, they became attached to some of the settlers, and showed them great kindness. Still incidents sometimes occurred which showed that they were not to be trifled with, and that their old savage nature might be revived. A young man, wrestling with an Indian, by a dextrous movement, which his opponent thought unfair, tripped him. The Indian was very cross, and sometime after, the young man going to the Middle River, where the former had his camp, his squaw came out and earnestly warned him away, saying that her husband would kill him if he found him there.

"During the summer months the settlers experienced little inconvenience from the weather, but they found the winter much more severe and of longer continuance than in their native clime. They were consequently ill prepared to meet its blasts, and suffered intensely from its inclemency.

"As their provisions diminished, they directed their enquiries to the internal resources of the country, and this investigation was amply recompensed by discovering the forest to be plentifully inhabited by different species of wild animals. In hunting these, the settler usually had the Indian for an associate, and his faithful dog for a follower. Among the several kinds of animals, none were more valuable or abundant than the moose. The hunter, in endeavoring to procure these, was subjected to much fatigue, having frequently to pursue one of them a whole day, with the probability of not overtaking it at the end.* If, however, he were fortunate enough to catch it, the quantity of excellent venison it produced might have been deemed an equivalent for the labor of the chase, but, besides, its skin, when properly prepared, was valued at ten shillings, and was advantageously bartered for necessaries to traders, who were accustomed to run into the harbour with small crafts.

"Necessity is truly the mother of art. Congregated as the early inhabitants of this district were, in a place which was devoid of every conveniency, where the most common and indispensible commodities were wanting, their creative powers were laid under heavy contribution, in order to provide for the deficiency, and their inventive genius was called into ceaseless operation in constructing articles for household use, in forming implements of husbandry, and making instruments for hunting. They thus became more ingenious and more fertile in resources—what, in America, is called more 'handy'—than if living in older inhabited places."

In the following spring they found it necessary to proceed to Truro for seed; the journey required three days to go and as many to return. They returned, bearing each a bag of seed potatoes on his back. The labour of such a journey through trackless and intricate forests, carrying a burden, we can scarcely estimate. That year they planted the seed thus brought, and succeeded in

* The late James Patterson told the writer of starting one back of the town, and killing it in descending the southern side of Green Hill.

raising a quantity of good potatoes, but not sufficient for their subsistence, so that winter had not much more than begun before their supply was exhausted.

The following year they again went to Truro for a supply, but this time they cut the eyes out with a penknife, by which they could carry a large supply, and that season they raised enough for their subsistence during the following winter.

We subjoin slightly abridged, a return of the population at the close of the year 1769, which is, probably, the first census taken in this place, from which it will be seen that besides the settlers we have mentioned there had arrived here, in 1767 or 1768, the families of Robert McFadden, the Rev. James Lyon, and Barnabas McGee, and that the first of these left in 1769. By this return it appears also that nine families arrived in that year, but only five remained. Some of these were from Truro and some were from Philadelphia. Of those from Truro William Kennedy deserves special notice. He was one of the grantees of Truro, but sold out there in 1768, and obtained a lot in Pictou, at the Saw Mill Brook, extending in front from the mouth of the brook up the harbour, now the McKenzies' property. Here he made a clearing long known as Kennedy's Clearing, or Kennedy's Hill. On that stream he, in 1769, erected the first saw mill in the county, which was the first frame building in Pictou. On the 28th of September, 1774, he deeds "half of saw mill now built on a stream now known by the name of the Mill Brook." Two years later he returned to Truro, and in 1780 settled in Stewiacke, where he was the first settler. Moses Blaisdell also came from Truro. He settled on the lot since occupied by the Becks, at the head of the harbour, but afterward removed to the eastern part of the Province.

Of those who came from Pennsylvania we are certain of the names of only two—Matthew Harris and Barnabas McGee. The former was an elder brother of the Doctor, and settled on the farm now owned by George Davidson, about five miles from town. The remains of his cellar may still be seen near the shore, and the apple trees which he planted are still growing, and, if properly cared for, might still yield good fruit. We may here mention that all the American settlers planted fruit trees, and originally of good quality, especially the apple trees. The settlers that arrived afterward, being chiefly from the Highlands of Scotland, paid no attention to the raising of fruit, and their children after them showed the same spirit, so that among them even yet but a few have orchards of any account. McGee was a native of the North of Ireland, who had emigrated to Pennsylvania or Maryland, and there married a London woman. He had his land on Rogers Hill, but being dissatisfied with the want of frontage on the shore, he gave it up and was afterward the first settler in Merigomish.

Thus during this year 67 souls had arrived and 4 children had been born, but 36 had removed, and one had died, leaving the net population 84.

The return of produce raised in that year exhibits 64 bushels of wheat, 60 of oats, 7 of rye, 8 of barley, 6 of pease, and some flax, potatoes not given. When we consider that this was the result of the labors of six families at most, in what we may regard as the second year of their labors in the forest, the progress made, we think, was creditable. Their show of cattle under the circumstances is very good, viz., 6 horses, 16 oxen, 16 cows, 16 young cattle, 37 sheep, and 10 swine. We also find thus early the commencement of our Marine, for Dr. Harris is credited with owning a fishing-boat and a small vessel, and Kennedy had a saw-mill.

Of those in the above list, Henry Cumminger, who had originally come in the Hope, afterward removed, and Nathan Smith, William Aiken and Thomas Skead seem scarcely to have made any settlement. Other settlers also

A RETURN OF THE TOWNSHIP OF DONEGAL OR PICTOU, THE FIRST OF JANUARY, 1770.

NAME OF MASTER OR MISTRESS OF FAMILY	NUMBER IN EACH FAMILY					RELIGION		ORIGIN					STOCK AND SUBSTANCE IN CATTLE						PRODUCE LAST YEAR IN BUSH		ARRIVED SINCE LAST YEAR			LEFT SINCE LAST YEAR		
	Men	Boys	Women	Girls	Total Persons	Protestant	Catholic	English	Scotch	Irish	American	Acadian	Horses	Oxen and Bulls	Cows	Young Cattle	Sheep	Swine	Wheat	Oats	Males	Females	Total	Males	Females	Total
John Harris	6	2	2	1	11	11				3	6	2	1	2	1	1	6	4	20	30				2		2
Robert Patterson	2	5	1	2	10	10				1	6	1	1	2	1	1	15		10							
Robert McFadden	1	3	1	2	7	7					6	1														
Henry Cumminger	1	3	1	3	8	8				2	8		1	2	3	4	6	3	6							
James McCabe	1	2	1	3	6	6					5	1	2	4	4	8	10	3	15	30	2		2	4	3	7
Nathan Smith	1		2		2	2					2								10							
Rev. James Lyon	3	2	1		7	7				1	6								3		7	5	7	5	4	
Barnabas McGee	1		1		2	2				2		1									10	2	10			
Wm. Kennedy	1		1	3	6	10					5															
Moses Blaisdell	1		1	4	6	6					9	1	2	2	3	1	6									
William Aiken	1			1	2		1						1													
George Oughterson	1	2			1	1					6															
Thomas F'head	1	4			1	14			1		12			2	2						14	4	14	5	7	
Matthew Harris	4	3	2	4	11	11				2	7		2	2	2	1	10				9	7	9	1	2	
Barnett McNutt	3		2	2	4	4				2	6		1		1						8	2	8	4	2	6
James Archibald	1	4	1	5	9	9				1	5	1		2	2						6		6	1	1	
Charles McKay	1	2	1	1	8	8		1		1	3								3		4	2	4	4	2	
Robert Dickey	1	3	1	1	6	6										1										
Total	31	36	20	23	120	119	1	2	2	18	93	5	6	18	16	16	37	10	64	60	35	32	67	18	18	36

arrived, among them, it is said, two or three from Cumberland. Of these, one was named James Fulton, and another, we believe, was named Watson. Fulton was born in Ireland in the year 1726, and with his wife and family emigrated to Halifax in 1761. He went first to Lahave and afterward to Cumberland, whence he removed to Pictou. His name appears in the list of town officers of the latter place in 1775, but he removed shortly after to the lower village of Truro. "In removing from Pictou to Truro," says Miller, "they underwent great hardships; they had then to travel through the woods without any roads, and carry their stuff and their children on their backs. This journey occupied the whole of the week, although they had the assistance of several men. While on their way there came on a snow storm which caused them much suffering, as they had to stop in the woods for five nights, and one night in particular, their fireworks being damp, they could get no fire for some time, and were in danger of perishing." Watson lived on the west side of West River, and died there, and his farm was afterward purchased by Robert Stewart. His family moved away.

Of the settlers who arrived, some took up land of the Philadelphia company, and occupied the west side of the West River, nearly up to the ten mile house; others had their lots assigned to them in the rear, but discouraged by their location in the woods, they either moved away or squatted in other places round the harbour, without titles. Besides those already mentioned, we find the name of Jonas Earl, who had his farm to the west of Watson's, already mentioned; and Isaiah Horton, who lived to the east of him, besides others, some of whose names are given in the following list of town officers, which may be inserted as a curiosity:

ONSLOW SESSIONS,
FEBRUARY TERM, 1775.

A list of Town Officers for the township of Pictou:
Clerk of the District............John Harris.
Overseers of the Poor............Robert Mersom, John Harris, James Fulton.
Overseers of the Road............Matthew Harris, William Kennedy.
Surveyors of Lumber.............Moses Blaisdell, William Aikin.
Constable......................William Aikin.
Clerk of the Market.............James Fulton.
Culler of Fish..................Abraham Slater.
Approved and established by the Sessions.

(Signed,) NOAH MILLER,
Clerk of the Peace.

One other settler is deserving of notice, viz., James Davidson. He was a native of Edinburgh, where he married and the first of his family was born. He emigrated from Scotland in the same vessel in which the Rev. Mr. Cock brought out his family. Soon after he came to Pictou. He took up a considerable quantity of land, but specially claims attention as being the first school-master in Pictou, the school-house being situated at Lyons Brook. He also deserves notice as being a pious man, who first cared for the spiritual interests of the settlers. He collected the children on the Sabbath day for religious instruction, so that Lyons Brook is known as the site of the first Sabbath school in the County, and probably in the Province, established even before Raikes began that movement, which made these institutions part of the regular machinery of the Christian Church. "Here," says the editor of the " Colonial Patriot," " this worthy man taught school seven days of the week, and, to our shame be it spoken, the Sabbath was more sanctified then, when there was no place of worship, except the school-house where James Davidson taught and prayed, than it is now, when churches are in abundance, even at our doors." Partly from want of a minister in Pictou, and partly from friendship for Mr. Cock, he removed to Truro about the year 1776, and settled at Old Barns, where he died, leaving no sons, but several daughters, whose descendants are still numerous in Colchester.

We must remark here, however, that though these first settlers had a number of hardships to endure, yet they never suffered actual want of the necessaries of life. This was owing partly to the arrangements of the Company, and partly to their own industry and skill. The Company, had sent a supply of provisions, we believe, for two years. The settlers were acquainted with American life, and soon learned to avail themselves of the resources of the forest. The coasts abounded, particularly in spring and fall, with fowl, so little disturbed by man, that they were shot or even snared with little trouble and in great numbers.* Fish were abundant, the most valuable of which was the salmon, which came into the rivers in great numbers, as one said to me, " as thick as the smelts do now." I have heard old people describe them even at a later period, coming in such numbers into the West River, that at a narrow place they would seem almost jammed together, so that one would think he could walk upon them. These they not only caught for their own use, but salted for exportation.

As they cleared the land, they were able to grow crops, potatoes never failing to yield a bountiful return ; if only as much of the potatoe as had a sprout were planted. But perhaps their chief resource was the wood of the forest. The pine they split into four feet clapboards, and they manufactured staves from the oak and ash, both of which found a market in the old colonies.

Squire Patterson had brought a large supply of goods, with which he traded with both the Indians and the settlers. The former he supplied with guns, ammunition,

---

* I have heard James Patterson tell of even sometime later, going to the Beaches in the month of March with his gun, and, after being away a day or two, sending home for a horse and sled, which he brought home loaded with wild geese, which they salted down for their summer provision.

clothing, &c., in exchange for furs, or sometimes for food, and the settlers he supplied with various articles, taking the produce of their labor in return. Small trading vessels from the old colonies, employed principally in fishing, brought them supplies, receiving in exchange their fish, fur, and lumber. With all the toil and hardship connected with their life, there must have been something fascinating about it. The father of the Harrises having visited his sons, endeavored to persuade them to return to Pennsylvania, but they refused.

Of their social life at this period we have little further information. The peculiar circumstances of the first birth have preserved the remembrance of it. We may add that Dr. Harris' daughter, afterward Mrs. Robert Cock, born in 1769, was the first female child born in Pictou of English parentage. Of marriages we have no record; but we find in the foregoing return mention of one death, in the family of Oughterson, probably his wife, as he is returned as alone in his family. Probably there were more, as before the arrival of the Hector passengers, there was a burying ground. This, which was the first in the county, was on the farm owned by John Patterson, Squire's son. It was situated to the west of his house, the same in which his son Charles lived. The ground has long since been ploughed over, and the spot cannot now be distinguished.*
We may say here that the people were generally serious and religiously inclined, most of them being Presbyterians, and the others, with one exception, New England Puritans.

It is said that there were sixteen families in Pictou on the arrival of the Hector in 1775, but of these only six remained, and we shall conclude this chapter with a more

---

* A woman who came to reside in that neighborhood, not knowing there had been a burying ground there, took hold of a stake and, working with it in the ground, struck something that sounded hollow. She ran home to tell that she had found what must be money. Further information, however, led to the conclusion that it must have been a coffin.

particular notice of these. Especially deserving of attention is Robert Patterson, made a magistrate in 1774, and hence long known as Squire Patterson. He was a native of Renfrew, in Scotland, but had for some time been in the Old Colonies, residing, at least part of the time, at a place called Cross Roads, in the State of Maryland, about 14 miles south of the Pennsylvania line, now called Churchville, a small place in the midst of a rich agricultural district. Here also the Harrises resided. He had for some time been employed as a pedler, and also, I have been informed, as a sutler to the army, previous to the peace of 1763.

He was for many years the leading man in Pictou, laid out all the first lots, surveyed all the early grants, and was prominent in all the public affairs of the place. " On account of his steady adhesion to the soil and interests of Pictou; on account of his disposition and ability at all times to relieve the distressed, and on various other accounts, he fairly earned the title of FATHER OF PICTOU. As such he was loved and esteemed by the inhabitants during his life." *

His first location was, as we have seen, about two miles from town; but he afterwards obtained from McNutt a claim, afterward confirmed by Governor Patterson, to a lot a little above Mortimers Point, where he built the first frame house in Pictou. There he continued to live till his death, which took place on the 20th September, 1808. His remains were interred in the old burying ground at Durham. We shall have occasion to refer to him again. Mrs. Patterson died March 6th, 1812.

Of his children who were with him in the Hope, his eldest, known afterward as John Patterson, second, lived about two miles from town. He was an Elder in the Presbyterian Church, and died 8th May, 1820. He left a

---

* Editor *Colonial Patriot.*—An eight-day clock, brought with him in the Hope, is in the possession of a great grand daughter, Mrs. A. P. Ross, and still marks the hours.

large family, but there are now few of his descendants living. The second James will still be remembered in Pictou. He settled to the west of the town, where the remains of his old orchard still exist. He also was for many years an Elder in the Church, and died May 14th, 1857, aged 96. The third, David, lived above Mortimers Point. He and his brother James usually worked together, and their houses were the next frame houses in Pictou after their father's. He died September 26th, 1844. The Squire had two daughters also on board the Hope, Sarah, afterward Mrs. Mortimer, and Margaret, afterward Mrs. Pagan. Of the children born after his arrival, Thomas settled on Carriboo Island, and George was one of the early settlers in Merigomish, where, and elsewhere, he has left numerous descendants.

The Harrises, Matthew and John, were of the Scotch-Irish race, their ancestors, Edward Harris and Flora Douglas, having left Ayrshire, in Scotland, in the reign of Charles II., or James II., losing a fine estate for their attachment to Presbyterian worship. They settled near Raphoe, in the County of Donegal, Ireland.

Thomas, grandson of Edward and father of Matthew and John, and an older son, Robert, were members of the Philadelphia company. He was then described as of the county of Baltimore, Maryland, and his son as Doctor of Medicine, Philadelphia. He died in Elizabethtown, Lancaster county, Pennsylvania, on the 4th December, 1801, at the age of 106, having seen three centuries.

John was the younger son, but had most to do with the settlement of Pictou. He was born July 16th, 1739. He acted as attorney of the Philadelphia company in disposing of their land, and a host of deeds are recorded from him, of lots not only along the West River and Rogers Hill, but at Carriboo, Cape John, River John, and other places. He was the first magistrate in the district, having been appointed in 1769. He was first registrar of deeds, and held other public offices. He lived near Browns Point, on a lot purchased from McNutt, and confirmed to him by Governor Patterson, the same which has since been occupied by his son Thomas and his descendants. He, however, removed to Onslow about the year 1778, where he was clerk of the peace for some years, represented Truro in the House of Assembly from 1779 to 1785, celebrated marriages and was otherwise a public man. He died in Truro April 9th, 1802, through a fall from his horse. His descendants are numerous, a considerable number being in Colchester and some in Pictou, his eldest son Thomas having settled at the Town Gut, and his youngest, John W., having been long High Sheriff of the county.

Matthew Harris was born on the 12th January, 1735, according to a statement received from the United States, though his age as published at the time of his death, would make the date of his birth four years earlier.*

Of his family, the eldest, Thomas, was a surveyor, and laid out much of the land in this and the neighboring county of Colchester. He divided the back lands of the township of Truro under the writ of partition and made a plan of them, dated August 12th, 1788, which is still in use, and was for twenty years Sheriff of Pictou. Another son died unmarried; a third was lost at sea coming round from Halifax; a fourth, Robert, studied medicine and lived on his father's place, but afterward removed to Philadelphia, where he died; a fifth son, James, settled on Carriboo Island, and has left numerous descendants, and his youngest son removed to Pennsyl-

---

* "Died, at Pictou, December 9, Mr. Matthew Harris, aged 88 years, the last head of a family of the first settlers from the State (province) of Pennsylvania. In the year 1763 (should be 1769) removed to Pictou, which at that period was a rude, uncultivated wild, inhabited only by a few wandering Indians and four families who arrived there shortly before from the same place. He has left 9 children, 40 grandchildren, and 30 great-grandchildren." *Newspaper*, 1829.

vania. Of his daughters, one was married to John Patterson (deacon), hereafter to be noticed. The others were also married, but their descendants have nearly all removed from this county.*

James McCabe was a native of Belfast, who emigrated to the old colonies. He there married Ann Pettigrew, a North of Ireland Presbyterian. He was a Roman Catholic himself, but not a very strict one. He was too fond of the good things of this life to regard Lent or the other fasts of that Church, and cared little for her holydays. He attended Dr. McGregor's preaching, but never became a decided convert. He had with him in the Hope two sons, John and James, and four daughters. The sons afterward married, and had, the one thirteen and the other eleven children, or two dozen between them, whose descendants are widely scattered. His four daughters were married, one to a Watson and another to a Snow, both of whom removed to the United States; a third to Robert Gerrard, but died when only 26 years of age, and the fourth to Owen McKowen, or McEwan.

John Rogers was a native of Scotland, brought up in Glasgow, where he married, his wife's name being Ritchie.

---

* We may here remark that the Harris family to which these two brothers belonged is very widely spread through the United States. A gentleman of the connection has sent me a genealogical chart containing the names of 425 persons, and adds, "leaving hundreds and even thousands of whom I know little or nothing." He further remarks, " Taken as a whole they represent a very respectable body of people, none of them very distinguished either for wealth or genius, but nearly all of good character and fair respectability. Of the 425 names given, about 40 are those of professional men, while several others have "General," "Colonel," "Judge," or other title attached, and quite a number have been liberally educated. One was a member of the Legislature of Pennsylvania at the time of the Revolution, an active participant in the affairs of church and state of the day, and one who had particularly distinguished himself by his efforts to abolish slavery in that State ; and left a name behind him for patriarchal wisdom and goodness. For many generations christian faith and life have been manifested among them, several being elders and pillars in the Presbyterian Church, from attachment to which their ancestors were driven from Scotland."

He thence removed to Maryland. He never assumed much prominence in the settlement of Pictou, but was a quiet, industrious and inoffensive, and we have reason to believe, a good man, as have been many of his descendants. He left one daughter and four sons, of whom all but the youngest son were in the Hope. His eldest son, James, lived on the farm, since owned by Alexander McKay and Rae, above the Town Gut. His second son, John, lived and died on his father's homestead. His third son, David, went to River John, where his descendants are numerous. His youngest son died comparatively early.

Barnabas McGee, we have already mentioned. He was lost, with his eldest son, going down to Newfoundland. His descendants are still in Merigomish.

The Rev. James Lyon appears as one of the Philadelphia company. In the petition from the inhabitants of Pictou, in the year 1784, they say, "The Philadelphia company made provision for, and sent, a minister, viz.: the Rev. James Lyon, at its first settlement, yet he did not continue among us, which very much discouraged the people and was exceedingly detrimental to the settling of the place." It would appear from this that the company had been mindful of the spiritual wants of the settlers. In fact we have reason to believe that the zeal manifested at that time in the old colonies in the settlement of the Province, was induced partly by motives of religion, particularly a desire that these regions where French Popery had hitherto prevailed, should now be occupied by sound Protestantism. Mr. Lyon was regularly ordained by the Presbytery of New Brunswick, in New Jersey, and came to the Province in the fall of 1764 or early in 1765, and was the first Presbyterian Minister in the Province of whom we have any account. In the latter year he was in Halifax. By the return which we have published, he appears to have been residing here with his family since 1769. He has given his name to Lyons Brook, about three miles from the town of Pictou. At different dates, from 1767 to 1772, we find his name in deeds as of Onslow, which was then the place where the public business of that district was transacted, and after that we find him described as of Machias, in the State of Maine. We have been informed that the lot of land about two miles from town, now occupied by Mr. Daniel McKenzie, had been set apart as a glebe, and that there was a burying-ground upon it. After the Philadelphia company's grant was escheated, it was granted by Sir John C. Sherbrooke to Dr. McCulloch.

We have thus, with considerable labor, gathered all the facts within our reach regarding this first attempt at the settlement of Pictou. The result is by no means proportionate to the effort. We have been minute in the details, as being the first attempt of the kind, and amid difficulties which might have appalled the stoutest heart, we deemed it proper to preserve everything we can learn of the early actors in these scenes. None of the settlements in Nova Scotia had such obstacles to encounter as that of Pictou. At Halifax and Lunenburg colonization began under the superintendence of Government, which also expended large sums in providing for the wants of the settlers. Those again who came to settle the townships along the Bay of Fundy, in Annapolis, Kings, Hants, Colchester, and Cumberland counties, entered at once upon the rich marshes prepared to their hands by the French Acadians. But the first settlers in Pictou came to a country covered with heavy forest, without an acre cleared, and after a little were thrown on their own resources. We must admire the heroism with which they entered upon their work, the energy with which they so bravely combatted the difficulties in their path, and the perseverance by which they at length happily surmounted them.

The results for the first six years, it will be seen, were

very small. The great cause of this was the unfortunate position of the Philadelphia grant, in having within its bounds no place in the harbour suitable for a town, and so little frontage on the shore. This spoiled the efforts honestly made for the settlement of the place, and frustrated, as we shall see, the next great effort made toward that end in 1773. In that year came the ship Hector with emigrants, mostly from the Highlands of Scotland. From that the effective settlement of the place may be dated. The event was important and deserves commemoration. But the first honor is due to the little company who arrived here previously, who cut down the first trees, erected the first huts, run the first lines, cleared the first land, and planted the first seed—the little band of pioneers who, in their little brig, with its well omened name, the Hope, first planted the standard of British colonization upon our northern coast; and the true natal day of Pictou is the 10th of June, when she first dropped anchor in our harbour, or the 11th, when her precious cargo first set foot on our shores.

> "What noble courage must their hearts have fired,
> How great the ardour which their souls inspired,
> Who leaving far beyond their native plain
> Have sought a home beyond the western main;
> And braved the perils of the stormy seas
> In search of wealth, of freedom, and of ease.
> Oh, none can tell, but they who sadly share,
> The bosom's anguish, and its wild despair,
> What dire distress awaits the hardy bands,
> That venture first on bleak and desert lands;
> How great the pain, the danger and the toil
> Which mark the first rude culture of the soil.
> When looking round, the lonely settler sees
> His home amid a wilderness of trees;
> How sinks his heart in those deep solitudes,
> Where not a voice upon his ear intrudes;
> Where solemn silence all the waste pervades,
> Heightening the horror of its gloomy shades;
> Save where the sturdy woodman's strokes resound
> That strew the fallen forest on the ground."
> *Rising Village*, By H. GOLDSMITH.

CHAPTER V.

ARRIVAL OF THE SHIP HECTOR AND SETTLEMENT OF
HER PASSENGERS—1773-1776.

Some of the shares of the Philadelphia company were transferred, so that the celebrated Dr. Witherspoon became one of the proprietors,* and John Pagan, a merchant of Greenock, became the purchaser of three undivided shares. They seem at that time to have been combined in promoting the settlement of the old colonies. We find the ship Hector, which was owned by Pagan, in the year 1770 arriving in Boston with Scottish emigrants, and there is a deed on record in the Pictou registry office, after the American Revolutionary war, from Witherspoon, conveying to Pagan all the land of the former in Pictou, in exchange for the lands of the latter in the United States. †

To carry out the original obligations of their grant, the proprietors offered liberal terms for the settlement of it. They employed an agent named John Ross, with whom they agreed to give each settler that he might bring from Scotland, a free passage, a farm lot, and a year's provisions. Ross went to the Highlands, and, drawing a glowing picture of the land and the advantages offered, many, knowing nothing of the difficulties of settling a new country, and allured by the prospect of *owning* a

---

* Dr. Witherspoon is sometimes represented as the projector of the Philadelphia company's scheme, but his name does not appear among the first members of the company.

† In a petition against the escheat of the grant, Pagan's son alleges that the father and sons sent out altogether about 800 souls; that they had spent £280 sterling in provisions, and altogether had expended about £600 sterling in settling the grant.

farm, eagerly embraced his proposals. The Hector was chartered to convey them to Pictou. She was under the command of John Spears as master, James Orr being first mate, and John Anderson second. Three families and five young men embarked in her at Greenock, whence she sailed for Loch Broom, in Ross-shire, where she received the rest of her passengers, amounting in all to thirty-three families and twenty-five unmarried men, beside the agent. The number of souls is stated in one account as 189, in another as 179, while Governor Legge, on their arrival, speaks of them in round numbers as 200.

In the beginning of July, 1773, they finally bade adieu to their native land.* As they were leaving, a piper came on board, who had not engaged his passage. The captain ordered him ashore, when the passengers interceded, offering to share their rations with him in exchange for his music. At their request, he was allowed to remain. There was not one person on board who had ever crossed the Atlantic, except one sailor. Though hearts doubtless were saddened as they parted from kindred, and as their native hills faded from their vision, yet hope beat high in every bosom, and for a time all went cheerily among the pilgrims. Song, music, wrestling, dancing, and other amusements relieved the tedium of a sea voyage. But the passage was destined to be a long and painful one. The Hector was an old Dutch ship, and a dull sailor. Passengers said that they could with their hands pick the rotten wood out of her sides. When they arrived off the coast of Newfoundland, they met with a severe gale, which drove them so far back that they were a fortnight before they were again as far forward. The accommodations on board were poor and the provisions of inferior quality, perhaps not worse than

---

* Philo Antiq. says on the 10th July, but the universal statement among the old settlers is, that they were eleven weeks on the passage, which would make the date of sailing 1st of July.

in emigrant vessels of the time. Small-pox and dysentery broke out on board, so that eighteen, most of them children, died on the passage and were committed to the deep. The former disease was brought on board by a mother and child, both of whom afterward lived to a great age. And one child was born, afterward the late Mrs. Page, of Truro. As the voyage was prolonged, their stock of provisions and water became low. For some time before arrival, they were put on an allowance of water, the scarcity of which, with the salt provisions, was a great privation. During the voyage the oatcake supplied to the passengers became mouldy, and the passengers often threw away pieces of it or other food. Hugh McLeod was in the habit of gathering up all these fragments and putting them into a large sack, and the last two days of their voyage, they were glad to avail themselves of this refuse food.

At length all the troubles and dangers of the voyage were surmounted, and on the 15th of September, this pioneer band of Scottish emigrants arrived in the harbour of Pictou, and the Hector dropped anchor opposite where the town of Pictou now stands. Previous to her arrival, as we have seen, the Indians had been somewhat troublesome to the settlers; if not positively dangerous, they at least gave annoyance, and the whites, from their small numbers, were kept in considerable alarm. It was even reported, that there was a plot among them at that time to cut off the whole settlement, which we have seen was the only one on the north shore of the Province. When the word was received of the coming of the Hector with Highland emigrants, the whites, in reply to threats of the Indians, told them that the Highlanders were coming— the same men they had seen in petticoats at the taking of Quebec. Sure enough the Hector appeared. Her sides being painted, according to the old fashion, in imitation of gunports, helped to induce the impression that she was a man-of-war. The Highland dress was then proscribed, but was carefully preserved and fondly cherished by the Highlanders, and in honour of the occasion the young men had arrayed themselves in their kilts, with *skein dhu*, and some with broadswords. As she dropped anchor the piper blew his pipes to their utmost power; its thrilling sounds then first startling the echoes among the silent solitudes of our forest. All the Micmacs fled in terror and were not seen for some time, so that trouble with the Indians was never heard of again.*

We may here remark the importance of the arrival of the Hector to these Lower Provinces. With her passengers may be said to have commenced the really effective settlement of Pictou. But this was not all: the Hector was the first emigrant vessel from Scotland to Pictou, or even these Lower Provinces. That stream of Scottish immigration which, in after years, flowed, not only over the county of Pictou, but over much of the eastern part of the Province, Cape Breton, Prince Edward Island, portions of New Brunswick, and even the Upper Provinces, began with this voyage, and even, in a large measure, originated with it, for it was by the representations of those on board to their friends, that others followed, and so the stream deepened and widened in succeeding years. We venture to say that there is no one element in the population of these Lower Provinces, upon which their social, moral and religious condition has depended more than upon its Scottish immigrants, and of these that band in the Hector were the pioneers and vanguard. We may mention here that after returning to Scotland from this voyage she was condemned, and went to sea no more. Truly her work was done.

The first care was to provide for the sick. One woman,

---

* We do not believe that there was really any plot of the kind. It was good fun for the Indians to frighten the white people, and we believe they raised the reports for that purpose.

wife of Hugh McLeod, afterward of West River, had just died of smallpox; the body was sent ashore in a boat and buried, we believe, at the burying ground already mentioned. Several were sick; some dying. The resident settlers did what was in their power to provide for their wants, and with the supply of fresh provisions most soon recovered their health, though some, I cannot learn how many, died on board the vessel.

If the expectations of these people had been excited by the prospect of an estate in America, their hopes were lowered by the sight that met their view as they crowded on the deck of the vessel to see their future home. One unbroken forest still covered the whole land, with the exception of a few patches on the shore between Browns Point and the head of the harbour.

But if the first view of matters was discouraging, worse was in store for them. Squire Patterson and Dr. Harris, the agents of the company, lived near Browns Point, nearly a mile above the town, and had erected a small store, in which they kept the supplies of the company, though even there the woods were scarcely broken. Here the immigrants were landed without provisions and without shelter, except as with the assistance of those here before them, they erected rude camps for themselves, their wives and their little ones. However glad to be relieved from the confinement of shipboard, bitter were the feelings of disappointment, with which they contrasted the expectations they had entertained of a free farm and plenty in America, with the reality before them. We need not wonder to hear of some sitting down and giving way to bitter weeping. The arrival of such a number swept the place like a torrent of all the provisions it contained and left it nearly destitute. The few settlers previously here could not have provided food for one-third of the number for any time, and it was too late in the season to raise any crops that year.

In the meantime they began to select their future homes. The company had the land laid out in regular blocks, named A, B, &c., which were subdivided into lots regularly numbered. But here the fact to which we have already adverted, of the small frontage of their grant on the harbour, spoiled the whole of this well meant and not ill contrived effort at settlement. When the Hector arrived, all the shore of their grant was occupied, and her passengers were taken back one, two or three miles,* and there, amid the primeval forest they were invited to settle. Never did there seem to be offered to men such an utter mockery. The gigantic trees would have seemed to any person a serious difficulty in their way, but to men unaccustomed to clearing the wood in America, and unskilled in the use of the axe, the work seemed hopeless. Without roads or even paths, and unprovided with compasses, they were liable to be lost in the forest and they were afraid of Indians and wild beasts. Even if these difficulties could have been removed, they saw that they would be shut out there, from what must hereafter form a large part of their subsistence, viz.: the fish in the harbour and rivers. We have heard of McCabe taking some of them back to where he promised to show them good land, which they might take up, when, looking round on the big trees, they only asked, with an air of helplessness, that he would take them back to the shore.

In consequence of these circumstances they all refused to settle on the company's land, and when a supply of provisions arrived the agents refused to give them any. A jealousy arose between them and the American settlers. Ross and the company quarrelled. They refused his demands and soon after he abandoned the passengers he had brought out. A few who had a little money,

---

* Each division of lots was a mile in length from the shore, and so the lots on the north side of the harbour still are, the side lines running north and south by compass.

bought provisions for a time or even exchanged clothes for food, but the majority had absolutely nothing to buy with; and the little that the others had was soon exhausted, so that they were left without provisions and entirely destitute of means to provide for themselves.

Driven to extremity they insisted on having the supplies sent by the company. On one occasion Donald McDonald and Colin Douglas were in the store claiming a supply; being rather pressing the agent ordered them out; they refused to go, when he threatened to lock them in. As they still refused he went out and attempted to lock the door, when Donald drew his dirk, an article which many of them then wore, and drove it in before the bolt. Finally they resolved to take the provisions by force. They seized both Squire Patterson and Dr. Harris, tied them and took their guns, which they hid at some distance, told them that they must have the food for their families, that they were willing to pay for them when they were able. They then proceeded to weigh and measure the various articles; they took account of what each man received, which they left. Roderick McKay, father of our late custos, a man of great energy and determination, and who in this and all the proceedings of the time, was recognized by the Highlanders as their leader, was left to release the prisoners. After a sufficient time had elapsed to enable the rest to get to a safe distance, he undid the ropes by which they were tied, and having informed them where their guns would be found, got out of the way himself.

Intelligence was despatched to Halifax, that the Highlanders were in rebellion, with a request for assistance. We may suppose that at a time when the scenes of "the forty-five" were still fresh in memory, this was heard with dismay. Report says that orders were despatched to one of the Archibalds of Truro, usually known as Captain Tom, or Uncle Tom, to march his company of militia to

Pictou to suppress the rebellion. He received the order with the most unmilitary reply, "I will do no such thing; I know the Highlanders, and if they are fairly treated there will be no trouble with them." Representations of the true state of the case were sent to Halifax. Lord William Campbell, whose term of service, as Governor, had just expired, was still there, and interested himself on behalf of the immigrants as his countrymen, so that orders came from the Government to let them have the provisions. We may add here, tnat Squire Patterson used to say afterward, that the Highlanders, who had arrived in poverty, had paid him every farthing that he had trusted them, but he had lost two hundred pounds by his good friend the Governor of Prince Edward Island.

Beset with such difficulties and with winter approaching, the majority of the immigrants removed to Truro and places adjacent, to obtain by their labour food for their families. A few settled at Londonderry, at a place which has since been known as Highland Village. Some went to Halifax, and some even to Windsor and Cornwallis. Not only men, but mothers of families, hired out, and their children, male and female, they bound out for service, till they should come of age. Some went that season, and others not till the spring following. One man stayed till the musquitoes made their appearance in the following summer, when, thinking it a judgment, he left. The majority of them, however, returned in subsequent years. / The number who remained is stated at seventy, and for a time, particularly during the following winter, they endured almost incredible hardships. Not having taken up land, they remained at Browns Point, with only rude huts, covered with branches or the bark of trees to shelter them from the cold, of the severity of which they had previously no conception. To obtain food for their families [they had to proceed to Truro, through a trackless forest and in deep snow, and there obtaining a bushel or two of potatoes, and perhaps a little flour, in exchange for their labour, they had to return, carrying this little supply on their backs or dragging it on a handsled.

The labor of this we can scarcely estimate. One bushel of potatoes was a sufficient load for a man to carry that distance. One who boasted of his strength undertook to carry two and started off with his load quite jauntily. The Highlanders have a Gaelic proverb, that a sheep the first mile will be a cow the second, meaning that a burden which a man can carry easily a short distance will be intolerable afterward; and so this man found, for before he had reached half way to Pictou he was glad to get quit of part of his load. Then there was the climbing of stiff braes or the descending steep banks, the crossing of brooks on a single tree, or the sinking in wet or boggy ground, or in winter in deep snow; this continuing for three days, involving two nights camping in the woods. Even the potatoes they did get were inferior, being of a kind known as Spanish potatoes, large and soft, like a kind known some years ago as yams, or like some of the coarse kinds still used for feeding cattle. Sometimes they froze on their backs, but even so, when roasted in the ashes or sliced and roasted on the coals, they were heartily relished. No wonder that some of those who had gone through these scenes, could not bear in after life to see even the peeling of a potatoe thrown into the fire.

Perhaps, however, a better idea of their privations may be gained by giving a few incidents of this period. Two young men set off for Halifax. They could get so little provisions when they left, and had so little on the way, that they were scarcely able to travel from weakness, and when they reached Gays River they were nearly ready to give up altogether. But there they saw a lot of fine trout, strung on a rod, hung on a bush. They hesitated whether to take them or not. They thought they belonged to the Indians, who they feared would come after them and kill them. They, therefore, left them and went on a short distance, when, finding that from their weakness they would not be able to prosecute their journey, they returned to where they had left the trout. Each put it upon the other to take them. At last the claims of hunger prevailed and they proceeded to make a meal of them. They afterward discovered that they had been caught by two sportsmen from Halifax, who had disputed who should carry them, and finally left them, where, in the kind providence of God, they afforded a meal for the hungry travellers.

The late Alex. Fraser, elder, of Middle River, when only a lad of about sixteen, carried a younger sister to Truro on his back, while the only food he had for the whole journey was the tail of an eel.

One or two incidents of this family, though at a somewhat later date, may be given. Hugh, a younger brother of the last, and who was one of the last survivors of the Hector, told the writer that on one occasion his father, having exhausted every other means of obtaining a supply of food for his family, cut down a birch tree and boiled the buds, which he gave them to eat. He then went to a heap where Horton, one of the old settlers, had buried some potatoes, and took out some. Before he could inform the owner of what he had done, some of his neighbors maliciously did so, when Horton merely replied that he thanked God he had them there for the poor old man's family.

On one occasion, when the husband and eldest son had gone to Truro for provisions, everything in the shape of food for the younger children was exhausted, except one hen, which the mother finally killed. She boiled it in salt water for the benefit of the salt, with a quantity of weeds or herbs, which she had collected, and of the nature of which she was entirely ignorant, which she served up for them with the flesh of the hen. But not long after the children found the hen's nest with ten eggs, some of which she cooked for their next meal and the rest she retained till her husband's return.

On another occasion, the men of the family had brought home a supply of potatoes, from Truro, for seed, but after planting them and enclosing the ground, they were so much in want before going back, that they had to dig up some of the splits to use for food. Some time after, having earned as much money as would buy a cow, Alexander was sent to Colchester to make the purchase; but having fallen in with his brother Simon, who had been bound out, and finding him dissatisfied, he applied the money to the purchase of his time. On arriving home, on his mother meeting him, her first enquiry was "Have you got the cow?" "No, but I have brought Simon instead," was the reply. "Well, poor as I am," said the mother, "I would rather see Simon than the cow." The girl whom this same Alexander afterward married, was bound out in Truro, and served till she was eighteen years of age.

These few incidents, most of them in the history of one family, and that one of the few which had arrived with some means, will give an idea of what they endured for the first few years. All were in the same condition, and none could help another. The remembrance of those days sunk deep into the minds of that generation, and long after, the narration, of the scenes through which they had passed, beguiled many a winter eve, as they sat by their, now, comfortable firesides.

To return to our narrative. That winter the first death occurred among the immigrants, a child of Donald McDonald, who was buried at John Patterson's (second) place, already mentioned; and the first birth occurred, a son of Alexander Fraser, afterward of Middle River, named David, afterward Captain Fraser, who lived at what is now Evans' place, about five miles from town.

In the following spring they applied themselves earnestly to provide for the wants of themselves and their families. Though unaccustomed to the use of the axe and the employments peculiar to a new country, yet, except in this respect, never were immigrants better adapted to the work of settling in the wilderness. They were children of the mountain and the flood. They were accustomed to coarse food, inured to hardship in its roughest form, and were not easily dismayed by difficulties. " They accordingly exerted every energy, and sought out suitable spots on which to settle. In their enquiries after these, they were enabled to judge of the virgin mould from the growth and species of wood. Where high and bulky black birch, ash, rock maple, elm or oak was discovered, the land was accounted to be of a strong and superior kind. They explored the different rivers, which abounded with fish; and finding the soil near their banks to be the most fertile, and capable of being more easily improved than that of higher lands, they seated themselves upon it." *

Difficulties were thrown in the way of their getting their grant, principally, we presume, through the opposition of the agents of the Philadelphia Company, by whom they had been brought out. The first grant was to Donald Cameron, who had been a soldier in the Fraser Highlanders at the taking of Quebec. His lot was situated at the Albion Mines, being the same lot afterward purchased by Dr. McGregor. It is dated 8th February, 1775, and beside the condition of payment of quit rent, as in the other grants, contains the following:—

"That the grantee, his heirs or assigns, shall clear and work, within three years, three acres for every fifty granted, in that part of the land which he shall judge most convenient and advantageous,[or clear and drain three acres

* Philo Antiquarius.

of swampy or sunken ground, or drain three acres of marsh, if any such be within the bounds of this grant, or put and keep on his lands, within three years from the date hereof, three neat cattle, to be continued upon the land until three acres for every fifty be fully cleared and improved.

"But if no part of the said tract be fit for present cultivation, without manuring and improving the same, then this grantee, his heirs and assigns shall be obliged, within three years from the date hereof, to erect on some part of said land a dwelling house, to contain twenty feet in length by sixteen feet in breadth, and to put on said land three neat cattle for every fifty acres, or if the said grantee, his heirs or assigns, shall, within three years after the passing of this grant, begin to employ thereon, and so to continue to work for three years then next ensuing, in digging any stone quarry or any other mine, one good and able hand for every 100 acres of such tract, it shall be accounted a sufficient seeding, planting, cultivation and improvement, and every three acres which shall be cleared and worked as aforesaid ; and every three acres which shall be cleared and drained as aforesaid, shall be accounted a sufficient seeding, planting, cultivation and improvement, to save for ever from forfeiture fifty acres in every part of the tract hereby granted."

The rest of the Hector passengers, who remained in Pictou, occupied land on the three rivers, especially the intervales, on what had been McNutt's grant, which was now escheated. They did not, however, obtain a title to it for some time. As late as the 22nd January, 1781, they complained, in a petition to the government, that a grant had been often promised but never received. At last it was issued on the 26th of August, 1783. It contains the names of forty-four persons, some of whom had arrived from other quarters after the Hector, conveying the lots on which they had been located, the size of the lots being regulated by the number of their families. The conditions were the same as in Cameron's grant, and the mines reserved are gold, silver, lead, copper and coals. We append a list of the grantees with the number of acres received by each and notices of the situation of their lots. (Appendix A.)

In the meantime they were energetically using the means in their power to supply the wants of their families. They learned to hunt moose. Timber of the finest quality abounded, and they soon could split staves or the long shingles formerly mentioned, with their neighbours. Small vessels came from the old colonies, which supplied them with necessaries in exchange for these articles.

Seeing the majestic trees on every side, and knowing the value of timber in Great Britain, they formed the idea of preparing a quantity for exportation. Unskilled in the use of the axe, they invited a company of hewers from Truro, and with their aid prepared, during the summer, a sufficient quantity of squared pine to load a vessel, which had been condemned in Prince Edward Island and purchased by Governor Patterson. This was the first timber ever shipped from Pictou, and the commencement of that wood trade afterward carried on so extensively from this port.

It is just to say that the Indians, as soon as the mutual terror had subsided, treated them with much kindness. From them they learned to make and use snow-shoes, to call moose, and other arts of forest life. From them they often received supplies of provisions. One old man used to say that the sweetest meal he ever ate was provided and prepared by them. Hunger, we presume, was the sauce. The Indians were indeed sometimes disposed to make use of the terror which they knew their name and appearance inspired, particularly among the weaker sex, to secure their object; but it is due to that unhappy race to say, that from the time of the arrival of the Hector, they never gave the settlers any serious molestation, and generally showed them real kindness, which, when the tables were turned, so that the whites had plenty and they were needy, has not always been reciprocated.

During that summer they also prepared to occupy the land which they had selected, but could get little, if any, ready for crop that season, and in the fall the majority, even of those who had remained, disheartened at the prospect of another such winter as the past, left for Colchester or other places. By a return made on the 1st

January following (1775), the following were the families and unmarried men on the settlement at that date :

FAMILIES.—John Rogers, Robert Patterson, William McKenzie, Alex. Ross, Kenneth McClutcheon, Wm. McCracken, Abram Slater, Moses Blaisdell, Wm. Kennedy, Colin McKenzie, James McCabe, James Davidson, Bar. McGee.

UNMARRIED MEN.—John Hall, John Patterson, George McConnell, Joseph Richards, James Hathorne, Thomas Troop.

The whole population consisted of 23 men, 14 women, 21 boys, and 20 girls; total, 78. The produce raised in that year was 269 bushels wheat, 13 of rye, 56 of peas, 36 of barley, 100 of oats, and 340 lbs. of flax. The farm stock consisted of 13 oxen, 13 cows, 15 young neat cattle, 25 sheep, and 1 swine. There were manufactured 17,000 feet of boards, and Squire Patterson was the owner of a sloop or schooner.

Of the above list only five or six were Hector passengers. The return seems imperfect. At all events, quite a number returned the following season (1775). As the law of the Province then allowed a representative to each township having 50 families, we find a return in that year by Dr. Harris, showing that Pictou contained the required number. (See Appendix B.)

That year their circumstances continued to improve and some crop was raised, though not sufficient for their subsistence; and still there were the same weary journeys to Truro for necessaries. They were, however, acquiring more skill in availing themselves of the resources around them. The moose afforded them a supply of meat for the winter, and the rivers plentifully supplied them with fish, and they learned to make sugar from the juice of the maple. One mode of laying up a supply of food for winter was, to dig a large quantity of clams in the autumn, pile them in a heap on the shore, and then cover them with sand, though they were sometimes in winter obliged to cut through ice a foot or more in thickness to get at them.

We give in the appendix a list of the Hector passengers, with notices of their places of settlement, and history, so far as known. (See Appendix C.)

Though still poor enough, they were provided with at least the necessaries of life, when they were again tried by the arrival of a class poorer than themselves. Inducements having been held out by some of the proprietors of Prince Edward Island (then called St. John) to parties in Scotland to settle their land, John Smith and Wellwood Waugh, then resident in Lockerby, in Dumfriesshire, sold out their property, and chartered a small vessel to carry thither their families and any others who might join them. They accordingly arrived at Georgetown, or Three Rivers, in the year 1774, and were followed by others a few months later.

They commenced a settlement with fair prospects of success, when their hopes were blighted by a remarkable visitation. Diereville, a French writer, in a work published in 1699, says :—" The Island of St. John is stated to be visited every seven years by swarms of locusts or field mice, alternately—never together. After they ravage the land, they precipitate themselves into the sea." There is no evidence of any such regularity in this visitation of mice, but later writers speak of it as recurring on the Island at longer or shorter intervals, and there was one of the kind some years later in Nova Scotia, though now it is unknown. At all events, it came upon the new settlers, to whom we have referred, in full force. These animals swarmed everywhere, and consumed everything eatable, even the potatoes in the ground.*

The new settlers would have had difficulties enough under any crcumstances, but this filled their cup to the brim, and during the eighteen months that they remained there, they endured all the miseries of famine. For three

---

* In some houses at West River are still preserved books of which the leather on the covers has been gnawed by them.

months in summer, they subsisted on lobsters and other shell fish, which they gathered on the shore. In the spring they had obtained from Tatamagouche a few potatoes for seed, but the mice devoured them in the ground, and everything else in the shape of crop, so that when winter came, they were on the verge of starvation. An old woman in my congregation, though a strong child and with a constitution which carried her to ninety years of age, told me that when she was two years of age, she was not able to walk from weakness, owing to want of food. One boy died, it is supposed from eating some herbs which were injurious or poisonous. Waugh had brought a supply of provisions and other articles, so that the first summer they did not suffer much, but at the end of the second season, he had all his goods in the store of a man named Brine, who traded with the small fishing vessels from the colonies. A number of these vessels happened to be in the harbour, and before returning home the crews came ashore for a carousal. The American Revolution was just commencing, and they were leaving with the idea of not returning, expecting when they reached their homes to serve either as soldiers or sailors. Before going on board they plundered Brine's warehouse of all it contained, carrying off all Waugh's property.

That winter they would have perished, were it not for a French settlement some miles distant, from which they received supplies, principally of potatoes, in exchange for the clothing they had brought with them from Scotland, until they scarcely retained sufficient to clothe themselves decently. From scarcity of food the men became reduced to such a state of weakness, and the snow was so deep, that they became at last scarcely able to carry back provisions for their families, and when, with slow steps and heavy labour, they brought them home, such was the state of weakness in which they had left their children, they trembled to enter their dwelling, lest they should find them dead, and sometimes waited at the door, listening for any sound that might indicate that they were alive.*

Having heard that there was food in Pictou, they, in spring (1776), sent one of their number (the late David Stewart) to enquire into the state of matters there. Some of the American settlers had brought slaves, one of whom had been sold in Truro by his owner, who brought home part at least of the proceeds in wheat, which he was consuming in his family when Stewart arrived and lodged in his house. The latter, amid all his troubles, retained some measure of cheerfulness, and on his return his friends gathered round him to hear his report. " Well, what sort of a place is Pictou ?" was the enquiry. " Oh, an awful place," was the reply, in a very solemn tone. " How ?" it was again asked. He replied, " I stayed with a man who was just eating the last of his nigger." Such was their own condition on the verge of starvation, that for a minute they actually supposed that the people of Pictou were reduced to such a state from hunger as to have devoured the flesh of their colored servants.

Having explained the true state of the case, his report was on the whole so favorable, that they were glad to exchange total want in Prince Edward Island for the partial supply to be found in Pictou. About fifteen families accordingly moved over, of whom seven settled on the West River. When they arrived, the only break in the woods on the west side of the West River was

---

* One old woman, living in 1831, used to tell that for three months her children had neither bread nor potatoes. During that time their food was principally shell-fish and boiled beech leaves. One calamity she described as having tried them severely. They had brought with them iron pots, but not knowing the severity of the frost in this country, had left water in them, by the freezing of which they were cracked. In their circumstances, believing that they could not obtain others nearer than Scotland, and seeing no hope of obtaining them there, she said that the loss was next to the loss of a child.

where the Rev. George Roddick now resides. Four settled on the Middle River and two on the East River. The John Smith who came with Waugh removed to Truro. He first visited that place to have his child baptized and to hear the gospel, camping in the woods between Pictou and that place. In the fall he brought over part of his movables, carrying a large two cwt. anvil to Truro on a horse, which he hired from Squire Patterson. Wellwood Waugh settled on what has since been known as Dunoons farm. He used to tell that he left the Island with only a bucket of clams for the support of himself and family; that the day after his arrival in Pictou he went to the woods to make staves, and was able to make a living for them ever after. His stepbrother, William Campbell, then a young man, who came with him, settled on the farm next above.

Though the Highlanders were ready to extend their wonted hospitality to the new comers, and did so, to the best of their ability, yet, having barely sufficient for the support of their own families, such an influx pressed heavily upon them. Though these people arrived here in such destitution, they were among the most valuable of the early settlers of this country, and their descendants to this day are among the most respectable members of the community.

We give in an appendix a list of these settlers, with notices of their places of settlement. (Appendix D.)

CHAPTER VI

PICTOU DURING AMERICAN REVOLUTIONARY WAR.
1776—1783.

The breaking out of the American Revolutionary War at first subjected the settlers to serious inconvenience. They had hitherto received most of their supplies by trading vessels from the Old Colonies, which received in exchange the proceeds of their labour, especially fish, fur and lumber. This trade, however, was now stopped, and the want of it was at first severely felt. Even salt could not be obtained, and in summer the settlers might be seen for days boiling down sea water to obtain a supply of this necessary. But the war soon had an enlivening influence upon the trade of the Province. Halifax was chosen as the chief depot for the British Navy in this Hemisphere. Large sums of money were expended on the dockyard; vessels of all classes were there annually refitted, and employment was given to artizans. A large military force was kept at Halifax, and there was, in consequence, a larger circulation of money, in the advantage of which the country districts shared.

The following is given as the price received by the settlers for their wood :—

	1775.	1776.
Squared Pine, per ton	9s.	12s. 6d.
Hardwood, "	18s.	20s.
Barrel staves, per M	25s.	50s.
Hhd. " "	35s.	70s.

The settlers in Pictou were for a time, however, still at a loss for British goods, but in the year 1779, John Patterson went to Scotland and brought a supply, and from that time continued to trade.

But the American war had another effect, in the division which it occasioned between the new and the old settlers. The Scotch were loyally attached to the British Government. But, with the exception of Squire Patterson, most of the American settlers strongly sympathised with the American cause. Murdoch, in his history of Nova Scotia, tries to make it appear that those who came to this Province from the Old Colonies, and settled various townships before the American Revolutionary War, were at this time loyal. From the facts that have come to our knowledge regarding these people in Colchester, and the few settlers in Pictou, we can assert most positively that they generally sympathized with the Americans, and that a number were ready to manifest their sympathy by taking arms, if there had appeared a favourable prospect of thereby serving the cause. And when this seemed hopeless, they manifested their spirit in more harmless ways, as in the refusal of tea, of which the good wives could sometimes only secretly brew a small quantity for private indulgence, and more permanently in the names, which their children have carried down to our own day—the Adamses, the Burkes, and the George Washingtons, the latter of which, however, it was found more convenient to change into John Washington or George William.

In Pictou, it will surprise many of the present generation to hear, the feeling was quite violent. A circular was addressed to the magistrates throughout the Province, requiring them to be "watchful and attentive to the behaviour of the people in your county, and that you will apprehend any person or persons who shall be guilty of any opposition to the King's authority and Government, and send them properly guarded to Halifax." The inhabitants were ordered to take the oath of allegiance, and magistrates were required to furnish lists of those who complied and those who did not.

Patterson, who had been made a magistrate in 1774, was

active and zealous, perhaps more sc than wise, in carrying out these instructions. He started for Halifax, intending to get copies of the oath required, for the purpose of imposing it upon the inhabitants. When he reached Truro, his purpose becoming known, one of the Archibalds invited him to his house, and took him to a private room where, drawing out a pistol, by its persuasive influence, he induced him to return home.

The squire also attempted to arrest some of the old settlers, who had openly declared their determination to swear no oath of allegiance, while the others endeavoured to conceal them. We have heard, for example, of Horton being obliged to hide under a haystack. On the other hand, their passions became so excited that they threatened to murder him. So serious did the danger become that his older sons were obliged several times to hide him in the woods, taking him over to Frasers Point for the purpose.* Matthew Harris, having had some dispute with Squire Patterson, regarding some business in which they had been engaged together in Maryland, started thither in the heat of the American war. While in Halifax the circumstance exciting suspicion, or perhaps, in consequence of his giving too free expression to his sympathy for the American cause, he was arrested as a spy, and placed in the care of a guard of soldiers, who went into an inner room of a tavern to drink, leaving him to move about on his parole. While he was calmly walking on the platform, a woman rushed in where they were, exclaiming, "Your prisoner is escaping." They rushed out, half intoxicated, and one of them struck him over the head with a weapon he had in his hands, cutting him very severely. He was detained in custody till evidence was obtained from Pictou that he was a peaceful

---

* We had heard of this, but regarded it as an exaggeration. The family of James, one of his sons referred to, asscits positively that their father froquently mentioned it to them as a fact.

resident. The old man was Christian enough to say, in after life, that he could forgive everybody except that woman.

A few incidents connected with the war, as affecting the County of Pictou, may here be given. The first was the capture of a vessel at Merigomish by an American privateer, which took place near the beginning of the war, probably in the spring of 1776. She was not a large vessel, but was loaded with a valuable cargo of West India produce. The previous fall she had been on her way to Quebec, but being too late to get up the St. Lawrence, she made Merigomish harbour, where she remained in the ice all winter. The captain and crew landed, and from the scarcity of provisions, some of the latter went to Truro or Halifax. One of the settlers, named Earl, went off, it was supposed to the States, and with the design of giving information which might lead to her capture. At all events, early in spring, as soon as the gulf was clear of ice from the Strait of Canso, a vessel appeared off Merigomish. Those in charge of the vessel in the harbour, suspected her purpose, and commenced conveying to the shore and hiding in the woods articles of value that could easily be removed. Soon, however, parties from the strange vessel came on board and took possession of her. James and David Patterson had been making oak staves on the land near where she lay. The captors, to prevent the word circulating, or any attempt to frustrate their purpose, sent a boat on shore, with a crew, who seized them and carried them on board their vessel, where they were put in irons. The captors then set to work to get the vessel to sea. When they got her well out into the gulf they released the two brothers. There was some difficulty in unloosing the handcuffs on David's hands, when one of the men struck it with a marlin spike to break it, and in so doing smashed his thumb, which bore evidence of the fact till his death. They then put the

two brothers into a small boat with a few biscuits and a
small earthen jar, called a coggie, of sugar, to find their
way back to port as best they might. In the meantime
word had circulated of the capture, and as it was expected
the privateer would come to the harbour, the inhabitants
collected with every old musket and fowling-piece, prepared to offer a sturdy resistance to the enemy. They
assembled at the Battery Hill and soon saw a small boat
coming up the harbour, which they eagerly watched, and
as it approached they saw in it two men, whom, as it
drew near, they recognized as the Pattersons, who had
thus made their way to port.

The next incident was the capture of Captain Lowden's
vessel in the harbour in 1777. Haliburton speaks of it as
effected by rebels from Machias, who came from Cumberland. The information I have gathered attributes the
work to the American settlers in Pictou, and some friends
in Truro. It is certain that they were in the plot. At all
events, the circumstances of the capture are as follows :—
The vessel was loading with timber for the British
market. A time was chosen when the crew were absent
with the boat for part of the cargo. The captain was
invited to the house of W. Waugh, where a number of
them were gathered. Waugh was an old Scotch
Covenanter, and from rigid adherence to the principles of
that body, would not swear allegiance to the British
Crown, and though afterward he was in the employment
of the Government, yet, at this time, seemingly from the
common fact of their not taking these oaths, sympathized
with the Americans. The Captain went without suspicion,
leaving the ship in charge of the mate. During his visit,
at a given signal, the company gathered round him,
informed him that he was a prisoner, and commanded him
to deliver up his arms. "Gentlemen," said he, "I am
very sorry to say I have no arms," was his reply, in a tone
of indignation at their treachery. In the meantime, a

strong party, fully armed, proceeded to the vessel, and finding scarcely any person on board, easily took possession of her, and made the mate a prisoner, confining him in the cabin. They then placed sentries on deck. Some time after, the rest of the crew came on board, and as they did so, they were made prisoners and confined in the forecastle.

Some of the captors then took a boat belonging to the ship and proceeded up the East River. On their way they met Roderick McKay and his brother Donald coming down the river with a boat-load of staves. They gave no hint of their object, but encouraged the McKays to proceed to the vessel. They then continued on their way to Roderick's place. He had erected a blacksmith's forge, and had it duly stocked. They plundered it of everything worth taking away, loading their boat with his tools, iron, &c. In the meantime, the McKays had proceeded to the vessel. As Roderick mounted the deck, he saw the sentries with their muskets on their shoulders, and before he could take in the situation, one of them tapped him on the shoulder, saying he was a prisoner. His reply was a tap on the face with the back of his hand. The sentry brought down his musket and told him he was serious. Roderick was obliged to yield, and both he and Donald were taken to the cabin as prisoners.

After some time the party who had gone up the East River returned, their boat laden with the plunder of Roderick's forge. They came on board, leaving the boat alongside, which afterward sank with its contents, and remains to this day beneath the waters of the harbor. They then proceeded to celebrate their success by a night of carousal. When they became pretty well under the influence of liquor, Roderick, with his usual determination, wished to take the ship and urged his brother Donald to join him in the attempt. His plan was that they should make a sudden rush up the cabin stairs to the deck; that he should seize the sentry and pitch him overboard, while Donald should with an axe stand over the companion and not allow any of them to come up. Donald, however, was a quiet, peaceable man and refused to join in a scheme involving the danger of bloodshed, and Roderick could not communicate with the mate. He was deeply disappointed and used to say that if the mate had had two words of Gaelic, they would have retaken the ship that night.

The McKays were soon set at liberty, and the captors, anxious at once to secure their prize, sailed as soon as they could for Bay Verte, where the Americans for a time had possession, taking Dr. Harris, under a certain kind of compulsion, with the mate and part of the crew, to navigate the vessel. Information of these proceedings was immediately sent to Halifax, the late John Crockett and Colin Douglas being the messengers. They proceeded on foot to Shubenacadie, and finding the rivers very high and difficult to cross, they employed an Indian to proceed by the lakes and deliver the letter, which he did.

After the sailing of the vessel, Capt. Lowden was released and started for Charlottetown in a canoe. He found there a man-of-war, under the command of Lieut. Keppel, which immediately started in pursuit. In the meantime, the captors had reached Bay Verte, but finding that the American invaders had retired, they, on the approach of the man-of-war, abandoned the vessel and took to the woods, where it is supposed many of them perished. One reached the settlements in Colchester, after having eaten the upper leather of his boots, and died soon after. The mate took charge of the vessel and hailed the man-of-war as she was about to fire, when Capt. Lowden, who was on board the latter, knew his voice. The vessel was taken charge of by the commander, who came into the harbor of Pictou, threatening vengeance on all who had any share in the affair. All

Waugh's goods were seized and sold,* and such was the feeling against him amongst the old settlers, that he left the place and afterward settled at Waugh's River, Tatamagouche, to which he gave his name. It may be mentioned here, however, that not only did he afterward act the part of a loyal subject, but the communication between Halifax and Prince Edward Island being through Tatamagouche, he was employed by Government as their courrier between that place and Truro.

The affair of Capt. Lowden's vessel, I have no doubt, made the place too hot for the settlers, who sympathized with the American cause, and was one reason for their removal. Some whom I have been able to trace, moved eastward *without selling their farms*, and we may here mention an incident which occurred at this time. Matthew Harris embarked with his family in a vessel intending to remove to Guysborough. But while on their passage thither, they fell in with an American privateer. Those on board were unwilling to lower the British flag, when the privateer fired a shot ahead and another astern of her. Upon this, one of the men hauled down their colors and the vessel was brought to. The captain of the privateer came on board in great wrath. An infant child of Harris was sick and laid upon the deck, wrapt in a blanket. The captain struck the bundle with his sword, not knowing what was in it. The mother sprang forward, saying, "You have killed my child." The captain immediately calmed down, asking what the child was doing there, and shortly after left, taking only a few tubs of butter that were on deck.

During the war American privateers were on the coast, but had very little effect on Pictou. One of the Hector passengers, who had moved to Halifax and there earned

---

* Another tradition says that this was done by the officers and crew of the Malignant when in Pictou as hereafter mentioned. This may be correct.

some money, married and came to Pictou by land, but put all his things into a vessel to come round by water. She was captured and he lost his little all. One came into the harbour, and the alarm was given, and the settlers began to gather to repel the intruder, when one of the American settlers went out to her and urged that there were only in the place a few Scottish settlers commencing in the woods, not having anything worth taking away, and that all they could do was to burn Squire Patterson's house. In consequence of his representations they sailed, taking only a boat belonging to Waugh.

What excited the greatest alarm, however, during the war was a large gathering of Indians, it is said, from Miramichi to Cape Breton, probably a grand council of the whole Micmac tribe, which took place at Frasers Point in 1779. In that year some Indians at the former place, in the American interest, having plundered the inhabitants, a British man-of-war seized sixteen of them, of whom twelve were carried to Quebec as hostages and afterward brought to Halifax. This led to a grand gathering of the Indians. For several days they were assembled to the number of several hundreds and the design of the meeting was believed to be, to consult on the question of joining in the war against the English. To this they were probably instigated by French agents. The settlers were much alarmed, but the Indians dispersed quietly.

Another incident which excited some attention in Pictou at this time was the wreck of the Malignant, which took place near the close of the war. She was a man-of-war, bound to Quebec, and was wrecked late in the fall, at a place ever since known as Malignant Cove. The crew came to Pictou and were provided for through the winter by the efforts of Squire Patterson, as far as circumstances would permit.

To finish what we have to say here regarding the settlers in Pictou from the old colonies, we may here advert to another circumstance in connection with them. Some of those who came to Pictou, as well as other parts of the Province, had brought slaves with them, and as a curiosity of the time we shall insert here a copy of a document, which is on record in the office of the Registrar of Deeds in Truro:

Be it known to all men, that I, Matthew Harris, of Pictou, in his Majesties' Province of Nova Scotia, yeoman, have bargained and sold unto Matthew Archibald, of Truro, within said Province, tanner, and I do by these presents bargain, sell, alien, and forever make over to him, the said Matthew Archibald, his heirs and assigns, all the right, property, title or interest, I now have, or at any time hereafter can pretend to have, to one Negroe boy, named Abram, now about twelve years of age, who was born of my Negro slave in my house in Maryland, for and in consideration of the sum of fifty pounds, currency, to me in hand paid by the said Matthew Archibald, or secured to be paid, and I do by these *presence*, for myself, my heirs, and assigns for ever, quit claim to my Negroe boy, now in possession of said Matthew Archibald. In testimony of which I have to this bill of sale set my hand and seal, this 29 day of July, Anno Dom., 1779, in the 19th year of his Majesties' reign. Truro, County of Halifax.

MATTW. HARRIS.

Signed, Sealed, and Delivered
in presence of
DAVID ARCHIBALD, Js. Peace.

The following, however, which we find in the records of Pictou, is still more curious:

Know all men by these presents that I, Archibald Allardice, of the Province of Nova Scotia, mariner, for and in consideration of the sum of forty pounds currency to me in hand paid by Dr. John Harris, of Truro, have made over, and sold, and bargained, and by these presents do bargain, make over, and sell to the aforesaid Dr. John Harris, *one negro man named Sambo, aged twenty-five years or thereabouts, and also one brown mare, and her colt now sucking.* To have and to hold *the said negro man and mare with her colt, as his property,* for and in security of the above sum of money until paid with lawful interest. And at the payment of the above mentioned sum with interest and expenses, the aforesaid Doctor John Harris is by these presents firmly bound to deliver up to the aforesaid Archibald Allardice, the said negro man, named Sambo, with the mare and colt (casualties excepted). But if the said negro man, mare or colt, should die before the said money should be paid, then in such proportion, I, the said Archibald Allardice, promise to make good the deficiency to the said Doctor John Harris. In witness whereof I have hereunto set my hand

and seal, this tenth day of August, in the year of our Lord, one thousand seven hundred and eighty-six, and in the twenty-sixth of our Sovereign Lord, George the Third's Reign.

Archibald Allardice, L. S.

Signed, sealed, and delivered in presence of
James Phillips,
Robert Dunn.

Truro, August 26th, 1786, Recorded on the oath of James Phillips.

John Harris, D. R.

Along the margin the following words were written: "Assignment to Thomas Harris, 20th day of April, 1791."

per John Harris, D. R.

We have not heard of any cases of those in Pictou who owned slaves ill-treating them. On the contrary, a poor woman who had belonged to Matthew Harris, and obtained her freedom, used to confess that her life had never been so free from anxiety as when living with him; but in other places tradition has preserved the remembrance of some cruel deeds, showing the character of the system. We have heard, for example, of a negro slave in Truro, who was so treated by his master, that several times he ran away, usually making for Pictou. On one occasion his master having caught him, cut a hole through the lower lobe of his ear through which he passed the end of a whip lash, and knotting it, he mounted his horse and rode off, dragging after him in that way the poor man, who shortly after died, it was believed, in a large measure through the treatment he had received.

At this time the first settlement was made in Merigomish by Barnabas McGee. As we have already mentioned, he had first taken up land on Rogers Hill, but dissatisfied with its distance from the shore, he removed to Barneys River, which took its name from him. Here he settled in the fall of 1776 or spring of 1777; his daughter Mary, afterward Mrs. Gillies, the first child of English descent born in Merigomish, being born in May of the latter year.* The harbour and coast then swarmed with fish, particularly the salmon. The islands were visited by great flocks of geese and other wild fowl, while moose were plenty in the woods, so that he had no difficulty in providing at least flesh for his family. The Indians were then numerous, their chief place of encampment being on the west side of his farm, and his children, from want of associates, made playmates of their little Micmac neighbours.

He was soon after joined by George Morrison, who settled on the adjoining lot to the west. He had originally come in the Hector. He was a strong and determined man. On one occasion, being from home, a number of Indians came to his house, made his wife cook whatever they saw in the house that they desired, would not allow his children to the fire, and otherwise frightened the family. On his return, hearing of their behaviour, he immediately started in a rage for the Indian encampment, and meeting some of the offenders, he attacked them, in detail, with his fists, giving them a hearty drubbing as a hint for better behaviour in the future. The next day the whole band had decamped.

They were joined soon after by Walter Murray. He had been originally a soldier and had served in India, but had emigrated to Nova Scotia in the Hector. He first settled on the East River, but now removed to Merigomish, where he took up land on the east side of Barneys River, McGee taking him, with his family and household goods, in a boat round the coast. In commencing their labours, Murray and Morrison each carried a bushel of potatoes on their backs from Truro. They took the eyes out of them, for seed, with a knife or a quill, retaining the rest for food, so that, as they used to say, each planted his bushel and ate it.

* We may add that his son Charles, born the 24th November, 1778, was the first English male child born there. He died in the autumn of 1876.

The Rev. Mr. Cock, on one of his visits to Pictou, extended his journey to Merigomish, and preached the first sermon in the settlement, in Morrison's house, either in 1783 or 1784, probably the latter, and at the same time baptized all the young children.

At the period at which we have now arrived, the following may be regarded as a view of Pictou : A few settlers were thinly scattered along the north side of the harbor, from below the town to the head of the harbor, and on both sides of the West River, as far up as the late Deacon McLean's place. There was one family on Rogers Hill, three or four on the Middle River, and some others on the intervale of the East River from Stellarton nearly up to Fish Pools, and there were three families in Merigomish. Altogether, the population might be from 200 to 250.

We append a return to Government of the men capable of bearing arms, made at this date. (Appendix E.)

We may here give some account of the social condition of the inhabitants at this time. " The society of Pictou," says Philo Antiquarius, "down to the moment of which we are now treating, might be viewed as one family, where the children were all under the immediate superintendence of a good parent. One venerable settler had heretofore presided over the others, advising them to discharge their various duties, and impressing upon them the necessity of honesty, unanimity and industry, while they, with confidence, looked to him as their best director, and yielded in most cases obedience to his counsels." Squire Patterson, referred to in this extract, is described as short and thick-set, one of those men sometimes said to be as broad as they are long, with a free and pleasant manner, and was highly esteemed. From his skill in business he was very influential, indeed, a sort of factotum for all the settlers, even celebrating their marriages, notices of the same being posted up for three weeks as a substitute for the proclamation of banns.

Along with him we must notice "John Patterson, commonly known as Deacon Patterson, and, after his death, as the old Deacon, from the circumstance of his eldest son of the same name being also an Elder in the Church. He has been called the Father of the Town of Pictou, from his having been the means of fixing the town on its present site. But the old Deacon merited the title of Father of Pictou on other accounts. For many years after he came, there was neither law nor lawyers. In those happy times men took the Scriptural mode of settling disputes. They were not afraid to leave the adjustment of "the things that pertain to this life" to their conscientious neighbors. These two old patriarchs, the Squire and the Deacon, famed as they were for integrity and sound sense, became the general peace-makers. None dared or wished to gainsay their decisions. Generally when two men in any place are upon an equality, the disposition to be first, so universally distributed among men, creates feuds between them, and the public good is left in the back ground, and the public peace disturbed. The two good men of whom we are speaking formed an honorable exception from that common occurrence. They lived together, not merely on good terms, but a pattern of warm and inflexible friendship."*

The most of the Highlanders were very ignorant. Very few of them could read, and books were unknown among them. The Dumfries settlers were much more intelligent in religion and everything else. They had brought with them a few religious books from Scotland, some of which were lost in Prince Edward Island, but the rest were carefully read. In the year 1779, John Patterson brought a supply of books from Scotland. Before leaving the old country, he had built a range of small houses for working

* Editor *Colonial Patriot*.

people, on what was called a thirty-nine year tack, that is, a lease for that period, the buildings at the end of the term reverting to the proprietor. When he returned, his rents had accumulated to about £80 sterling, a good portion of which he laid out in books, among which was a plentiful supply of the New England primer, which was distributed among the young, and the contents of which they soon learned. Of teachers, I have not heard the names of any, after James Davidson left, about the year 1776.

The people, however, were all religiously educated and desirous of religious ordinances, and some of them decidedly pious. They met together on the Sabbath day, Robert Marshall, known afterward as Deacon Marshall, holding what was called a reading for the English, and Colin Douglass doing the same in Gaelic. The exercises at these meetings consisted of praise and prayer, and especially, as their name indicated, the reading of the Scriptures and religious books. Marshall was a man of strong powers of mind, well informed, especially in theology, and particularly distinguished by the boldness with which he rebuked sin. He is said some years later to have reproved the Governor for travelling on Sabbath.

They also received occasional ministerial service. The Rev. Daniel Cock, of Truro, and the Rev. David Smith, of Londonderry, visited them, Mr. Smith only once or twice, but Mr. Cock several times. We cannot tell the date of the first visit of either of them, but know that the latter visited them each summer for several years, spending a week or two among them preaching in private houses, or in the open air, and baptizing their children. The people considered themselves under his ministry, and went on foot to Truro to be present at his communions, and some of them carrying their children through on their backs to be baptized by him there. This was done by a people who had so little English that they could scarcely have understood any sermon in that language.

This may be judged from an incident that occurred some years later. A Highlander, living in Truro, attended Mr. Cock's preaching. The latter one day took as his text the words "Fools make a mock of sin." The former bore the sermon patiently, but said afterward, "Mr. Cock needn't have talked so about moccasins; Mr. McGregor wore them many a time."

They were also visited by travelling preachers, the most important of whom was Henry Alline, so noted in the early religious history of the western part of the Province. In his journal he says, under date July 25th, 1782: "Got to a place called *Picto*, where I had no thought of making any stay, but finding the Spirit to attend my preaching, I staid there thirteen days and preached in all the different parts of the settlement. I found four Christians in this place, who were greatly revived and rejoiced that the Gospel was sent among them."

The Rev. James Bennet, itinerant missionary of the Church of England, also visited this place. We have never heard his name mentioned by the old settlers; but Mr. Aikin, in his sketch of the rise and progress of the Church of England in British North America, says that in 1775 he visited the eastern harbors of the Province, and at Tatamagouche administered the Lord's Supper to 28 communicants; that in 1780 he again visited Pictou and Tatamagouche, and on his return lost his way in the woods.

During the war the price of timber rose, and the trade in it from Pictou increased. During each year three or four cargoes were shipped to Great Britain. It was at this time that Capt. Lowden, afterward an active man in the county, first commenced trading to this port.

CHAPTER VII.

FROM THE CLOSE OF THE AMERICAN WAR TILL THE ARRIVAL OF DR. M'GREGOR.—1783-1786.

The next accession of settlers to the county, and the largest it had yet received, was at the peace of 1783. These, however, were not loyalists from the revolted colonies, as in some other counties. It might have been well for them and for the county, had they occupied such a rich district as Pictou, instead of the rocky shores of our southern coast. The most who came here were disbanded soldiers, with a few families who had emigrated from the old country about that time. The largest body of them were of the 82nd, or Hamilton Regiment, as it was called. The main body had been employed in garrison duty in Halifax, with the exception of an expedition to Casco Bay, in the State of Maine, under General McLean. Another portion were employed in the Southern States, at least some of the men saw severe service there. The most important event, however, which befell the regiment was, the loss by shipwreck of a transport on the coast, near New York, when, of three hundred men on board, only eighteen were saved, who were taken off the rigging, to which they had clung for some time.

Being disbanded in Halifax at the close of the war, a large tract of land was set apart for them in Pictou, principally of the grants of Fisher and others, which had just been escheated, in spite of the efforts of Wentworth, who strove to maintain the titles of the old grantees, but only succeeded in upholding his own. This tract, which has since been so well known in the county as the 82nd grant, embraced the shore on the south side of the harbour, at Frasers Point, and from the upper part of Fishers Grant around the coast, almost to the eastern extremity of the county, including Fishers Grant, Chance Harbour, Little Harbour and Merigomish, with the exception of the Wentworth grant and of some smaller grants previously made at Barneys River, and extending into the interior to the depth of three or four miles. It was said to "contain in the whole 26,030 acres, allowance being made for a town plot, common, glebe and schools, and for other public uses." It was divided as follows: to the Colonel, (Robertson, of Struan, in Perthshire,) the Big Island, hence often known as Robertsons Island, estimated at 1,500 acres, though in reality containing considerably more; to Capt. Fraser, 700 acres at Frasers Point, which obtained its name from him; to four other officers, 500 acres each; to another, 300 acres; to thirty-two non-commissioned officers, 200 acres each; to two others, 150 acres each, and to 120 privates, 100 acres each. The following is the description of the grant, which is dated 15th February, 1785:

"Six certain several lots or tracts of land, containing on the whole 22,600 acres.

"One tract beginning at west boundary of land granted to Robert Patterson, near the head of Merigomish harbour, thence to run south by the magnet 373 chains of 4 rods each, thence west 120 chains, thence south 38 chains, thence west 109 chains, thence south 26 chains, thence west 187 chains, 50 links, or until it comes to Wentworth grant; thence north 276 chains to the harbour aforesaid; thence by the several courses of the said harbour, running east up to the grant made to Robert Patterson as aforesaid; thence crossing on that line to the west side said harbour, and running west down the harbour and round the sea-coast, running east to the bound first-mentioned, containing 11,388 acres. Also, one other tract beginning at a stake and stones on the west point of the entrance into the harbour of Merigomish, thence to run south 48° west 300 chains, thence north 78° west 107 chains, thence north 12 chains, thence north 45° west 48 chains, to the harbour of Pictou; thence bounded by the several courses of the said harbour and sea-shore, running east to the bounds first mentioned, containing 12,000 acres. Also, one other tract beginning at the point between East and Middle Rivers in Pictou harbour, thence to run south 65° west 43 chains, thence south 67 chains, thence east till it comes to the East River aforesaid, thence bounded by the several courses of the shore to the bound first mentioned, containing 500 acres, hereby granted wholly to the said Colin McDonald. Also, one other tract, beginning at the first mentioned bound of the last described tract, thence to run south 65 ° west 43 chains, thence south 104 chains, thence north 85 ° west till it comes to the harbour of Pictou, thence bounded by the several courses of the said harbour to the bounds first mentioned, containing 700 acres, hereby granted wholly to the said John Fraser. Also, one other tract, beginning at the northern bound of lands granted Archibald Allardyce, on the aforesaid harbour of Pictou, thence to run south 85 ° east 158 chains, or until it comes to lands granted to Rod. McKay, thence north 48 chains, thence west 34 chains, thence south 3 chains 50 links, thence north 85 ° west to the harbour, and by the several courses of the said harbour to the bound first mentioned, containing 500 acres, hereby granted to the said Donnet Fenucane. Also, one other tract beginning 28 chains to the east of land granted to R. Patterson aforesaid, thence to run south 140 chains, thence east 75 chains, thence north to the sea-shore, thence running westwardly by the several courses of the sea-shore to the bound first mentioned, containing 950 acres, and containing in the whole of the aforesaid tract of land, 26,080 acres of land, allowance being made for a town plot, common, glebe and school, and for other public uses, excepting always the land marked on the plan as reserved, and being all wilderness land."

The ground reserved for a town was at Fishers Grant, which was laid out to contain every public convenience, and was duly named Walmsley, but as our readers are aware, it was a town only in name, and we venture to say that even its name is entirely unknown to the majority of the young generation of Pictonians.

By the tenure of the grant, "all mines of gold, silver, copper, lead, and coals" were reserved, and also an "entire right to all His Majesty's subjects to fish on the coasts of the tract hereby granted, where it butts upon the sea-shore." We hope, therefore, none of those resident on the grant will attempt to hinder any of the lieges of Queen Victoria from this privilege. It proceeds: "also saving and reserving to His Majesty, his heirs and successors, all white or other pine trees of the growth of twenty-five inches diameter and upwards, at twelve inches off the earth, and if such trees shall be so cut or felled without license for so doing, either from the Surveyor-General of the Woods or his deputy, or from the Governor of the Province for the time being, the lot or share of land on which said trees shall be so cut, shall be forfeited and the lands revert to His Majesty, his heirs and successors." Surely with such a penalty nobody on that grant has ever cut down any big pine trees on his lot, and surely Her Majesty, as the lawful heir of George the Third, must have a large reserve of masts for her navy. But somehow we don't see them there nowadays.

The grantees were also required to pay a quit rent of two shillings sterling for every hundred acres on the Feast of St. Michael (which we may inform our ignorant readers is on the 29th September) in every year, the first payment on the first term after the expiration of ten years from the date hereof. There were also the same conditions as to working as in Cameron's grant.

We have given the conditions of the various early grants, as curious exhibitions of the ideas of the time, but it is to be observed that those issued at the same period in the other counties were in the same terms. At this time, however, a condition was inserted, not in previous grants, as follows :—" If the land hereby given and granted shall, at any time or times hereafter, come into the possession and tenure of any person or persons whatever, inhabitants of our said Province of Nova Scotia, either by virtue of any deed of sale, conveyance, enfeoffment, or exchange, or by gift, inheritance, descent, or marriage," (most likely, we think, it would at some time in some one of these ways), " such person or persons being inhabitants, as aforesaid, shall, within twelve months after his, her, or their entry, take the oaths prescribed by law, and make and subscribe the following declaration, that is to say : — ' I, ——— ———, do promise and declare that I will maintain and defend, to the utmost of my power, the authority of the King in his Parliament, as the Supreme Legislature of this Province,' before some of the magistrates of the said

Province, and such declaration and certificate of the magistrate that such oaths have been taken, being recorded in the Secretary's office of the said Province, the person or persons so taking the oaths aforesaid, and making and subscribing the said declaration, shall be deemed the lawful possessor or possessors of the land hereby granted." If this were not done within twelve months, the grant was to be void. We hope all our friends in Merigomish and the other settlements upon this grant, have carefully attended to this, otherwise, they see what a certainty there is of their losing their lands.

The land was surveyed by Squire Patterson and a son of the Surveyor-General, and divided into lots, which were duly numbered. The men were drawn up on the Barrack Square in Halifax, and each man drew his lot by number.

The attempts made in the Colonies to form settlements by disbanded soldiers have not generally been very successful. Governor Lawrence, writing sometime previously to the Lords of Trade and Plantations on this subject, says:—" According to my ideas of the military, which I offer with all possible deference and submission, they are the least qualified, from their occupation as soldiers, of any men living, to establish a new country, where they must encounter difficulties with which they are altogether unacquainted." This was soon realized in the present attempt. Some came and looked at the land they had drawn, and without cutting a tree upon it, returned to Halifax and re-enlisted. Others sold out for trifling sums. The county records, for 1785 particularly, contain a number of transfers of their lots, sometimes " with their right to a town lot and their share of provisions," for sums of four and five pounds. A number never sold, and their land has since been unoccupied, or occupied without title.

Still a good number came to settle their grants, some

arriving in the fall of 1783, others in the spring of 1784, and thus the whole shore of the eastern part of the county was in some measure occupied. A large proportion of these settlers were reckless and profligate, but a number proved steady and industrious, and from them are descended many of the most prominent and useful members of society in the county at the present day. The most of them were Scotch, and of these the majority were Lowlanders; but a number were Highlanders, of whom a considerable proportion were Catholics, principally from the Island of Barra, who had enlisted under that persuasive influence which Highland Lairds were accustomed to exercise over their dependants. A number of these afterward removed into the neighboring county of Antigonish. We give in the appendix a list of grantees, with such notices of them as we have been able to obtain. (Appendix F.)

We must now notice another band who arrived—some of them at this time, and others not till a little later, who first occupied the upper settlement of the East River. These belonged to the 84th Regiment, known at that time as the Royal Highland emigrants. It consisted of two battalions, originally embodied in the year 1775, though not numbered as the 84th till the year 1778, when each battalion was raised to 1,000 men. Their uniform was the full Highland garb, with purses made of raccoon instead of badgers' skins. The officers wore the broadsword and dirk, and the men a half basket sword. The first battalion was raised among the discharged men of the 42nd, Fraser's, and Montgomery's Highlanders, who had settled in Canada or the old colonies at the peace of 1763. It was stationed at Quebec, under the command of Col. Allan McLean, where it did good service in defence of that post, and was thus the principal means of preserving the Province to the British crown. The other battalion was raised principally among immigrants arriving in the United States or Nova Scotia. At the time the war broke out, a large number were on their way from Scotland to settle in various parts of the old colonies. In some instances the vessels were boarded from a man-of-war before arrival.* After arrival they were induced partly by threats and partly by persuasion, to enlist for the war, which was expected to be of short duration. They were not only in poverty, but many were in debt for their passage, and they were now told that, by enlisting, they would have their debts paid, have plenty food as well as full pay, and would receive for each head of a family 200 acres of land and 50 more for each child, "as soon as the present unnatural rebellion is suppressed," while, in the event of refusal, there was presented the alternative of going to jail to pay their debts. Under these circumstances, most of the able-bodied enlisted, in some instances fathers and sons serving together. Their wives and children were brought to Halifax, hearing the cannon of Bunker Hill on the passage.

This battalion was under the command of Col. Small. Stewart, in his history of the Highland clans and regiments, says: "No chief of former days ever more firmly secured the attachment of his clan, and no chief certainly ever deserved it better. With an enthusiastic and almost romantic love of his country and countrymen, it seemed as if the principal object of his life had been to serve them and promote their prosperity. Equally brave in leading them in the field, and kind, just, and conciliating in quarters, they would have indeed been ungrateful, if they had regarded him otherwise than they did. There was not an instance of desertion in

---

* The tradition in several families is, that they were captured by a British man-of-war. I do not understand how this could be. A number were on their way to Virginia.

the battalion. Five companies remained in Nova Scotia during the war. The other five joined General Clinton and Lord Cornwallis to the southward. At Eutaw Springs, the Grenadier company was in the battalion, which, as Col. Alex. Stewart, of the 3rd Regiment, states in his despatches, drove all before them."

That portion of the regiment which remained in Nova Scotia, was stationed at Halifax, Windsor and Cumberland, and the men were distinguished by their good behaviour, in which they presented a remarkable contrast to the rest of the army at that time.

At the close of the war, both battalions were disbanded. The first battalion settled in Canada, the second in Nova Scotia. The transports, with the flank companies, from the Southern army were ordered to Halifax, where the men were to be discharged, but owing to the violence of the weather, and a consequent loss of reckoning, they made the Islands of St. Nevis and St. Kitts, which delayed their final reduction till 1784. The largest portion of the battalion obtained their land in Hants County, where they formed the township of Douglass, but a number of them settled on the upper settlement of the East River. The first who came was James Fraser (Big) who, in company with Donald McKay, the elder, followed the river up till he reached the intervale, a little below St. Paul's Church, where his grandson Donald resides, which he chose as his future home. We may mention that intervale land was then eagerly sought, and that it was this that principally attracted the settlers to such a distance from the shore. Accordingly he and fifteen of his comrades took up a tract of 3,400 acres, extending along both sides of the river,* on the east side, from Finlay

---

* The discharges we have seen are dated 10th April, 1784, but the grant is dated 3rd November, 1785. There was, however, always delay in the issue of the old grants. Curiously enough I could find no record of this in the Crown Land office at Halifax. It was surveyed and the lots laid off by Squire Patterson.

Fraser's, a little below Springville, to a little above Samuel Cameron, Esq.'s house, and on the west side, from John Forbes' to James Fraser's (Culloden). This grant, usually known on the East River, as "the soldiers' grant," is in the same terms with that of the 82nd. The settlers also received a town lot at Fishers Grant and a supply of provisions.

They made a beginning of settlement here, as near as we can ascertain, in the summer of 1784. We may say of these as well as those of the same class who settled on the West Branch, that they were very different from disbanded soldiers in general, being sober and industrious, and many of them serious. But they had for some time great hardships to endure. Till they made a blaze, there was no path to the neighbouring settlement. All their seed and supplies for their families, they were obliged to carry on their backs, or in winter, to drag on handsleds. And from their seclusion from the rest of the settlement, they were for a time exposed to peculiar privations. We append a list of these grantees with notices of them. (Appendix G.)

About the same time with the occupation of the East Branch, or a little after, the West Branch was occupied, principally by men of the same regiment. The first to make their way thither were David McLean afterwards Esquire, and John Fraser, who made their way along the bed of the river to the falls, at Gray's Mills, where they spent the night sleeping in the open air. Their grant however did not come out for some time after. The one in use is dated 13th December 1797, and includes a number of other parties, who had settled on other parts of the river. This grant is not on record, but there is one registered for the same quantity of land with nearly the same names, dated 1st April 1793. We subjoin a list of the grantees with brief notices of them. (Appendix H.)

The same summer (1784) there arrived at Halifax a

vessel with immigrants, of whom some eight families, all Highlanders, removed to Pictou, all of whom, so far as we know, settled on the East River. Perhaps the most noticeable of these were Thomas Fraser, who settled nearly opposite where New Glasgow now stands, and who was long known as Deacon Thomas, and John Robertson who was the first settler at Churchville. A list of these settlers so far as we have been able to obtain, we subjoin.(Appendix I.)

About the same time a number of others, who had served during the war, settled in various parts of the county. Thus we find Lieutenant Gordon, who lived at Mortimers Point, hence for some time after known as Gordons Point, Henry Burnside, a native of Glasgow, who had served ten years in the 42nd Regiment, who joined the settlers on West Branch, taking up the farm since occupied by Peter Ross, Esq., Robert and Joseph McDonald, brothers, who settled on Barneys River; and William McKay, Archibald Gray and Dalgliesh, who had served in one of the Highland Regiments, and Donald McGillivray, who had been a dragoon, all of whom settled on the lower part of the Eight Mile Brook; Samuel Cameron, who had been a light horseman, and who settled at Merigomish, and Gregor McGregor, a native of Perthshire, who had also served in the army and now settled at Barneys River. There came also others of those whom we have mentioned as having been on their way to the United States when the war broke out, among whom may be named Alpin Grant, who, we believe, belonged to the 84th, and who settled below the town, where Capt. Foote now lives, the McMillans, (though one son William served his full time as a soldier in the Cavalry and did not arrive for some time after) and James McDonald and John McKenzie, of West River, who had been employed in Halifax during the war, the former as a tailor to the troops, and the latter as a carpenter. To these we may add James Chisholm, a son of a parish minister in the North of Scotland, who had been on the staff of Gen. Washington, but finding his Highland countrymen generally taking the other side, deserted and had a price set on his head.

About the same time, Governor Wentworth made efforts to settle his grant. In a letter written in the year 1783, to prevent process of escheat against him, he says, " I had made an agreement with agents of 120 families in Connecticut, all loyalists, and churchmen, with their missionaries, to remove upon our lands in the spring next, to give them alternate lots of 100 acres, and something more to the missionary and one or two principals among them. They are dissatisfied with their present Government, are well recommended, and determined to sell their present possessions."

This scheme was never carried out, but at this time to secure his grant from forfeiture, he offered liberal terms to those who should occupy it. Several embraced his proposals, the first of whom was Mr. John Sutherland. He had immigrated a young man in the Hector, and removed to Windsor, where he married. After this he returned to Pictou, he and his wife travelling on foot, and carrying that distance an iron pot, as the beginning of house furnishing. After being a short time on the East River, he removed about the year 1785, and settled at the mouth of Sutherland's River, which received its name from him, on the farm still occupied by his descendants. Among others, who settled on the Wentworth Grant about this time, we may mention Alex. McDonald, (Garty) who had been a soldier in one of the Highland Regiments during the war, and three brothers, George, Charles and Joseph Roy, who had just emigrated from Scotland.

The most important accession to that settlement at that time was Nicholas Purdy (properly Purdue, that being his mother's name), Olding. Both by his father and mother's side he belonged to families of some rank in the County of Kent, England. He was well educated, had studied for the bar, and commenced practising in the State of New York, where he had married, when the war broke out. He took the British side, though his father-in-law took the opposite, and joined one of the loyal American regiments, and served throughout the war as a chasseur, or light horseman, with great credit. At the end of the war he removed to Halifax, and commenced practising law, and might have attained to the highest honors of his profession. But he had received a wound in the head, which had been trepanned, and rendered him unfit for the excitement of the bar and the social habits of the time. He had drawn his land at Sheet Harbor, but, not liking the situation, he, at the solicitation of Governor Wentworth, removed to Point Betty, where he spent the rest of his life. He was for many years a magistrate, and on the list of lawyers, though he did not practise much, and, in his old age, recognized as the father of the Bar of Nova Scotia.

The large influx of settlers produced important changes upon the state of the community. Perhaps the most important was the injury done to its morality, by the large number of drunken profligates, discharged from the army, a fact which will come under our notice hereafter. Another circumstance must here be mentioned. From the large influx of male settlers, there was such a scarcity of the gentler sex, as we now hear of at the first settlement of some of the Western States. An old woman in the author's congregation used to say that she recollected the time when there was "only one girl in all Pictou"—marriageable, of course, she meant. Why that one remained in the market, we regret that we omitted to enquire, but presume that it was because she was "owre young to marry yet." But extreme youth was not always a protection. A case is well authenticated of a woman who was married when she was fourteen years of age, and had six children before she was twenty. What a contrast to our present degenerate race! Men then travelled, like Jacob, long distances for wives, and married them with as little previous acquaintance as an Oriental. A vessel having arrived in Halifax with immigrants, three young men on the East River set off through the woods to the city. On their arrival, they went among the newly arrived, and each selecting the girl whose appearance caught his fancy, at once made proposals to her. We suppose it must have been through the persuasive influence of the Gaelic language, but, at all events, the fair ones yielded, went home with men they had never seen before, and proved faithful and, we have no doubt, happy wives. One reason given for so many of the 82nd men leaving was that they could not get wives in Pictou.

We must here notice another class of settlers who, about this time, commenced in River John, and as their history is somewhat peculiar and interesting, we shall give it at some length. They were originally from the town of Montbeliard (pronounced Mong bilyar) which formerly formed part of the dominions of the Duke of Wurtemburg, but which was annexed to France by the ambition and treachery of Louis XIV, after the treaty of Nimeguen, in 1679.*

In the third volume of D'Aubigne's History of the Reformation will be found an interesting account of the introduction of the Reformation into this place, by Farel, in the year 1524. His labours were successful; a large number embraced that system which also spread through

---

* "The late-treaties had ceded to France several important cities and districts, 'with the dependencies belonging to them.' This vague expression opened a wide field to the grasping ambition of Louis. He proceeded to institute courts called *Chambres de reunion*, for the purpose of ascertaining what dependencies had appertained at any former period to the territories now annexed to France, and by this ingenious device he soon added to his dominions no less than twenty towns, wrested from neighbouring princes, including Saarbruck, Luxemburg, Deux-ponts and Montbeliard."—*Smith's History of France.*

the surrounding districts. Soon after their annexation to France, came the revocation of the Edict of Nantes in the year 1685, which let loose the floodgates of suffering upon the Reformed Church of France. But in that act the districts referred to, were excepted, and the Protestants for some time suffered no molestation, owing to the stipulations of the Treaty by which they were annexed. But after a time the Lutherans in the annexed Provinces were exposed to the same sufferings as their Reformed neighbours, and the remembrance of these is still handed down among their descendants in this country. One incident connected with their emigration may be mentioned. Orders had been given that one of their chapels should be taken away from them and handed over to the Romanists. Fifty young men, among whom were George Tattrie and Peter Millard, assembled at it, armed only with stones, prepared to resist. A detachment of troops was sent against them, with a priest at their head. He warned the party gathered of the uselessness of resistance. They, however, refused to yield, when a section of the troops were ordered to fire, which they did, killing two and wounding others, among whom was George Tattrie, who received a ball in the fleshy part of the leg. The order to fire was answered by a volley of stones, by which some of the soldiers were badly injured, and, it was said, one killed. The Protestants were again summoned to surrender, but refused, until the priest called on the whole detachment to fire, when they submitted, and saw the house where their fathers had worshipped given to their enemies.

As soon as his wound was healed, George Tattrie,* who we may here mention, had previously been a French soldier, and fought at the battle of Fontenoy Millard, and most of the party joined a body of their fellow-coun-

---

* In 1873 I conversed with a son of his, over ninety years of age, from whom I received these particulars, as I had received them some years before.

trymen, who were preparing to emigrate to Nova Scotia, in response to the invitations, which had been addressed to Protestants on the Continent by the British Government, offering liberal terms to those who should settle in this Province. They came down the Rhine and took shipping at Rotterdam for England, in the year 1752. They landed at Portsmouth, and for a time were left in destitution by those who had brought them there. But their case was taken up by the British Government, by whom they were despatched in four vessels, two for South Carolina and two for Halifax. Those who came in the latter reached their destination in the following spring, and were landed at Georges Island, to the number of 224. From Halifax they proceeded to Lunenburg, where they endured the hardships and dangers of the first settlers.

After the peace of 1763, Col. DesBarres, a countryman of theirs, and a son of one of their old Protestant ministers, who had entered the British military service, and who had served at the taking of Louisburg and Quebec, being, it is said, one of the officers in whose arms Wolfe fell, and who was afterward successively Governor of Cape Breton and Prince Edward Island,* had obtained a grant of a large tract of land at Tatamagouche, extending from Point Brulé along the shore westwardly, some distance beyond the present village. By his persuasion, a number of them were induced to settle there. Accordingly, eleven removed from Lunenburg with their families in or about the year 1771. The names of these were George Tattrie, who settled on the French River at what is now Donaldsons place ; George Gratto ; David Langill, who settled on what has since been known as Lombards place, and his son John

---

* Old Mrs. Mattitall, who lived where the village of Tatamagouche is now situated, had been his nurse. When he was Governor of Prince Edward Island she went to see him. He took her to Government House and showed her every kindness.

James Langill, then married, who also settled on the French River; Matthew Langill, his brother, James Bigney and George Mattitall, who located themselves where the village now is, and at the same time, or perhaps a little later, Peter Millard, who settled at the point below Mrs. Campbell's, and John Millard, who settled next him to the west. There were three other settlers who did not remain; Ledurney, who settled where Waugh afterward lived, John Lowe and John Buckler.

When they arrived they found the indications of what had once been a flourishing French settlement. A considerable extent of land on the shores of the bay and harbor, from the church to McCulley's, had been cleared by them, and their furrows were still visible. The intervales both on French and Waughs Rivers had also been cultivated, and on the former they had been extracting and attempting to smelt the copper ore. The remains of no less than five mills were found; one on Mill Brook, one at Blockhouse Bridge, one at Murdochs, one on the main French River, and one at Goosar. Traces also of a graveyard, with crosses still standing at the head of the graves, and of a Romish chapel, were to be seen between what is now Mr. Wm. Campbell's field and the schoolhouse. The first settlers for a time endured great hardships. A supply of implements and provisions was to have been sent round to them in a vessel, but she never arrived. They had to carry wheat and potatoes on their backs from Truro, the former article costing them twelve shillings per bushel and the price of the latter being proportional. They frequently resorted to a plant growing on the marsh, which, when boiled, made a palatable sort of greens. But they had the benefit of the clearings made by the Acadian French, those on the intervales being particularly rich, from which they soon derived a comfortable subsistence, and they were soon joined by others of their countrymen from Lunenburg.

But Des Barres was unwilling to sell his land and wished to keep the settlers as tenants. This, of course, they did not like, when there was so much land around, which might be had in full ownership. Accordingly, as the young men of these families grew up they began to look elsewhere, and were attracted by the land in River John, then known as Deception River, and, as near as we can ascertain, in the year 1785, four young families, viz.: George Patriquin, John Patriquin, James Gratto and George Langill, son of John James, above mentioned, took up their abode there.* Following the plan which had been adopted at Lunenburg, especially from fear of the Indians, of living in a town and having their farm lots outside, they laid out for themselves small lots at what is called Smiths Point, and where they intended also building a blockhouse. They took up land for farming purposes, but for several years continued to live together at Smiths Point, thus strengthening each other's hands, and overcoming the feeling of loneliness, incident to the situation of the new settler.

John Patriquin took up and occupied for a number of years the farm owned afterward by the late Alexander McKenzie, Esq., and from him the point where the shipyard is, now occupied by Charles McLennan, Esq., was known as Johns Point. It is supposed by some that the river received its present name from him. It was for a length of time known as "Johns River," and on the communion cups of the Presbyterian congregation of River John, it is so named; but before the English settle-

---

* The land here belonged to the Philadelphia Company, and the first deed on record is dated 25th July, 1786, and is to James Gratto, (John) James Langill, George (Frederick) Langill, John (Frederick) Patriquin, George Patriquin and George Tattrie, of land described as "lying on a river and bay known by the name of Deception River, near Cape Jean, beginning about a mile north from the entrance of said river on west side," &c.

ment the cape, at the entrance, was known as Cap Jean, or Cape John and in Des Barres' chart published in 1770, the estuary of the river is called "Harbour or River John." This John Patriquin again removed to Tatamagouche, but several of his family afterward returned to River John.

George Patriquin took up his farm adjoining John's to the north, where Thomas Mitchell now resides. He had four sons, James, who settled in New Annan, David and George, both of whom settled on the road leading to Earltown, and Frederick, who, when five years of age, was stolen by the Indians. This was a very severe trial to the afflicted parents, who mourned over their loss during their lifetime, between hope and despair, the feeling still lingering in their minds that, probably by some event in Providence, their darling might be restored to them again. Whenever they heard of any white person among the Indians, they would make enquiry, but no tidings of the lost were ever received. His daughter Phebe, universally known as aunt Phebe, wife of Joseph Langill, Brook, was the first white child born in River John. James Gratto, a son of George Gratto, one of the first settlers in Tatamagouche, took up his farm beyond Smiths Point. He left two sons, George and Matthew, and his descendants still occupy part of the old property.

The last of the four first settlers of River John was George Langill. He was a grandson of David, or John David Langill, already mentioned, as one of the first settlers of Tatamagouche, whose widely scattered family we shall presently notice.

Five years later or about the year 1790, the settlers in River John were joined by the families of George Joudry, who settled next above him, by George Bigney, son of James mentioned as one of the first settlers of Tatamagouche, who settled where Thomas Bigney now resides, and Mattitall and George Langill, only son of Matthew

Langill, already mentioned, as one of the first settlers of Tatamagouche. He obtained John Patriquin's place by exchange. His father Matthew, who had been a light horseman in the French army, came afterward to River John, and died there at the age of 76, and was the first person buried in the old graveyard, his tombstone bearing the date 1800.

At a later date, Louis Tattrie, son of George already mentioned, settled on the Tatamagouche road, and others from Tatamagouche, and two brothers Perrin, Christopher and George, came direct from Lunenburg.

We must now notice, however, the family of (John) David Langill. By his first marriage, he had one son, John James, who, as we have mentioned, settled with him on the French River. He had five sons, George, David, James, Joseph and Frederick. This George was, as we have seen, one of the first settlers of River John, but he remained only a few years, and then removed to New Annan, and the Langills, in that quarter, are descended from him. Frederick removed to the United States, and the other three settled in River John. David is the ancestor of the Miller Langills, and James and Joseph settled on the Mill Brook, where their descendants are still numerous.

(John) David Langill, by his second marriage, had no children. He was married a third time to a woman, who had a son previously, who assumed the name of Langill, and who settled at Point Brulé, where he became the ancestor of the Langills in that quarter.

By this third marriage, David Langill had (1) Nicholas, who went to the United States and was never heard from, (2) John David, (3) John George, (4) John Frederick, and (5) John Louis.* These four all settled on the shore from

---

* Among the Germans in Lunenburg, it was a practice not uncommon till recently, and perhaps still existing, to give each son in a family the same first name and to distinguish them by their second names.

River John toward Tatamagouche, about the year 1792. John George was long an elder in the Presbyterian Church, and was regarded as a man that feared God with all his house. His son Ephraim was also an elder for 36 years, and now his grandson Ephraim is also an elder. One son of John Louis, David, has also long been an elder in the Presbyterian Church, and is now the oldest member of Session.

The old people spoke a corrupt dialect of French, but with a German tone and accent, as the inhabitants of Alsace and neighboring districts do till the present day. But they understood pure French; some of them could read it fluently, and they could also understand the *patois* of the Acadians. Some of them also had Bibles and other books in that language. But among the present generation it has nearly died out. Indeed, in their general character, they show more of the staidness of the German than the vivacity of the Frenchman.

The large accessions to the population of the district induced in 1783 and 1784 an effort to obtain the services of a settled clergyman. A meeting of the inhabitants was accordingly held in the fall of the latter year with that view, when it was agreed to apply to Scotland for the services of a Presbyterian minister. For his support they agreed to "raise £80 per annum for the first and second years, £90 per annum for the third and fourth years, and thereafter £100 currency, that is, £90 sterling annually,—one-half thereof in cash and the other in produce; and if Providence should smile upon the settlement and their industry, to make additions to that sum." They also agreed to build a house and barn for their minister, and that he should have a glebe, and that they should clear so much of it, from time to time, for his encouragement. A committee was appointed, consisting of Robert Patterson (the Squire), John Patterson (deacon), of the harbor, as it was then called,

William Smith, of West River, Robert Marshall, of Middle River, and Donald McKay, of the East, to act for them in obtaining a minister. A petition, drawn up by Mr. Cock, was accordingly signed by them and entrusted to John Buchanan, Sen., and John Pagan, respectable inhabitants of Greenock, with authority to present it to any Presbyterian Church Court with which they were likely to be successful.

It was accordingly presented to the General Associate Synod of Scotland, then usually known as the Antiburgher Synod, at their meeting in Spring, 1786, and as the result, the Rev. James, afterward Dr., McGregor was appointed to proceed to Pictou. He accordingly, on the 4th June, set sail from Greenock, in the brig Lily, for Halifax, where he arrived on the 11th July, and the same week came to Truro, travelling on horseback. There was something like a road for eleven miles from Halifax, but beyond that there was only a narrow avenue through the woods, on which the trees had been cut down and sometimes cut across and rolled to one side. The ground was generally so soft that even at midsummer, as it was then, the horses sank to their knees in mud and water, and as each horse put his foot where his predecessor had, the path became a regular succession of deep holes, such as one may see in a road recently made in deep snow.

From Truro there was only a blaze, but men were then employed in opening the road, which, however, consisted only in cutting down the trees along the line of travel. On the 21st he left Truro, and arrived at George McConnell's, now the ten-mile farm, and then the nearest clearing to Truro. On the following morning he was taken in a canoe to the harbour. His impression he thus describes:—"When I looked round the shores of the harbour I was greatly disappointed and cast down, for there was scarcely anything to be seen but woods growing down to the water's edge. Here and there a mean timber hut was visible in a small clearing, which appeared no bigger than a garden, compared to the woods. Nowhere could I see two houses without some wood between them."

On the following day he commenced his labors by preaching in Squire Patterson's barn. He thus describes the event :—

"The Squire gave orders to lay slabs and planks in his barn for seats to the congregation; and before eleven o'clock next morning I saw the people gathering to hear the Gospel from the lips of a stranger, and a stranger who felt few of its consolations, and had but little hope of communicating them to his hearers. None came by land except certain families who lived a few miles to the right and left of Squire Patterson's. Those who came from the south side of the harbor and from the river, had to come in boats or canoes, containing from one to seven or eight persons. The congregation, however, was not large; for numbers could not get ready their craft, the notice was so short. I observed that the conduct of some of them, coming from the shore to the barn, was as if they had never heard of a Sabbath. I heard loud talking and laughing, and singing and whistling, even before they reached the shore. They behaved, however, with decency so long as I continued to speak, and some of them were evidently much affected. I endeavored to explain to them in the forenoon in English, 'This is a faithful saying, and worthy of all acceptation, that Christ Jesus came into the world to save sinners;' and in the afternoon, in Gaelic, 'The Son of Man is come to seek and to save that which is lost.' The first words which I heard, after pronouncing the blessing, were from a gentleman of the army calling to his companions, 'Come, come, let us go to the grog shop;' but instead of going with him, they came toward me to bid me welcome to the settlement, and he came himself at last."

In the same vessel with Dr. McGregor arrived one who was afterward well known in the county, viz., William Fraser, surveyor. Having traversed the eastern part of the Province about this time, he says :—" In 1787 there were only four or five houses from Salmon River to Antigonish. To the eastward of the East River there was not even a blaze on a tree. There was not one inhabitant on the Cape Breton side of the Gut of Canso, and but one on the Nova Scotia side. In 1788 there was one house at Ship Harbour. I may add that from Pictou to Cocaigne, there were but four or five families at River John; a few more at Tatamagouche; some refugees at Wallace, and but one at Bay Verte. At Miramichi there were but five families."

We may add here, that by a return to Government, signed by John Fraser and Robert Patterson, dated 8th June, 1786, the following was the amount of farm stock in Pictou and Merigomish:—

Oxen.	Cows.	Small Cattle.	Sheep.
230	356	450	1500

CHAPTER VIII.

DR. M'GREGOR'S EARLY LABOURS.—1786–1788

Hitherto we have had great difficulty in getting anything like clear information regarding the state of society among the early settlers, but Dr. McGregor, in his autobiography, has given a very vivid picture of the condition of the country at this period. We have already published this in another form, but it is necessary, for the completeness of our present work, to give, at least, a condensed view of the state of matters as he found them, and of his early labours.

" As for our population," he says, " Pictou did not contain 500 souls; if Merigomish be included, I suppose they would amount to a few more." These were settled principally along the intervales of the three rivers, and thinly along the shore from Fishers Grant to Merigomish. The site of the town was still covered with woods. The majority of the settlers, having commenced within the previous two years, were in extreme poverty. Squire Patterson's house was the only framed one.

Of the rest, but seven or eight had two fire-places. The most were of round logs, with moss stuffed in between them, and plastered with clay, while the roof was formed of the bark of trees cut in pieces of equal length, disposed in regular tiers, the ends and edges overlapping, and kept in position by poles running the whole length of the building, placed on the ends of each range of bark, and fastened at the ends to the building by means of withs. Their furniture was of the rudest description, a block of wood, or a rude bench, serving for chair or table. Food was commonly served up in wooden dishes or in wooden plates, except when discarding such luxuries, they gathered round the pot of potatoes on the middle of the floor. Among the new comers, at least, straw formed the only bed. Money was scarcely seen, and almost all trade was by barter, wheat and maple sugar being the principal circulating medium.

"Not a loaf could be afforded of our own wheat. There was no mill to grind it. We had an imitation flour by the hand-mill, but of oatmills we had not a semblance." These hand-mills, or querns, were in almost every house. They consisted of two stones, about two feet in diameter; the lower was fixed, and the upper surface "picked," as millers say, and a pintle of iron inserted in the centre. The upper stone was heavier, being about ten inches thick, with a hole in the centre through which the pintle in the lower stone passed, and by which, also, the grain to be ground was introduced. The lower end of an upright pole was fastened to the upper surface of this stone, near its outer edge, while the upper end was fixed in a cross piece of wood between the upper beams of the house. The operator seizing the upright pole in one hand, whirled the upper stone by means of it with a rapidity according to his strength, at the same time, with the other hand, putting the grain into the hole in the centre. John Patterson made an improvement upon them.

by putting a rim round them and a spout at one side, so that the flour might come off the stone at one place. This was afterward sifted and made wholesome bread. " Grinding on the hand-mill was so laborious that it was let alone till necessity impelled to it. This was the occasion of saving much wheat, for many a meal was made without bread on account of the trouble of grinding."*

" There was not a merchant in the district, nor any who commonly kept goods for sale, or made the third of his living by the sale of goods. Little schooners came round in the summer with some necessary articles, to which the people repaired in their canoes, and got a few things, for which they exchanged a little produce. Sometimes John Patterson got a few pounds' worth more than he needed, and afterward sold them. We had scarcely any tradesmen of any kind."

" There was not a foot of road in the district. There was a path from the West to the Middle River, and from the Middle River to the East, but no path from any of the rivers to the harbor. We had not a dozen of horses,† and

---

* Mention is made of a grist mill of Kennedy's in the census of 1774, but it must have worked but for a short time, if it ever worked at all, as he returned to Truro in 1776. Certainly, at this time, there was none in operation, and the tradition is, that the first application of any other power than the hand-mill to grinding was by the erection of a windmill by John Patterson, at Norway Point, where he then lived. The wheat which the settlers exchanged with the merchants for goods, was shipped to Halifax and ground at a mill at Dartmouth.

† William Fraser, surveyor, says that in 1787 there was only four or five horses between Salmon River and Antigonish. Though there were horses in the settlement, they were still rare enough to be objects of wonder and dread to the rising generation in some places. We have heard John McLean, of West River, tell of the first horse he ever saw. He had heard of a man in the neighborhood having got such an animal, and not long after, being down on the intervale, he was struck with terror at the sudden appearance of a huge beast, which he concluded must be the aforesaid horse. He retained his faculties sufficiently to consider whether it would be better immediately to take to flight, but concluded that if he did so it might lead the animal to pursue him. He, therefore, glided away quietly, till he got some bushes

for carriage, neither sleighs nor gigs." These roads, if they can be called such, served scarcely any other purpose than to prevent the traveller going astray. Though there were inconveniences in travelling by them from their roughness owing to stones or roots, from the branches of trees which crossed the path, or from wet and boggy ground, yet old people have assured me, that from the soft moss, with which the ground was covered, walking on them was easier and less fatiguing than on an ordinary road. Over most of the district, however, there was not even this convenience. The most there was to direct the traveller between one settlement and another was a "blaze," which we presume all our readers know means a chip taken off the sides of the trees along the line of travel. The chief of the travel on land was along the shore or the banks of the rivers, which were often incumbered with trees and stones, and at other places presented bogs, the crossing of which was most inconvenient, or creeks which required a long circuit round, or brooks, which it was necessary to ascend for some distance, to a convenient places of crossing. In winter, the regular mode of conveyance was the snow shoe. It is certain, that whether more snow fell then than now, it lay more continuously through the winter, and as winter advanced, was of greater depth than is now commonly seen, rendering this at that time the only possible way of travelling. In summer, canoes were largely used in crossing harbours and streams on passing along these shores. These were what are in the west called dugouts, being constructed of a section of a large pine tree, hol-

---

between himself and the horse, when he took to his heels and ran with all his might till he reached home. The late John Douglass, of Middle River, used to tell, with equal interest, of the first horse he saw. It belonged to a man from Truro, who called at his father's house. John, returning home from a short absence, was surprised at seeing such an animal tied to a tree. While peeping curiously at it from behind another tree, he was still more surprised to see a strange man, who came out of the house, mount upon his back and ride away.

lowed out and properly shaped. These were large and capable of carrying four or five persons with perfect safety, but from the unskilfulness of many of the settlers, who had not been accustomed to them, accidents frequently happened.*

"It was no little discouragement to me that I saw scarcely any books among the people. Those who spoke English had indeed a few, which they brought with them from their former abodes; but scarcely one of them had got any addition to his stock since. Almost all of them had a Bible, and it was to be seen with some of the Highlanders who could not read. Few of them indeed could read a word. There was no school in the place. Squire Patterson had built a small house and hired a teacher for a few months now and then for his own children. In three, or perhaps four, other places three or four of the nearest neighbors had united and hired a teacher for a few months at different times, and this

---

* I have been surprised in tracing the history of families of the early inhabitants, to find so many cases of drowning by the upsetting of canoes or falling through the ice. On one occasion at the Middle River three men were drowned in attempting to save a woman. She was a Mrs. Cumminger, who lived on the east side. She was on her way to town by the ice on foot, when, for some reason unknown, she took a wrong course and went over to the opposite side, where the ice was bad. When opposite what is now Blairs place, it gave way under her; her cries attracted the attention of three men, who were working in the woods, two named Ross and one named McLean, who immediately proceeded to her assistance. They cut poles which they laid upon the ice, and on which they walked out towards her, but when close to her, stepping on the outer ends of the poles, on the edge of the water, the ice gave way, and they were plunged in. Their cries in turn attracted the attention of two other men, the late Samuel Archibald and a Mr. Hingley, who were going to town on skates. Owing to the state of the ice on that side the river, they had to make a considerable detour to reach the place, so that when they arrived the woman and two of the men had sunk. The third was supporting himself on his pole which he held in an upright position, but just as they approached he let go and disappeared. The next day the whole four bodies were taken out of the one hole. The three men were all young, had not been long in this country; two were brothers, the third a cousin, and one of them had only been about six weeks married.

was a great exertion. What was more discouraging, I could not see a situation in Pictou where a school could be maintained for a year, so thin and scattered was the population. Besides, many of the Highlanders were perfectly indifferent about education, for neither themselves nor any of their ancestors had ever tasted its pleasure or its profit. But afterwards I found that children made quicker progress in the small and temporary schools, with which the people were obliged to content themselves, than they did at home in large and stationary schools; and I found it easier than I had thought, to rouse the Highlanders to attend to the education of their children, so far as to read the Bible."

Dr. McGregor, we may here observe, was a native of Perthshire, born at what is now the village of St. Fillans, at the foot of the romantic Loch Earne, in December, 1759. His father had been brought to the knowledge of the Saviour under the celebrated Ebenezer Erskine, when a young man laboring near Stirling. He returned to his native parish, to be an earnest friend of the Gospel, and continued active in promoting its extension during the rest of his life. His son was early devoted to the ministry, and, possessing strong natural powers, an earnest spirit and active habits, he passed through his college curriculum at Edinburgh with credit, studied theology under William Moncrieff, at Alloa, then professor in the Antiburgher branch of the Secession, and was in due course licensed to preach the Gospel. Believing that duty called him to preach the Gospel to his Highland countrymen, he gave himself to the study of their language, and became a most accomplished Gaelic scholar. He not only spoke it with ease and fluency, but wrote it with precision and elegance, so that before leaving Scotland he had been employed in preparing a corrected version of the Gaelic Scriptures. We may here observe that he had somewhat of a poetical genius, and in his later years, with a desire to benefit his countrymen, he prepared a small volume of Gaelic poems, in which he exhibited the doctrines of the Gospel, in verse, adapted to the sweetest melodies of his native land. The work is still popular in many parts of the Highlands.

Now commenced that course of protracted and energetic labors, which endeared him to the hearts of the people of that generation, which established the moral and spiritual character of the county and built up the Presbyterian cause through the eastern parts of the Province and in the other Maritime colonies. We have in another work described these labors at some length, but as his history is for some time the history of Pictou, a brief account of them is necessary in this place.

On the second Sabbath after his arrival (July 30) preaching was at the East River, at the head of the tide, a little below the present Albion Mines, and he complains that the conduct of those in attendance was as disorderly as before. "Their singing and whistling, and laughing and bawling, filled my mind with amazement and perplexity. I took occasion to warn them of the sin and danger of such conduct." During the service one man stood up and, in a loud and angry voice, told him that he was good for nothing and did not deserve the name of a minister, and that he would never pay him a shilling, as he had refused to baptize his child.

The following Sabbath he preached at the lower part of the Middle River, at what was then Alexander Frasers place, near where Samuel Fraser's house stands. Service was under the shade of a large elm tree. At first it was contemplated to erect one church here, as being central for the whole district. This idea, however, was soon abandoned, and it was resolved to erect two, one on the East River and the other on the West.

During the summer preaching continued thus alternately, with some improvement in the conduct of the people, but not very decided, till the cold weather led the gentlemen of the army to dispense with their presence. He remarks that though public worship was conducted in the open air till they were compelled by cold to go into houses, they were never disturbed by a shower.

Early in October he first visited the upper settlement of the East River. The only mode of travelling to this quarter was by walking along the edge of the river till they came to a brook, and then ascending it till they reached a place where it could be crossed. His first sermon was preached at James McDonald's intervale, under the shade of a large tree. "On Sabbath," he says, "they came all to hear me with wonder and joy; for they had not indulged the hope of ever seeing a minister in their settlement. They had very poor accommodations. I had to sleep on a little straw on the floor."

A little before winter set in he paid his first visit to Merigomish, preaching and visiting. The people solicited a share of services, and for about thirty years he continued to give a portion of his labors to that settlement. To attend his ministry, a number were in the habit of travelling to his church at the lower settlement of the East River, going in canoes to the head of the harbor, and thence on foot through the woods to the church. It was not, however, till two years after, that they were fully organized as a congregation, by the ordination of Elders, the first being Walter Murray, John Small, and George Roy.

The winter following was the severest known among the early settlers for many years. It set in on the 15th November. There had been snow previous, which had melted, but what fell on that day remained till the middle of April, and some of it till the month of May. Before the end of the month, the harbour was sufficiently frozen for persons to cross on the ice. "When winter came on," he continues, "preaching was in private houses. People could not sit in a house without fire, and they could not travel far. It was therefore agreed, that I should preach two sabbaths at the East River, two upon the Harbour, two upon the West River, and two upon the Middle River, and then renew the circle till the warm weather should return. The Upper Settlement of the East River, being unprovided with snow-shoes, were excluded through the whole winter from all communication with the rest of the people, as effectually as if they had bolonged to another world, excepting one visit by two young men, who made a sort of snow-shoes of small tough withes, plaited and interwoven in snow-shoe frames. This circulating plan of preaching was no little inconvenience to me. For six weeks in eight, I was from home almost totally deprived of my books and all accommodation for study, often changing my lodging and exposed to frequent and excessive cold. But it had this advantage, that it gave me an easier opportunity of visiting and examining the congregation, than I could otherwise have had, for I got these duties performed in each portion between the two Sabbaths on which I was there."

With this winter began his regular course of family visitation, and catechetical instruction. "I resolved not to confine my visitations to Presbyterians, but to include all of every denomination, who would make me welcome; for I viewed them as sheep without a shepherd. The purport of my visitations was to awaken them to a sight of their sinful and dangerous state, to direct them to Christ, to exhort them to be diligent to grow in religious knowledge, and to set up the worship of God in the family and closet, morning and evening. I did not pass a house, and although I was not cordially welcomed by all, my visits were productive of more good than I expected; and I trust they were the means of bringing to Christ several who were not Presbyterians." He also, annually held meetings in each section of the congregation, at which young and old were duly catechised according to the old Scottish mode.

This course of labour, both in preaching, visiting and catechising, he regularly fulfilled over the whole district during the nine years he was sole minister of Pictou. With the state of traveling as we have described it, it may be understood that this involved a large amount of toil. "I had to learn," he says, "to walk on snow-shoes in winter, and to paddle a canoe in summer, and to cross brooks and swamps upon trees overturned or broken by the wind, and to camp in the woods all night—for there is no travelling the woods at night, where there is no road." But he possessed an ardent temperament, and an active, wiry frame. People have said, that they never saw one brought up in the old country, become so good a traveller on snow-shoes, and such were his powers of endurance, that he outdid many who were accustomed to labour and travelling in the forest.

He was also subjected to serious privations. For weeks he was from home, and in the poor huts of the settlers, he suffered extreme cold and had to partake of the poorest fare. Often the plank was his only bed, and a potatoe his fare, but never did he complain. Cheerfully he went in and out among them, cheering them with the message of life.

The effects of his labors soon began to appear. The people generally began to awake to the subject of religion, many were found turning to the Lord, and a great change in their religious habits passed over the whole population. Family worship, and family religious instruction became almost universal, and people flocked from all quarters to attend on the preaching of the word, young women even walking in summer from the West to the East River, a distance of ten miles or more, for that purpose. When the settlers thus became in earnest on the subject of religion, a most bitter feeling of opposition was raised against him by a set of profligates, at the head of whom were the retired army officers. These men were living in drunkenness and disregard of the marriage tie. The Dr. as in duty bound, spoke to them about their conduct, and induced one to reform, but the rest were hardened.* As the influence of his labours was felt, the people reprobated their conduct in the plainest language. Besides they had hoped to exercise over the soldiers the same authority as previously, but now as he says, "time, intercourse with the other settlers and doubtless an increase of knowledge, induced them to withdraw their subjection." Of this he had to bear the blame, and their animosity against him was excited to such a pitch, that before the end of the first winter they threatened to shoot him, and burn the house in which he lodged. The following winter they held a meeting with a view to send him bound to the Governor, hoping that their mere word would be sufficient to procure his banishment. But, as he says, they went fast to destruction, and on the breaking out of the French war in 1793, all the drunken among the old soldiers enlisted, so that he could look upon Pictou as purged.

He also labored to promote the educational and social interests of the community. Parents receiving Baptism for their children were put under pledges, to give them as good an education as their circumstances would permit. He encouraged them in establishing schools, and when established, frequently visited them. And though

---

* One of them, who lived on Robertson's Island, had bought a soldier's wife from her husband (for selling wives was one of the venerated institutions of the olden time), and was so jealous, that when he left home, he was in the habit of taking her out in his boat and leaving her on a small island off the main one. Dr. McGregor urged upon him the duty of separating from her. "But what will become of the children." "Oh" said the Doctor, "You should do your duty, and leave them to the care of Providence." "They would be the better of my help."

for a length of time they were poor enough, they were the means of giving the young at least the elements of learning.

Among the settlers there were three, who had been ordained elders before leaving Scotland, Thomas Fraser and Simon Fraser, in the parish of Kirkhill, and Alexander Fraser (McAndrew), in the parish of Kilmorack. These were called to exercise their office here, and soon after, the following were elected, and on the 6th of May, of the following year, were ordained, viz.: Donald McKay and Peter Grant for the East River; Robert Marshall and Kenneth Fraser for the Middle River; John McLean and Hugh Fraser for the West, and John Patterson for the Harbor. These first elders, from all we have heard of them, were men eminent in godliness, and a large proportion of their descendants now occupy positions of usefulness in the community and are active members of the church.

This summer (1787) were built the first two churches in the country. "During this month" (July), he says, "the men were chiefly engaged in building the two meeting houses; but, instead of employing contractors to build them, they agreed to divide the work into a number of lots, and appointed a party of themselves to every lot. One party cut the logs and hauled them to the site; another hewed them and laid them in their place; a third provided boards for the roof and floors; a fourth provided the shingles; those who were joiners were appointed to make the doors and windows, and those who did not care to work provided the glass and nails. Moss was stuffed between the logs to keep out the wind and rain; but neither was one of them seated otherwise than by logs laid where seats should be. Public worship was conducted in the open air all this summer, and part of harvest, till the churches were finished, and we had the same kind Providence preserving us from wind and rain and tempest as we had last year; but no sooner were the houses built than great rain came on the Sabbath."

"Such were the first two churches of Pictou, and for a while they had no pulpits, purely because they could make a shift without them, and when they were made they were not of mahogany, but of the white pine of Pictou." These two churches were some thirty-five or forty feet long, by twenty-five or thirty wide. The only seats in them at first were logs of wood or slabs supported on blocks; there was a gallery, or rather, an upper story, with a floor seated in a similar manner, to which the young went up by a ladder. The one on the East River was situated on the west side of the river, a short distance above New Glasgow, on a rising ground between the old burying ground and the line of the present railroad. The one at Loch Broom was situated near the head of the harbour, on the farm then owned by William McKenzie, still held by his descendants. It was situated near the shore, close by the brook that there enters the harbour.

"As soon as the meeting-houses were built, the people set themselves to make roads to them, that they might be as accessible as possible by land; but these roads were nothing more than very narrow openings through the woods, by cutting down the bushes and trees that lay in their line of direction, and laying logs, with the upper side hewed, along swampy places and over brooks which could not be passed dry, by way of a bridge. The stumps and roots, the heights and hollows, were left as they had been. The chief advantage of this was that it prevented people from going astray in the woods. During winter the roads and meeting houses both were totally useless, for the preaching was in dwelling-houses where there was a fire."

It was in November following that he received the first payment of stipend. He should have received £40 in cash and as much more in produce ; but he actually received only £27 of the former, and about £30 of the latter. And yet, of this £27, about £20 was expended in an act of charity, which, we venture to say, has rarely been surpassed, and which, as connected with the early social state of the Province, we must here notice. As already mentioned, some of the settlers from the old colonies had brought with them slaves, and retained them as such for a number of years. Among others, the late Matthew Harris was the owner of a colored girl, afterward known as Die Mingo, and a mulatto man, named Martin. The question of the slave trade had, just previous to the Doctor's leaving Scotland, begun to agitate the public mind of Britain. His feelings had been warmly enlisted on the subject, and he now interested himself in securing their freedom. For this purpose, he actually agreed to pay £50 for the redemption of Die, and of the £27 he received in money the first year, £20 was paid toward this object, and for a year or two a portion of his produce receipts went to pay the balance. He also persuaded Harris to give Martin his pardon, after a period of good service. He also relieved a woman, who was in bondage for a term of years, paying £9 or £10 for her freedom, and, in addition, aided in supporting and educating her daughter.

Fired with zeal on this subject, he soon after wrote a severe letter to a clergyman in a neighboring district, who held a black woman as a slave. This letter, enlarged, was published in pamphlet form in the following year, and led to an epistolary controversy.

In the year following (1788,) the Sacrament of the Lord's Supper was dispensed for the first time in Pictou. The spot selected was a beautiful piece of intervale on the Middle River, partially shaded by an overhanging bank. This was chosen however as central for the whole district, and as accessible by boats from the harbour and coast.

Here the ordinance was observed with all the services then customary in Scotland. These were all performed by Dr. McGregor himself, who preached every day from Thursday till Monday, both in Gaelic and English. The number of communicants was 130. In this spot the supper was observed annually and with the same solemnities, during the whole time, that he was sole minister in Pictou, and people came from long distances, even from the County of Hants, to attend them.

The same summer he commenced that series of missionary labors, which rendered his name so venerated and beloved among the older settlements throughout these Lower Provinces. There was not at that time another minister of any denomination on the north shore of Nova Scotia or Cape Breton, and not a Presbyterian minister, and scarcely any other in New Brunswick or P. E. Island, and from this date, for a period of forty years, he employed a portion of every summer, and even of winter, in visiting the settlements throughout these regions. For this work he possessed the highest qualifications. From the first loving the gospel, all the energies of a very ardent nature were aroused, as he saw the destitute condition of those who dwelt solitarily in the woods, and his preaching grew in impressiveness and power, so that it would be impossible to convince the old settlers that there ever came to America one so eloquent. He possessed the special gift of directing conversation into religious channels, so that whatever subject was started, he gave it a pious turn. When he visited a settlement all gathered, and days and nights were spent for weeks together, in preaching, praying, religious conversation, and travelling from place to place. In this way he traversed the eastern part of Nova Scotia, Prince Edward Island, Cape Breton, and New Brunswick, and was the means of founding or cherishing in their infancy all the older Presbyterian congregations throughout this widely extended territory. In this work he spared no fatigue, and readily endured hardship, finding pleasure in the work, and the richest reward in the joyful reception of the gospel by the solitary dweller in the wood.

The Highlanders having now surmounted the first difficulty of settlement and, above all, having the gospel preached in their native tongue, now invited their relations over from Scotland, and they continued to arrive in greater or less numbers till all those portions of the county most desirable for settlement were occupied. Others who had settled in other parts of the Province were so attracted by his preaching, that they sold their farms and removed to Pictou to enjoy his ministry.

CHAPTER IX.

FROM THE COMMENCEMENT OF THE TOWN TILL THE BEGINNING OF THE FRENCH REVOLUTIONARY WAR, 1789—1793.

We must now, however, give some account of the origin of the town. We have already mentioned that on the arrival of the first settlers, a town was laid out above the Town Gut, and another afterward at Fishers Grant; but at neither of these places was one ever built. Its being located on its present site, was owing to old John Patterson. We have already mentioned that at the time of the issue of the first grants, the block on which the town stands, embracing the shore from Browns Point round to Carriboo, had been reserved out of the Philadelphia Company's Grant, and given to Col. McNutt and his associates. The grant of the latter, however, was escheated in the year 1770, and an order of survey issued in favor of Richard Williams, described as "late lieutenant in the 80th Regiment," and said to have served at the taking of Louisburg and Quebec, for this block and another on the Middle River, extending from Middle River Point up to Kerrs, a distance of ten miles along its banks, and one mile back, the rear line to run parallel with the river, the first said to contain 5,000 acres, but in reality containing 10,000; the second estimated at the same amount, but also considerably in excess.

The grant is dated 21st January, 1776. The conditions as to pine trees, fisheries, quit rents, and reserves of minerals, are the same as in the 82nd Grant, and the conditions as to working are the same as in Camerons. But it contains the following additional clause, that the grantee shall " settle one person upon it for every hundred acres, and the settlers to be introduced to be Protestants, from such parts of Europe as are not in His Majesty's Dominions, or such persons as have resided within His Majesty's Dominions in America for two years, antecedent to the date of the grant." This exhibits an idea that was prevalent at this time, of the danger of the British Isles being depopulated by emigration, in consequence of which parties were bound to settle their grants with Protestants from abroad.

This grant was transferred to Walter Patterson, Esq., who came out in 1770 as Governor of Prince Edward Island, but who had previously been a judge in the province of Maryland, and was, we believe, originally from the North of Ireland. According to tradition, which there is every reason to believe well founded, the title passed to him in the following manner: Riding one day he accidentally met Williams, with whom he entered into conversation, when the latter said, " I have a lot of land in Pictou; what will you give me for it?" " I'll give you this horse," was the half joking reply. "Will you give saddle and bridle?" was the next enquiry. "Yes." "Then jump off." He did so, when the other mounted, and thus the bargain was completed. At all events, Governor Patterson became the possessor, as it was found afterward, without any proper conveyance, and at the date of our history began selling portions of his land. He also laid out a town, which he called Coleraine, to the eastward of what is still known as Coleraine street. According to his plan, it was to contain various public conveniences, the ground at the foot of that street being reserved for a public landing, and a lot near set apart for a market. Hence, the wharf there has been till recently, and we suppose is by some yet, known as the Market Slip. This, however, like previous efforts, was for some time at least a town only on paper.*

On the 10th September, 1787, for the sum of £62 10s., he deeds to John Patterson 150 acres, afterward purchased and occupied by Mortimer, at the Point; and another lot

---

* We may here give the subsequent history of this grant: Patterson, while on the island, sold portions both of the part on the Middle River and at the town, by his agent, Lieutenant Gordon, already referred to; but his title being doubtful, persons squatted on various portions of his land, particularly on the Middle River, and some who purchased one lot took possession of another. We have seen deeds running in something like the following strain: "Whereas I have made improvements on the land of Governor Patterson, at such a place, I hereby give all my right, title and interest in the said lot of land," &c.

Patterson became involved in debt to Messrs. Cochrane, then leading merchants in Halifax, doing business at what has since been known as Cochrane's Corner, where the Dominion public offices now stand. In consequence, after he left the Province, they sold his property under a judgment, and one of them became the purchaser, and hence this grant has since been usually known as the Cochrane Grant. But there was, for some time, a difficulty about the title, partly in consequence of want of proof of a transfer from Williams to Patterson, but partly also, we believe, from some irregularity in the legal proceedings of Cochrane against the latter, and squatters resisted attempts to eject them. One of Patterson's heirs came some years after, claiming the land, with whom Cochrane compromised. After this his title was generally conceded, and he sold the rest of the land, but a good deal having [previously been occupied, the parties held it by possession.

of 100 acres, described as follows : " Beginning at a stake and stones at the west corner of a town lot, thence running north 320 perches, from thence west 50 perches, thence south 320 perches, thence east the different courses of the harbour to the place of beginning." The front of this lot embraced the central portions of the town, extending from Coleraine street westwardly to the market. A settler named, we believe, Kennedy, had previously squatted on a portion of this, having made a small chopping, and erected a shanty on the face of the hill, below Dr. Johnston's house, so that the deacon had to buy out his claim.

By the terms of the deed, the ground along the shore, to the width of forty feet, was reserved for a highway, and this became the front or lower street of the town, from the west end as far as Glennies. This will explain its sinuous course. To this point the south side of the street was originally water lots, and all the buildings now upon it are upon made ground. The ground near Meagher's slip was regular bog, through which logs were laid to form a pathway. At Glennies the coast took a turn to the southward, with a somewhat high bank, forming a cove, which was long used for the reception of timber. From this point the deacon run a street in a straight line to Coleraine street, although there was a road round the shore to the Battery Point, within the memory of many still living.

In the previous winter (1787) the wood had been cut on part of the ground where the town now stands, and Patterson removed from Mortimer's Point to where his grandson, A. J. Patterson, now resides. He next commenced dividing the front of his land into small building lots, which he sold at low rates. These, however, were conveyed upon the condition that the purchasers should build upon them, and under a penalty for every year in which this was not done. Thus, in his ledger we find, in

1805, one man charged, "To 7½ years' damages for not building one house;" and in another case we find the following entry, "To 5 years you did not build on the lot, at 5s., £1 5s." These lots are described as in the town of New Paisley,* doubtless, so named after his native town. Among the first lots, of which the deed is recorded, is one to James Dunn, of the lot on which the Royal Oak Hotel lately stood, on which he erected the first tavern in the town,† just behind where that building lately stood.

The Deacon having been originally a carpenter, commenced now the erection of small buildings to sell or let to tradesmen, and it is in this way that the town was fixed on its present site. We have been unable to ascertain where the first house was built, but believe that it was on the street leading down to Messrs. Yorston's wharf. At all events, one that still stands there was among the first. It was originally occupied by him as a store, afterward by his sons, John and Abraham, in the same way; again, for a length of time as a cooper's shop, and now as a sailmakers. The timber was cut on the site of the lower part of the town. This is, without doubt, the oldest house in town. On repairing it lately, the carpenters found that the frame was entirely of hard wood, oak, ash, or beech. The date of erection of the first house was either 1789 or 1790. Dr. McGregor says in one place, "There was not a single house for years after I came here. The town was for some years without a single inhabitant; then there was a shed with one

---

* We may here notice the variety of names selected for the town or district. Besides the deacon, we find in old documents, the township of Alexandria, the township of Donegall, Teignmouth, Southampton, Walmsley and Coleraine.

† We read, however, the name of Francis Hogg, tavern keeper, before this date, and, as we have seen, the grog shop was in full blast, before Dr. McGregor's arrival. The site, however, of these establishments we do not know.

family; then another with it, and so on, till it became what we see it now." In his narrative he says:—"I think it was in this year (1790) that the first house in Pictou was built. It was some years without a second." Philo Antiquarius, speaking of the year 1789, says:— " Two or three houses were also erected about this period on the spot which was cleared the preceding year. Families were soon attracted, buildings were run up, and, ere the lapse of much time, a respectable hamlet rose into view."

He also built a wharf, the first in the town, on the site of what, afterward coming into possession of his son, also an Elder in the Church, long continued to be known as the Deacons Wharf, now Yorstons. It was described as consisting at first of three logs, but afterward a more respectable structure was erected. We may add, that he always showed an earnest and hopeful interest in the progress of the place, and a most enterprising spirit in undertaking measures for this end.

Thus it was to the sagacity and public spirit of John Patterson that the town rose upon its present site. The question has been raised where it ought to have been. Richard Smith maintained that it should have been on the south side of the harbor, at Abercrombie or Frasers Point. With the knowledge we now have of the mineral wealth of the East River, much may be said in favor of that locality. But the fact that after attempts to build a town elsewhere, the deacons site held its own, even though opposed by Mortimer, shows that something may be said in its favor. Under any circumstances, its nearness to the entrance of the harbour would have almost necessitated the erection of a town somewhere in that neighborhood. One unfortunate mistake he committed, though a natural one in the circumstances; that is, in taking the road originally laid out round the shore as the basis from which to lay off lots and other streets. The side lines of the original lots run from the harbor on a due north and south course, and had the cross streets been laid off at right angles to these, the inconvenience which the subsequent inhabitants have suffered from the want of square corners, would have been avoided.

We may add, that the first teacher in town, so far as we have been able to ascertain, was Peter Grant. When his father, Alpin Grant, settled in Pictou in 1784, he remained with a friend in Halifax, where he was educated ; and on coming to Pictou, about the year 1793, he opened a school and continued to teach for six years. In the year 1800 he took up land at Scotch Hill and removed thither the following season, where he resided till his death. In the year 1802 we find S. L. Newcomb teaching, though probably for a short time.* Then, for some time, George Glennie occupied his place. He was a graduate of Aberdeen College, a superior scholar, and did much to form the minds of the youth of that generation.

Philo Antiquarius says, " a vessel was launched from the point above the Gut in the year 1788," he does not say by whom, but the manner in which he speaks seems to imply, that it was the first built in the harbour, but the tradition is that the first in the harbour, was built by Thomas Copeland, on the site of what is now Dr. Johnstons wharf — that she was what was called a snow, that is a two-masted vessel, with another small mast abaft the mainmast, to which a trysail was attached, and that the first built in the county, a schooner called the Ann, was built previously at Merigomish, for John Patterson. The date we cannot ascertain, but it must

---

* Mr. Newcomb married a daughter of Matthew Harris, and had a family, several of whom were once well known in Pictou and Cumberland. One son, Simon, after long sailing out of Pictou to the West Indies and elsewhere, as a captain, served in the Texan war of independence, and obtained a large grant of land there. He persuaded his brother Thomas, then a promising lawyer in Cumberland, to remove thither, where he died. and his son was lately Secretary of that State.

have been about this time. In a ledger of his in our possession, which however was not his first, we have "an account of things sent to the vessel, November 21st, 1789," so that it must have been built before that time, though probably not long. Farther on she is called "the schooner" and Jerry Palmer appears to have been the name of her first commander. Small vessels however were owned previously. In the year 1784, we find recorded a bill of sale to Hugh Dunoon from Barnabas McGee, of "the good shallop Nancy," and by the returns already quoted, Dr. Harris appears as owning a schooner or sloop in 1769, and Squire Patterson another in 1774.

About the same time arrived a number of persons, who occupied prominent places in connection with the early progress of Pictou, the most important of whom was Edward Mortimer, who is said to have arrived in 1788, but to whom we shall have occasion to refer fully in a subsequent chapter. Captain Lowden also at this time removed with his family to this country.

At this time also came the Copelands. Alexander and Thomas were brothers, natives of Castle Douglass, in the County of Dumfries, and for some time did business in Halifax. They brought out two cousins, Samuel and Nathaniel, also cousins of each other, whom they had in their employment. A story has been commonly told of their making money by purchasing at the sale of an American prize some kegs of nails, which, however, were found to contain dollars in the centre. Their descendants discredit this, and tell a story of an opposite character, viz., that during the American war, Halifax being menaced by some American armed vessels, which had plundered other places, they had buried their money in their garden, but afterward could not find it, and circumstances led them to suspect a neighboring family of having discovered their secret and stolen their treasure. This, and some other losses, led them to remove to Pictou, where they obtained the farm lot west of John Patterson's, including what has since been the western part of Abraham Patterson's farm and the lot of Matthew Patterson. They built a wharf at what has since been known as Smiths Wharf, the second in the town, and commenced business there. Alexander and Samuel (father of Robert S.) died in Pictou, the latter being the first buried in the Pictou cemetery, about the year 1807, the cemetery at Durham being previously used by all on the north side of the harbour. Their families removed to Merigomish, as did also Thomas and Nathaniel. They all settled on lands, which they had purchased from 82nd men, in most instances for a mere trifle.

About the same time came Robert Pagan. He was a son of John Pagan, already mentioned as one of the proprietors of the Philadelphia grant, and owner of the Hector. He was in Pictou as early as 1789. In that year we find a deed to himself and his brother Thomas, described as merchants, from Walter Patterson, Governor of P. E. Island, of "Lots 2, 3, 4 and 5 in the town of Coleraine," described as "beginning at a stake and pile of stones upon the shore or bank of the harbour of Pictou aforesaid, at the south-east corner of lot No. 1, in said town, and 117 feet east from the boundary line between lands granted to John Patterson, senr. of Pictou aforesaid, and the said town of Coleraine." This was at the head of the Mining Companys wharf and extending westerly to Crichton's wharf, and his house was on the site of that recently occupied by Mr. Boggs. He married a daughter of Squire Patterson and engaged in business, but was not very successful. He was a man of excellent education, and filled several offices with credit, the highest being that of Judge of the Court of Common Pleas. He died 3rd December, 1812. None of his descendants are in the county.

Hugh Dunoon, after doing business in Halifax, removed

to Pictou about the same time. He was a native of the parish of Killearnan, in the county of Inverness, of which parish his father and brother were successively ministers. He took up land on the East River, as early as 1784, and we find him soon after living in Merigomish, where he had bought out the rights of some of the 82nd men. He built a mill on Hogans brook, the first in that part of the county, and carried on business there. He afterward removed to Pictou, where he lived about a mile below the town, where Mr. Fogo now resides. Subsequently he occupied a prominent place in the public affairs of the County, filling a number of offices. He was for many years a Judge of the Inferior Court, Deputy Registrar of Deeds, Collector of Customs and Custos of the District. He died the 24th March, 1836, aged 74.

John Dawson was here as early as 1791, his marriage having taken place in that year, but he will be more appropriately noticed elsewhere.

We give a plan of the town at this time, copied from one in the possession of the late Peter Crerar, Esq., which will exhibit at a glance its progress to this period.*

In the meantime the rural settlements were receiving settlers, especially from the Highlands of Scotland, among whom we may mention Martin McDonald, who arrived in 1787, and settled at Knoydart, to which he gave the name from his native place; and Alex. McKenzie from near Inverness, who, after serving eight years in the army, settled with Archibald Cameron, on Irish Mountain, where they had been preceded by Patrick Finner, an Irishman.

In the year 1788 or 1789 came a number of Lowland Scotch, principally from Dumfries, who settled in various parts of the country. Among these may be mentioned John Cassidy, who settled on McLennan's Brook; William

---

* For this plan we are indebted to the kindness of J. S. Arnison, Esq., Sandyford House, Newcastle-upon-Tyne. The date of it is not given, but it is before 1794 and after 1790.

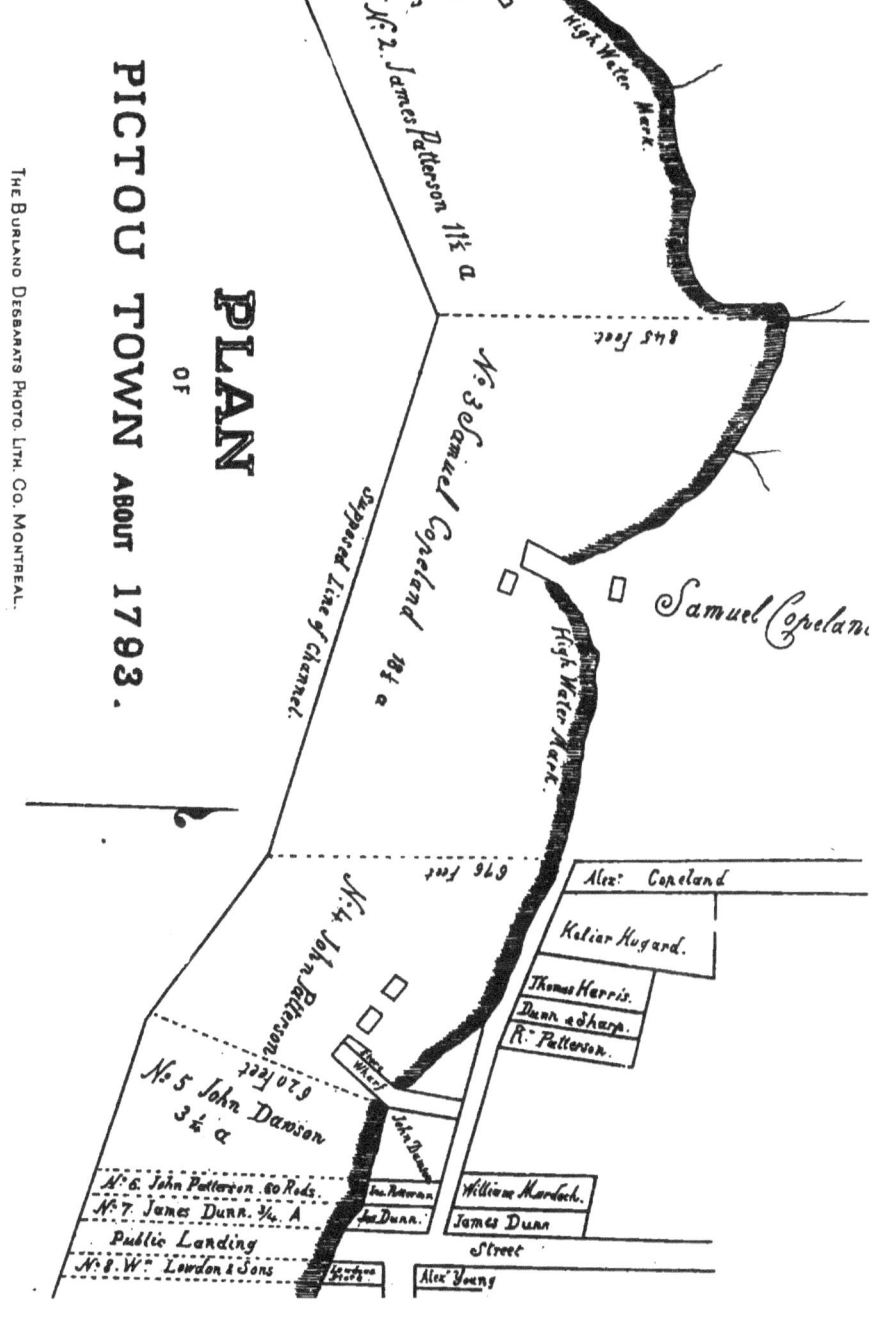

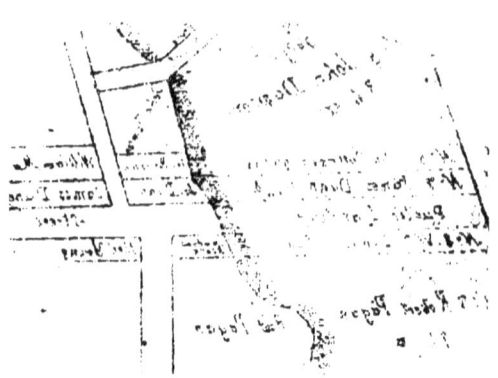

Munsie, who was the first settler on Green Hill, on the farm still occupied by his descendants; Robert Sturgeon, who settled on the south side of the hill; William Porter, who settled on the Middle River; John McGill, afterward of West River, and David McCoull, of Green Hill.

We may here give a notice of a visit to Pictou in the summer of 1787, of Sir John Wentworth, afterward Governor of the Province, and at that time Surveyor General of Woods and Forests in British America, as given in a letter of his, dated 19th January, 1788:

"Continued toward Miramichi, in the Gulf of St. Lawrence. Being overtaken with a hard gale of wind, and my boat, under 13 tons, unable to live in such cross-combing seas, we attempted Merigomish harbour in the night, and were soon on the sands; the tide rising we got off and rode out the night in a little pool between two reefs. The next day, at high water, we went over all into the harbour, which is very safe, and, with a pilot, of easy access for a ship of four or five hundred tons. Contrary winds detained me here and at Pictou (which is still a deeper harbour) for fourteen days, during which time I travelled into the woods from ten to twelve miles round, surveyed timber for the settlers and that proper to be reserved for His Majesty's service. In this district is some good pine timber, but the longest and best has been cut away by the Acadians and first English settlers, to whom that part of the country has been granted for some time. On the harbours and rivers of Merigomish and Pictou there was, formerly, the best growth of oak and pine; some still remains, and I have taken every measure to preserve the best of the latter, and have no doubt it will come into His Majesty's service, should any contract be extended to these Provinces.

" We sailed from Pictou the first possible moment, and met a very heavy sea, with which we contended until sunset, when the wind arose suddenly at southeast and obliged us to take shelter in Johns Harbour, to leeward of a rocky reef. We rode out the night, but at daylight were obliged to slip our cable, and run up the river, where the tide suddenly neaped on the change of the wind, and we were five days unloading provisions and ballast, perching out a channel, near five miles, and recovering the anchor. In this time I explored the adjacent woods, and found some good pine timber. Those unfit for the King's service I have granted license for to the settlers, who have promised to preserve those marked for reservation."

We have given the above for the purpose specially of noticing the claims, which it had long been attempted to maintain, of reserving the large timber for the use of the British Government. Wentworth had been Surveyor General of Woods for all the British Dominions in America, and, till the commencement of the American Revolutionary War, residing in New Hampshire; after that he retained the same office in the remaining British Provinces. The attempts to enforce this claim had produced great disturbance in the old colonies, and was one of the causes of the discontent, which led to their separation from the mother country. In this Province it had been a hindrance to its settlement, as it had been necessary to communicate with him in New Hampshire, as Governor Legge complained, before grants were issued to settlers. Yet, notwithstanding these things, and the utter absurdity of expecting settlers to clear their land and leave the big trees standing, Wentworth came to Nova Scotia, seeming to regard it almost as the chief end of his creation, to enforce this and every other arbitrary claim, that had ever been set up on behalf of the crown. As late as the year 1804, we find timber seized as cut in violation of this claim, and only released on the parties paying a certain sum as damages, and instructions issued that " the timber be surveyed on their lots and such as are fit for masts, yards or bowsprits marked ' I. ∧ W.,' also some of the best to cut into planks for the King's service. These to be faithfully preserved, then the rest to be free to the party."

Again, in the year 1806, we find timber seized, which, however, Wentworth instructed his agents to release on the party paying the sum of £12, "which sum is to be applied, under your care and direction, toward building a bridge at the head of Carriboo River, on the new road now to be made leading from Pictou to Carriboo Harbor."

We do not know that this claim was ever directly abandoned. To a very late period surveyors, in making their return of the survey of Crown lands, were accustomed to report that there was no timber upon it fit for the use of His Majesty's navy. But no such old fossil as Wentworth ever governed us again, and it died a natural death.

In the summer of 1791 arrived two vessels loaded with emigrants, mostly Roman Catholics, from the Western Islands of Scotland. They arrived so late in the season that few could provide houses for their families before winter. The old settlers, at the solicitation of their pastor, received them with the most open-hearted hospitality. Hundreds of them received the best shelter that the inhabitants could afford—such as could pay, at a very moderate price, and they that could not, for nothing. For a time they commenced settling in Pictou, and a number of them went to hear Dr. McGregor preach, but Priest McEachran came over from Prince Edward Island and persuaded them to go beyond the reach of Protestant influence. They accordingly left their settlements in Pictou, some of them with great reluctance, and took up land along the Gulf shore, a few in Pictou county, but the majority in Antigonish county. Some, however, went to Cape Breton. Among this class there is cherished to this day a fervent and grateful recollection of the kindness of Dr. McGregor and his people. But the Dr. complained that they proved dangerous guests, by foolish and profane conversation, and particularly by their tales of superstition, and that the evil influence of such close intercourse with them was felt upon his congregation long after.

A few of these, however, were Protestants, or became so, and settled permanently in the county. Among these may be mentioned John McKinnon, Lauchlan McLean, Angus McQuarrie, and Alexander McMillan, and others, who settled on the east branch of East River. All was woods above what is now Samuel Camerons place, except where Charles McIntosh, who had settled where David McIntosh now lives, and on the west side, where John Fraser, who had come from Strathspey in 1788, had commenced. There was only a blaze to the lower settlement. The salmon in the river were still exceedingly plentiful, and gaspereaux so abundant that as many as fifteen barrels were taken in a single night, which they often were not able to use for want of salt. Such was the difficulty of obtaining other conveniences, that they were obliged in summer to carry their produce on their backs to New Glasgow, and thence transport it by means of canoes to Pictou; or, in winter, drag it on hand sleds the whole distance. There they sometimes had to give a bushel of wheat in exchange for a bushel of salt. They adopted another plan, however, of getting their pork to market. They drove the animals to Fishers Grant, where they obtained the use of a slaughter house, and then butchering them, they disposed of the meat to the merchants for such goods as they required.

James Grant, who had originally settled at what is now called Grahams Pond, Carriboo, was induced by the settlers, who wanted a grist mill, to move up. He erected the first grist mill in the settlement, a little below the site of the mills, since occupied by his sons and grandsons, near Springville. The first mill above Springville was erected at Sunny Brae, by Hugh Fraser, about 1805, where the mill still is.

There was then no church, and preaching was, in winter time, in houses, particularly Charles McIntosh's, and in summer, when the weather was wet, in barns, but when fine, by the river side. The first church was built at Grants Lake, to accommodate both branches of the river, and was of logs.

Such was the progress of settlement that in that year the population of Pictou was estimated at 1,300 souls. In the first settlement every man was obliged to act as artizan for his own family. Now, however, mechanics were attracted, who attended to their own employments. The increased population created a demand for various articles which led to trade. Roads, as they were called, were opened, though as yet none of them were fit for a wheel carriage, being mere bridle paths. The price of land increased, and farms partially made brought what would now be considered fair prices.

In May, 1792, Wentworth became Lieut.-Governor of the Province, and that summer visited various portions of the province. His visit to Pictou is thus noticed in Murdoch's history : " Friday, Sept. 21, Lieut.-Governor Wentworth sailed in the armed schooner Diligent, for Pictou, having gone on board under a salute of thirteen guns. He arrived at Pictou, on the evening of Wednesday, the 3rd of October, and landed on the 4th. He received an address, signed by 30 persons. They stated that 500 men there had agreed to work on the road to Halifax. He returned by land by way of Musquodoboit, and it was stated that the road from Pictou to Musquodoboit was now open for horses and cattle." William Fraser (surveyor) says of this visit, that he came to Pictou, bringing plenty of provisions, axes and hoes, bought with the proceeds of a Government lottery, and that the new settlers were called to meet at William McKay's, for the purpose of receiving them according to their need.

There seems to have been at this time much ado about this road. After His Excellency's return to Halifax he and his lady gave a ball and supper, described in the gazette of the day, as " altogether the most brilliant and sumptuous entertainment ever given in this country." The writer adds, "Among other ornaments, which were altogether superb, there were exact representations of Messrs. Hartshorne & Tremaine's new *Flour-mill* and of the *Wind-mill* on the common. The model of the new *Light-house*, at Shelburne, was incomparable, and *the tract of the new road from Pictou was delineated in the most ingenious and surprising manner*, as was the representation of our fisheries," &c. Haliburton in his history also says, under the year 1792, " Great Pictou road opened," and has been followed by others.

The idea of a shorter route from Halifax to Pictou than by Truro has been often entertained. At a later period Sir James Kempt projected one by Stewiacke, Middle River, and Loch Broom. It seems that at this time the plan of a road by Musquodoboit had been so far carried out, that cattle and horses could pass over it. But, strange to say, for a length of time we could not find, even among our oldest inhabitants, any person who knew anything about this great road. It is, however, noticed in the early county records as the Governors Road. The fact is, that notwithstanding the great flourish of trumpets about the opening of it, nothing more was done to it, and the old route by Truro continued to be the one generally used.

Hitherto Pictou formed a part of the district of Colchester or Cobyquid, as it was commonly called, and all its municipal business was transacted at Onslow, where the Court of Sessions met. There also was the place of holding the poll, for the election of representatives. The increase of population, and the inconvenience of attending such a distance, led the inhabitants of Pictou, to petition to be relieved from attendance there. Accordingly in this year an act of the Legislature was passed, forming Pictou into a separate district, defining its boundaries as we have already given them, and appointing the Court of General Sessions of the Peace and Inferior Court of Common Pleas, to be held at Walmsley, on the 3rd Tuesday of January and 3rd Tuesday of July. It was also ordered, that a poll should be opened here, on the occasion of the election of representatives for the County of Halifax.

The changes which this introduced we shall notice hereafter. In the meantime we give a copy of Deacon Patterson's account of charges and receipts for the poor of the district at this period:

The Poor Dr. to John Patterson, Senr.

1787—To 1 blanket to a sick man at James Carmichael's........	£ 0	10	0
To rum and other things by Mr. Abercromby for burien of the above man at James Carmichael's	1	0	0
1789.—Nov. 16.—To 8s. payd to James Dun by order of the——..	0	8	0
To stocks by order of do.....................	0	15	0
To ½ gallon rum by order of the Justisses, omitted January, 1789.......................	0	4	0
1791—Dec. 19.—To 2 yds. blue cloth at 8s. and 2½ lbs blister steel, at 1s. 2d.............................	0	18	7
To 1 pair blankets by Pagan................	1	0	0
To 1 coverlet by Mr. Mortimer...............	0	12	6
To 1 cape 2s. and 1 lb—— by do 1s..........	0	3	0
To 1 pair shoes by do.....................	0	10	0
1792—To Butting, survayin, colecting and comission on timber this year.................	0	10	0
1793—To a weedow woman W R............................	0	15	10
To Lachlan McDonald, gulf..................	0	11	8
To Mrs. Patterson Dowry for gaill lot.........	1	10	0
	£8	19	3

Contra, Credit, July, 1792.

To ¼ of 1 log by James McKay, Mr. Pagan.....................	0	4	6
To ½ of log by James Reid for Mr. Mortimer.....................	0	2	0
To ¼ of log Frank Carmichael & Campbell for do...............	0	3	6
To drift timber by Mr. Pagan...........................	0	7	0
To do do delivered to Mr. Copeland, Amount..............	1	13	7
To 1 log McQuarry for do...........................	0	3	0
Neat proceds of the above............................ £	2	4	7

The spelling of our worthy grandfather is not always in accordance with Johnston or Webster, but as we like originality, we have given it in its primitive form. The logs of timber used in payment of poor tax would seem rather a cumbrous currency. When the poor asked for bread, to give them a stick of wood, scarcely seemed Christian charity. It does not appear that the district was very prompt in meeting his expenditures, but he has left a name behind him, especially for kindness to the poor.

CHAPTER X.

FROM THE COMMENCEMENT OF THE FRENCH REVOLU-
TIONARY WAR TO THE FIRST CONTESTED
ELECTION, 1793—1799.

In the year 1793, as we presume all our readers know, commenced the French Revolutionary War. One of the first effects of this upon the County of Pictou was, that the Governor raising a regiment, a number of the disbanded soldiers who had settled in Pictou, took the opportunity of enlisting. As Dr. McGregor describes them, " All the drunken and profligate," while none of the sober and industrious, either of the soldiers or the other settlers, followed their example.

Before this time the timber trade had been carried on, and was of some importance to the infant settlement. The first effect of the war was a slight reduction in the price of timber; but this was soon succeeded by its rising to an unprecedented height, and with this came a rapid increase in the trade from Pictou, which was at its height, as we shall see presently, from about the year 1800 to the year 1820.

At this time, too, ship building was being carried on to some extent, Captain Lowden's efforts in that respect being specially worthy of notice. Indeed, he may be considered the father of the ship building art in Pictou. He was a native of the south of Scotland, and had commenced trading to Pictou during the American Revolutionary War. Previous to this, he had been fifteen years in Russia, and also employed in carrying convicts to Virginia. In the year 1788, he removed with his family to Pictou. He first located himself near the Narrows, at the East River, where he erected a windmill at what has since been known as Windmill Point, and commenced ship building there. But soon after he removed to town, where he occupied a two-storey building of John Patterson's, on the site of Messrs. Yorston's store, the lower as a dwelling house, and the upper with goods, which he exchanged for timber. He also built a wharf, on the site of what has since been known as the Mining Companys Wharf, and commenced ship building there. The whole eastern part of the town, from Ives' store to the Battery Hill, was covered with a fine growth of hardwood, and the timber necessary for the work was cut close by his yard, or, afterward, on the top of the Deacons Hill, whence it was slid down on the snow to the shore, and, when once set in motion, it may be supposed, went with terrific rapidity. He erected a building on the east side of Coleraine street, which he used for boarding his men, but which was commonly known as the Salt House. Some years later he erected a windmill on a round hill near the head of the wharf, long after known as Windmill Hill, but which has now been carried away in levelling the ground near the Custom House. This mill was well constructed, had a large amount of machinery in her, and for some time did a large amount of work, both in sawing and grinding.

He continued for a number of years the business of ship building, his vessels being sent to Britain for a market, and was rather noted in the Province for his skill. Of one of his vessels we copy from Murdoch's History the following notice :

"Pictou, October 25, 1798.

"Yesterday was launched here, by Messrs. Lowdens, the ship Harriet, burthen 600 tons. She is pierced for 24 guns, and supposed to be the largest and finest ship built in this Province. Her bottom is composed of oak and black birch timber, and her upper works, beams, &c., totally of pitch pine; on account of which mode of construction, she is said to be little inferior in quality to British built ships; and does peculiar credit, not only to this growing settlement, but to the Province at large."

This we presume was the vessel known as Capt. Lowdens "big ship." She was commanded by his son David. She was mounted with four real guns, the rest being what were called "quakers." On her first voyage, she fell in with what was supposed to have been an enemy's privateer. The captain, backed by a determined Scotch crew, determined to fight rather than be taken. The other vessel, however, kept shy of them, and at night disappeared.

Another of his vessels he called the "Prince Edward" after the Duke of Kent, then in Halifax, who sent a sum of £50 to purchase a set of colours for her.

He had four sons engaged in business, first with him and afterward on their own account, Robert, who afterward removed to Merigomish, where his descendants still are, David who afterward lived at the Beaches, where his sons still reside, Thomas, whose house still stands near the head of the public wharf, and William, usually known as Bishop Lowden, long regarded as one of the characters of the place. He was a splendid scholar, knowing the classics and several modern languages, even acquiring the Gaelic. But owing, it was said to his being crossed in love in early life. he became partially insane. For years he never washed, and went about in a greasy coat, which made him the object of sport to the young. In his lodgings he pored over his books, and in later years, gave himself to the composition of an English grammar, which he succeeded in getting printed in the United States.

The year 1795 was noted among the early settlers as the date of the arrival of a second minister to share the toils of Dr. McGregor, the late Rev. Duncan Ross. The settlement of a single minister would not now be regarded as involving very important results to a county ; but at the time it was regarded as of sufficient interest to call forth rejoicing, and in many devout thanksgiving—in fact, to form an era in the history of the settlers.

Mr. Ross was a native of the parish of Tarbert, Rossshire, but at an early period of life, he removed with his parents to Alyth, in Forfarshire. He received his Latin education at the parochial school of that town, after which he passed through the usual curriculum at Edinburgh University. He studied theology under Prefessor Bruce at Whitburn, and was, on the 20th January, 1795, ordained by the Presbytery of Forfar. In June he arrived in Pictou, by way of New York and Halifax, along with the Rev. John Brown, afterward of Londonderry. They assisted Mr. McGregor at the dispensation of the Lord's Supper, and thereafter these ministers formed the first Presbytery of Pictou, under the name of "The Associate Presbytery of Pictou." Their first meeting was held in Robert Marshall's barn, as central for the whole district. It stood near the road from New Glasgow to Middle River, on the ascent of the hill to the west of the bridge across McCulloch's Brook, and on the left-hand side of the road as you go westwardly.

Immediately after, Mr. Ross was called as assistant to Dr. McGregor, and until the year 1801 they were jointly ministers of all Pictou, though Dr. McGregor labored principally on the East River, and Mr. Ross on the West. In that year it was agreed to divide the congregation into three, the East River, with Merigomish, under the charge of Dr. McGregor; the West River, with Middle River and Rogers Hill, under Mr. Ross; and the Harbour and Fishers Grant to form a third, to be supplied by the ministers of the other congregations, till they should obtain one of their own. This arrangement continued till the arrival of Dr. McCulloch.

Mr. Ross was a man of a very clear and logical mind and strong natural powers; he could scarcely be called a popular preacher, but by intelligent persons, his pulpit ministrations were highly relished for their clearness, variety and solidity of matter, and oftentimes ingenious and striking illustrations, while, among his ministerial brethren, for soundness of judgment, knowledge of Church matters, and intellectual capacity, he took rank among the "first three."

In private pastoral work he was laborious and faithful, visiting and catechising over the whole of his extensive charge to the end of his life. He also laboured for the advancement of the general interests of his people, especially by encouraging education, and promoting agricultural improvement among them. The influence of his example and recommendations was, in a variety of forms, perceptible among them, so that they became distinguished among our rural population for their intelligence and public spirit. As we shall see, he was also the first in the Province to found and support temperance societies.

He did not write much for the press. His principal publications were on the Baptist controversy, in which he showed himself a vigorous thinker and acute controversialist. He also contributed to the newspaper press, especially the *Acadian Recorder* and *Colonial Patriot*. He was a man of much quiet humour, one or two specimens of which may be given. Mr. Mortimer once meeting him riding on horseback, with a spur on one foot, said, "Mr. Ross, is one side of your horse slower than the other, that you have a spur only on one foot?" "Oh, yes," said Mr. Ross, "one side will get along without spurs as fast as the other will with all the spurring I can give it." Meeting the late Jotham Blanchard, the latter began playfully to tease him about his hat, which was of rather more than the usual breadth of brim. Mr. Ross replied, "Oh, Mr. Blanchard, you need not be so hard, it is only an error of the head, not of the heart." Some of his sayings of mingled wit and wisdom still float among the people of that part of the country, of which the following may serve as examples: Hearing a man described as "hard and honest," he said "that generally meant hardly honest." Again, he was accustomed to say, "that he had tried three ways of living; the first was to buy just what he wanted, but he found that would not answer; he then tried only buying what he could not do without, but did not find that to answer either. He then tried only buying what he could pay for, and that he found to answer well." These may serve to illustrate a wit which, if not sparkling, was genuine, and which, combined with his affability and intelligence, rendered him a genial companion.

In bodily stature he was below the middle size, broad and strongly made, and during the latter years of his life. inclining to corpulency. His appearance in the pulpit, especially at that period, was particularly clerical, his long white hair contributing not a little to the effect. He died on the 25th October, 1834, after a short illness.

As society was now fully organized, and the community had assumed a settled form, we may here pause to give a brief view of the social and material condition of the population at this time.

As to origin, the large majority were from the Highlands of Scotland. On the East River, when Dr. McGregor came, there were only two settlers who were not, and these were Lowland Scotch. In the other settlements there was a larger infusion of other nationalities; but in all, with the exception of River John, the majority were Scotch. The Gaelic language was everywhere heard; the customs of their fatherland everywhere seen, and its memories and traditions—in some instances, even its superstitions—fondly cherished. Some had been old enough to have been "out" in the Forty-five; many, at least, remembered Culloden; the sympathies of the majority were with Bonnie Prince Charlie, while all the older generation had their reminiscences of the scenes of that day. A few others had served under Wolfe, and had their tales of Louisburg and Quebec, while many more had been in the service of the British Government during the American Revolutionary War, and were full of hatred of "the Rebels."

The Highlanders, as settlers, have been pronounced unsurpassed for encountering the first difficulties of a settlement in a new country, but inferior to some other people in progressiveness. Accustomed to extreme poverty, they readily endure hardship; but it is said that they are apt to be content with a condition, but little beyond what they had previously enjoyed, and do not show the same eagerness for farther progress that others do. This has, to some extent, been the case where they have settled by themselves, but where they have been mixed with others, there is so much of the spirit of emulation in them, that they will soon compete with their neighbors in almost anything.

Physically the inhabitants were generally a superior class. An unusual proportion, both of the Highlanders and Lowlanders, were remarkably stout, strong men. This was no doubt in part owing to the fact, that it is the most adventurous who first emigrate, and they generally possess a good measure of physical vigour. But we cannot help thinking, that the tremendous drafts on the able bodied in the Highlands, to supply the British army, for more than fifty years previous to 1815, so much greater in proportion than in any other part of the empire, materially weakened the vigour of the race. Certain it is that the late immigration from the Highlands would not compare physically, with the first settlers of this county from the same quarter.

At this time, the population were scattered principally along the shores of the harbour, and the coast thence to the eastern extremity of the county, and along the banks of the rivers, wherever there was intervale, there the settler being attracted, as the eagle to the carcass. Only in a few instances, had settlers gone back from the rivers. William Matheson had settled on Rogers Hill, John Rogers farther up, William Munsie and John Blaikie were on Green Hill, and on the East River two or three settlers were on McLellans brook, Angus Campbell and perhaps one or two others were on Scotch Hill, but we know of none others, who at this time lived away from the shore or the banks of the rivers. And so scattered were they, that distances of from half a mile to three or four, commonly intervened between their residences.

They had now, however, reached that position in which they had plenty to eat. The lands chosen were good, and when the wood was burned, produced plentifully; but from the large size of the trees, the clearing involved much labor, and the stumps were left, so that there was yet but little ploughing. The most of the crop was still covered with the hoe, but even with such husbandry, potatoes, wheat and other crops never failed to yield an abundant return. Fish in the river were still abundant. A net set at the end of the Deacons Wharf has been found in the morning sunk to the bottom with the multitude of fish, and salmon and gaspereaux thronged the rivers, so that, even without the produce of hunting, to which we shall presently refer, the inhabitants were abundantly provided with the means of subsistence. But in regard to other conveniences, they were still deficient. Their houses were still generally of logs, small, and containing few elegances. Some British cloths were imported, but generally people were clothed in what they manufactured from their own wool or flax. As to their feet, all ages and sexes carried them a great part of the year in a state bare enough to appear in before an Eastern king. When the severity of the weather rendered necessary some additional covering, it was generally a raw-hide moccasin.* Store luxuries were little used. What would the present generation think, of Mr. Mortimer bringing home in a small green bag all the tea needed for a seasons trade in Pictou! There was consequently an ignorance of the proper mode of using it, which sometimes led to amusing mistakes. A party of men who had gone from home on some work on which they were engaged, took about half a pound with them. Delivering it to the woman with whom they were staying, to prepare the beverage for them, they were surprised to be treated to a black and nauseous draught, which they were unable to drink, and, on enquiry, found that she had boiled the whole at once. And there is an instance well established of a woman just arrived from the Highlands, who, wishing to show her gratitude to a person who had kindly entertained her on arrival, boiled a half-pound of green tea, which she had bought before leaving Scotland, as a great rarity, and, throwing away the liquor, served up the leaves, as a special entertainment. Her chagrin on learning her mistake may be imagined. Much later, a man said, "We bought a pound of tea; it cost eight shillings, but it did us eight years."

Horses were still few, so that the most of the travelling was on foot. But those who were becoming more independent in their circumstances, were beginning to use them more for travelling to any distance, and "riding double," or pillion riding, as it is called in some places, for business of more family interest was becoming an institution. And a cozy way it was, for the good man and his wife to proceed thus to kirk or market, or the lad with his

---

* A woman who arrived here in 1795, told me that in attending the Sacrament, wearing a good pair of shoes, she was told to take good care of them, as she would never see another pair. John McCabe, coming home from Halifax in the month of October, staid all night in a hut, which had been erected at Mount Thom, at what is now McKay's place. On getting up in the morning, he found the ground covered with snow, while his feet were bare, and his legs covered only by pants, made from flax spun and woven at home.

lass, to rural merrymaking. Years were yet to pass before there would be a single carriage in the district, and long after that, this was the common mode of travelling. We doubt not, many of the older generation still retain some pleasant associations of " riding double."

We may remark, that with all the cold and even hardship endured, the people were generally remarkably healthy and vigorous. In the country, consumption was almost unknown. Persons have told us of growing up to twenty years of age, without knowing a case in their neighbourhood. Infant mortality was rare, compared with what it is now. In settlements with which we are familiar, large families were reared on almost every farm, it being quite common to find cases of ten or twelve children, all growing up to maturity. Their little houses, comparatively open, and heated by large open fire places, had the benefit of the purest air, and were much more conducive to health, than more comfortable but ill ventilated dwellings, heated by close stoves; while in summer, women as well as men working much in the open air, they were just the class to rear a stalwart race.

Hunting was still largely followed, particularly by the young men brought up in the country, some of whom equalled the Micmacs in skill and endurance. The moose was the chief object of pursuit. Two modes of hunting were principally followed, the one was in September by calling, that is, imitating the cry of the female, so as to attract the male within gunshot; the other, and that chiefly adopted, was by running them down on snow-shoes in the months of February and March. Their pace is a trot peculiar to the animal. It is said that they neither gallop nor leap, but the disproportionate length of the forelegs, enables them to step with the greater ease over fallen trees or other obstacles. When the snow is light, they sweep through it without difficulty, and as their power of endurance is great, it was at such times no easy matter to run them down, but when the snow became deep, and especially, in the month of March, when, by the sun thawing the surface of it by day, and this freezing by night, a crust was formed, they were readily overtaken, and afforded a good supply of coarse but well flavoured meat. When it was inconvenient to remove it at the time, the hunters were in the habit of making a trough, in which they would deposit it, and putting a cover upon it, to preserve it from bears leave it till they found it convenient to bring it home.

Among the men of this period, there are two specially worthy of notice as hunters. The first is John McCabe. In the chase, particularly in the pursuit of the moose, he manifested both the skill and the enthusiasm of the children of the forest. Catching sight of a recent track, he became all excitement; his bundle, and perhaps his coat, was thrown away, and even with the thermometer near zero, freely perspiring with his efforts, he pressed on till within gunshot. One or two incidents which befell him may here be given.

On one occasion he had shot a moose, which lay apparently helpless, though not dead; his companion urged him to kill it outright; he refused, saying that it would get cold too soon for skinning. They began digging a hole with their snow-shoes to prepare a place for a fire, when suddenly the moose sprang to his feet and rushed at him with the utmost fury. They had not reloaded their guns, and he had only time to take refuge behind a tree. The moose pursued him, and for some time he kept running round the tree. He could make a quicker turn than the animal, and was thus enabled to baffle him, till his companion got his gun loaded and shot the creature.

On another occasion, having shot a large moose, he, as night came on, wrapped himself carefully in the skin and laid down to rest. In the night the fire went out and he slept the sleep of the wearied hunter. On awakening in the morning he found the skin frozen solid, and so tightly round him, that he could move neither hand nor foot. He rolled about for some time, helplessly struggling to get one hand freed sufficiently to get hold of his knife, which he managed to do, but only after considerable effort, and then cut himself free.

On another occasion, returning to his camp, he stood his gun alongside of it, and stooping down to enter by a low door, out came a bear, snuffing. The two were frightened about equally. He ran back, forgetting his gun, while the bear took to his heels, and before he recovered his presence of mind, was beyond his reach.

On another occasion, a whole flock of wild cats came round his camp. They climbed upon it and ran round it, making a continual howling through the night. There seemed to him, from their noise, to be as many as twenty of them. He kept his fire burning, and watched till dawn. Toward morning they all left, and on examination he found the snow round his camp beaten, as it would have been by a flock of sheep round a barn.

The other was Simon Fraser of Middle River. He would sometimes spend weeks in the woods, usually with the Indians, who regarded him as their equal, if not their superior, in all the arts of forest life. He killed twenty-seven moose in one year, so that he earned the title, which he sometimes received, of Nimrod. He was the first to make his way through the woods from Middle River to Stewiacke, and blazed the first path between these places. He moved to Port Hood, where he took up a large grant of land. Having quarrelled with his wife, he left in a small vessel, in which he had been trading, professedly for Newfoundland, but privately declaring to some of his most intimate friends that he would never return. He was accompanied by his son, a young lad, and by a Cape Breton Frenchman. They were never heard of by the public, but I am assured that the family received communications, which led them to believe that he was living at the Northwest, and it is believed that he is the same Simon Fraser who, in the year 1804, first explored the country from the Saskatchewan to the Fraser River, to which he gave his name, and who established, on behalf of the Hudson Bay Company, the first trading post in British Columbia, a short distance from the great bend of that river.

There was a smaller kind of hunters, of which, almost every settlement had its specimen, viz.: men who spent much of their time in hunting or snaring the smaller kinds of animals, especially for their fur, and to whom the catching of a black or grey fox was one of the prizes of life.

The bears were numerous, and gave great trouble, as the settlers began to keep farm stock, carrying off pigs, sheep and calves. Some were large enough, however, to attack cattle. Indeed, the settlers regarded them as of two kinds, which they distinguished as cow bears and sheep bears, the only distinction between them, except their size, being that the smaller had a stripe around the nose of a grey color, which, in the larger, was black; but we presume that they were the same species, differing only in age or size. At all events, the larger were able to carry off cows. I have heard of one springing across a brook, carrying a good-sized heifer. When killed, the meat dressed would sometimes weigh between five and six hundred pounds. The cattle, however, learned to resist their attacks. On being alarmed by the sight or sound of one, they would run together and take position like the spokes of a wheel, with faces outward, and heads in the attitude of attack. In some places near the woods in the interior, where these animals were numerous, by imitating the noise of a bear, the cattle would immediately run together and assume this position. McGregor says that the largest and most spirited bull is soon vanquished and killed by a full-grown bear ; but I have heard instances of their maintaining successfully a single combat with bruin. A man on the East River had brought home from Shubenacadie a very large animal of this kind. Missing him one day, he went in search of him, and found that he had maintained a fierce conflict with a bear, and finally had killed his foe. Few of the old settlers, and, indeed, few of the old men of the present generation, but could tell of bear hunts, of losing pigs or sheep, or even cows, by them, or of catching them in traps. These, at first, were constructed of logs, so placed, that when the animal pulled at the bait, they fell across his back. Afterward large spring traps were used, and we may observe, that the settlers regarded bears as superior in intelligence to most wild animals.

It was rarely that they attacked a man. The only instance we have heard of, was in the case of the late Alexander Cameron, Loch Broom. He had gone out to look after his sheep, among which bruin had been regaling himself, when he met a very large bear, which immediately ran at him. To escape he commenced climbing a tree. It was a good sized spruce, straight and clear of limbs for about forty feet. The bear followed, and overtaking him, caught his heel in his mouth and commenced dragging him down the tree. When about twelve feet from the bottom of it, the buckle, by which, according to the fashion of the times, the shoe was fastened, gave way, and the bear was precipitated to the ground. Enraged, he tore the shoe to atoms, and again climbed the tree in pursuit of Cameron. The latter succeeded in getting among the branches, and having broken off a dead limb, struck with it at the bear's eyes as he came near. Between this and the approach of others at his call bruin retired. But one was caught in a trap a day or two after, with the eye badly torn, supposed to have been the same.

The most amusing bear story we have heard, however, was an incident that occurred to the late William Clark, whose family arrangements were considerably disturbed one morning by a bear, suddenly and without notice or invitation, coming down the chimney, as his wife was preparing breakfast. His house was built against a bank, with a shed roof. The bear had been attracted by the smell of the viands cooking, and came upon the roof, making his way toward the chimney, which according to the fashion in those days, was very large. Just as he came to it, David Stewart, who lived close by, and had been watching him with his gun fired. The ball went right through his heart, and making one jump, he came flop down the chimney to the no small surprise of the inmates of the household. "Fac, it gave us a start," said the old man, as he told the story, which we can readily believe.

Strange as it may appear, the pigs became so fierce as in some instances to maintain a conflict with a bear They were allowed to roam the woods during the summer, feeding as they could. After the nuts began to fall in the autumn, they became in a short time very fat, and by that time they were so wild, that it was dangerous to approach them. They would turn on a man. Even a good sized dog could do little or nothing with them. If he attempted to seize them by the ear, he was in danger of being ripped up by their tusks, with which they would inflict very severe wounds, so that the common way of slaughtering them was by shooting them. Of their maintaining a conflict with a bear, the following instances we believe well authenticated. On one occasion, a bear came into a pasture of James McCabe, West River, where there was a very large pig, which gave him battle and maintained the conflict gallantly for some time. The contest issued, like many an engagement among wiser beings, in great loss to both parties. The pig was so badly injured, that he slowly made his way up to the front of the house and there laid down and died, while the bear was either unable or afraid to pursue. On another occasion, near the same place, a bear came out of the woods into a field, where were a number of pigs, who combined their forces to give him battle. They were literally too many for him. Finding himself unable to cope with their superior numbers, he sprang on to a large stump. One of the boldest of his enemies, however, executing a dexterous flank movement, charged upon him in that position, and dislodged him, when being attacked by the whole pack, he was eventually torn to pieces.

We must, however, notice the Aborigines in their relations to the settlers. We have already mentioned that they sometimes gave trouble to the first inhabitants. That they ever contemplated any serious injury to the English is not probable. Their boast is that when they made a treaty of peace by burying the hatchet, it has never been broken by them. Knowing the terror which their name and appearance inspired, they took pleasure in frightening people who showed any fear of them. The late Deacon McLean, of West River, used to tell that when a little boy, going down to an encampment of Indians, near his father's house, one of them came up to him, and assuming a fierce look, said :—" Supposum me killum you —scalpum," and, taking out his knife, he brandished it over his head, dancing round him, sounding his warwhoop, and frightening him generally. In the same way, they would take advantage of the absence of the men, to extort from the fears of the women what supplies they fancied, though in this, sometimes the mother wit of our good foremothers proved a match for their cunning. On one occasion, an Indian having entered a house where a woman was alone, and being rather threatening, she immediately went to another part of the house, and calling her husband loudly by name, let a roll of leather, which happened to be on a bench, fall on the floor. Before she returned, the Indian was off. Mrs. Roderick McKay was a woman of great firmness and strength of mind. She never yielded to them, and when they came to her house, she would order them round, and scold them if they attempted too much freedom. On one occasion, some of them coming in, asked her, "What news?" She replied, "Ah, ha, great news ; there is another regiment of soldiers arrived at Halifax, and Indians must now behave themselves." They were quiet, and soon went away, and shortly there came an invitation to the whites, to attend a feast provided for them by their red brethren. It was accepted, and on going to the place appointed, they found provided every variety of provision, which the sea or the forest afforded, fish, flesh and fowl, which they requested the whites to cook in their own way. This was intended as a grand peace offering, and as such was accepted.

The Government did what it could to gain their good will by annual supplies, particularly of blankets and a kind of coarse blue cloth, usually known as Indian cloth; but they doubtless retained a lingering attachment to the old French, and, there is reason to believe, that had there been a prospect of the French again having possession of these regions, they might have been roused to renewed hostilities. At times, too, particularly about the years 1804–8, there was considerable agitation among them; gatherings of different tribes at Quebec, at which Micmacs were present, and among those returning, hopes were freely expressed of a French invasion. In 1808, Judge Monk, Superintendent of Indian affairs, writes on the report of Mr. Dunoon and others, through the Province:

"That the Indians expect the Province will be invaded, and that it appears generally to be their intention, in case of such an event, to remain neutral until they can form an opinion of the strength of the enemy, and then (in their own words) to join the strongest party.

"That several Indians went last autumn from Pictou to Quebec, as it was understood, to establish a communication with the Indians of Canada,—that two Nova Scotia Indians, who had been for some time in Canada, had returned to this Province last summer, and informed a man employed in the Indian Department, that there were many Indians from the United States with the Canada Indians, and much talk of war with them,—that in the district of Pictou some Indians have declared they will not accept of anything from Government, as they expect the country will soon be invaded and conquered; and one of them was heard to say, that in case of war he and a few others would scalp all Pictou in two nights."

The last was, of course, bravado, yet the circumstances were sufficient to excite apprehensions, and rendered it prudent for the Government to take precautions against hostilities. As a general rule, however, they were kind to the whites; oftentimes they supplied the first settlers with food, and frequently I have heard old people speak gratefully of their kindness. It was their characteristic, to manifest strong feelings of gratitude to those who treated them well, but equally strong feelings of animosity against those who treated them with injustice or harshness, though a sense of the superior power of the white man kept them from any violent acts of revenge.* Indeed, when we consider the manner in which they were deprived of their lands, and the unfeeling manner in which they have often been treated, it is wonderful that they have been so quiet and free from deeds of violence.

It is, perhaps, more surprising that they have been so honest. How easy it would be for them to steal our sheep or cattle as they wander in the woods, or to purloin articles

---

* The only instance of this kind which we have heard, which threatened serious consequences, was an affair between Lulan and Rod. McKay. Soon after the arrival of the latter, he had in some way seriously offended the former, who came all the way from Merigomish to the East River to shoot him. It was night when he arrived, and McKay was at work in his forge. Lulan looked in but as he saw the glare of the fire on his face, and the sparks flying from the anvil, and heard the reverberation of his blows, he became scared, and his hands could not perform their enterprise. For long after the two were good friends, and Lulan used to tell the story, graphically describing his feelings, "Sartin, me taut you debbil."

from our barn yards; and yet an act of theft is rare among them ; and when such has occurred, the others have generally surrendered the guilty to justice. Squire Patterson and others of the first comers gained their confidence, and as an instance of his tact and their sense of justice we may give the following incident: On one occasion Simon Fraser going down the ice met Patlass, who pointed his gun at him. Fraser immediately went to him, took the gun out of his hand, and dashed it to pieces on the ice. Patlass went to the Squire for redress, who issued a summons, and directed him to bring as many Indians as he could to the trial. They came, and the Squire heard the whole case,—Simon defending himself on the ground, that he thought Patlass was going to shoot him, and the latter maintaining that what he did was in fun. The Squire said he had no jury, and must have one. He, therefore, selected five or six Indians to whose judgment he committed the case. The result was, that they gave a verdict against Patlass.

When the English came, the Indians had several places, where they had clearings and cultivated a few vegetables, as beans and Indian corn, but the Government, in granting the land, made no reserve of such rights. We do find it ordered, on the 18th December, 1783, that " a license be granted to Paul Chackegonouet, Chief of the tribe of Pictou Indians, for them to occupy the land they have settled upon, on the south-east branch of the harbor or River Merigomish or Port Luttrell, with liberty of hunting and fishing in the woods, rivers and lakes of that district." But the land was not reserved in giving grants to settlers, and the Indians were gradually worked out of all their claims. Some settlers honestly bought out all their rights, and had no trouble with them afterward. Walter Murray, in Merigomish, finding them coming and planting corn, even where he had planted his potatoes, finally agreed to pay them five pounds to relinquish all claims, which they accepted, and never troubled him again. Donald Fraser, McLellans Brook, paid Lulan a bushel of wheat annually, and was accustomed to speak of him as his landlord. Thus they were gradually deprived of all places of this kind. The last we know of was the front of the farm at Middle River Point, since owned by William McKay. This had been one of their places of encampment, from the time of the arrival of the first English settlers. It was on Cochrane's grant, the title to which was in dispute. Here they had some five or six acres cleared, each having his own patch, on which he raised potatoes and beans, and with fish offal and the refuse of their camps, it was very rich. Within the memory of persons living, they even raised a little wheat. Parties who attempted to settle, they drove off. On one occasion, a person came and built a house in their absence. Having left to bring his family, the Indians returned, and when they saw the intrusive dwelling, they gathered brush round it, which they set on fire, causing a great conflagration, around which they danced and yelled as long as it lasted. Through Mortimer's influence, McKay was allowed to remain on the land, and gradually obtained possession of their little plots.

The only land in the county, so far as we have been able to ascertain, reserved for them in Government grants is a small lot at their burying ground, at the mouth of the East River, but this they sold to the late James Carmichael, with the exception of the burying ground itself. The Sessions, however, have purchased a lot for them inside the beaches, which they now occupy. The only other land to which they lay claim are two islands in Merigomish Harbour, one, on which their chapel and burying ground are, known as Indian Island, and another claimed by Peter Toney, both of which are said to have been given them by Governor Wentworth, but attempts have been made to dispossess them of the latter.

Every year, usually in the month of September, they assembled in large numbers from Prince Edward Island, Antigonish and other places, their usual place of rendezvous being either Frasers Point or Middle River Point. A person brought up at the latter place, has told me that he has counted one hundred canoes at one time drawn up on the shore, and it was said that they would sometimes number one hundred and fifty. Sometimes two days would be spent in racing or similar amusements. At night came feasting. My informant on one occasion, when a boy, spent an evening at one of these entertainments. He says they had twelve barrels of porridge prepared, which the squaws served out to the men, ladling it into dishes that, he supposed, would hold near a peck each. Two moose were also served up on the occasion, and also a quantity of boiled barley. Afterward they had various plays and games, but the last night they spent in singing and praying. These gatherings continued yearly till a vessel with small pox was sent to quarantine at the mouth of the Middle River, about the year 1838. They have now a similar gathering annually, in the month of July, on Indian Island, Merigomish. All assemble in their best attire, and after mass, and the celebration of any marriages that may be coming off, the rest of the day is spent in feasting and dancing. As to the latter we are informed, that they have adopted the common Scotch figure of eight reel, in which men and women join.

A common amusement was to get the Indians to dance the war dance. At weddings or other occasions, where they might be present in numbers, for a share of the good things going, they would go through all the scenes of war, even to the scalping and torturing their prisoners. They acted the whole so perfectly, and their appearance in doing so was so frightful, that women and timorous persons would sometimes get thoroughly scared.

Among the Indians two were particularly noted, and

are still remembered by the older generation, viz. : Patlass
and Lulan. The former was particularly distinguished
for his skill in draughts, so that his death was announced
in a Halifax paper, as that of "the celebrated draughts
player." At this game, it is a question whether he was ever
beaten. When he met a stranger, he would allow him to
win the first game, but then he would induce him to play
for a wager, which was all he wanted to show his skill.

He was also noted for that grim humour, characteristic
of the red man. On one occasion a sea captain had
brought a fighting cock ashore, and set it fighting with
one belonging to the town. Patlass came along, where
a number of persons were standing looking on. After
looking at the scene for a few minutes, he seized one of
the combatants, and walked off with it. The captain
called out angrily after him to come back, asking him
what he was about. "Take him to jail, fightin' on the
streets," was Patlass' reply. On another occasion Mortimer
met him on the wharf, smoking one of the long clay pipes
then in use. The former being disposed to cultivate
familiarity with all classes, asked him for a smoke.
Patlass handed him the pipe, when Mr. M., taking a
silk handkerchief from his pocket, wiped the stem carefully,
before putting it into his mouth. When he had finished
smoking, he returned it to Patlass who, holding it up,
immediately broke two or three inches off the stem,
saying, "Dat more better, Missa Mortimer." The following joke is often told, but we believe that Patlass was the
real author of it. He was coming from Halifax and by
the time he arrived at Shubenacadie, his supply of rum
was exhausted. Applying at a tavern there for a supply,
he was charged at a much higher rate than in Halifax.
He grumbled at the price, when the tavern keeper said,
alluding to the cost of license, it costs me as much to sell
a puncheon of rum as to keep a cow. "No eatum as
much hay, but sartin drinkum more water," was the

Indians rejoinder. He was drowned near Middle River Point, on the 1st September 1827.

Lulan was of a milder disposition, though reputed to have been a great warrior in his youth. He used to boast that he had scalped ninety-nine persons, though there was probably some bounce in this. He was rather below the middle height, but straight and broad-shouldered, and in his later years corpulent for an Indian. He was the means at one time of saving the life of the writer's grandfather, old John Patterson. The latter was crossing the ice, when it gave way, and he fell into the water. The Indians put out to his help, and succeeded in rescuing him, but he was insensible for a time, and when he recovered, he found himself in a large tub in Lulan's camp. Lulan was ever after freely entertained at my grandfather's house, of which he did not fail to take advantage. As an instance of the attachment induced by kindness, we may mention an incident that occurred at my grandfather's funeral. It being customary then to hand liquor round to all present, some was offered to Lulan, who replied: "Me no drinkem long time, but bleev take some to-day; me most dead grief my friend." After my grandfather's death, however, he continued to expect from my grandmother the same attention as in his life time. "Me save your husband's life," was the appeal which he supposed would never lose its efficacy, which he rendered more impressive by adding particulars: "Walk out on thin boards; only head and arms out of water; most lose my own life save his." And after her death, their sons' store was laid under contribution on the same ground.

He died about the year 1827, when he was said to have been in his 97th year, so that he must have nearly reached manhood when Halifax was founded, and been in full vigor when Louisburg and Quebec surrendered.

His son is still well remembered as Jim Lulan. He had

somewhat of the dry humor of some of the race. Mr. Carmichael had built a vessel, which, in honor of the old Chief, he called the Lulan. Some persons, teasing his son, said to him that he ought to make her a present of a set of colors. "Ugh!" said Jim, "me build big canoe, call it Old Carmichael."

Speaking of the wit of the Indians, we may give an instance, which was long a standing joke of Pictonians against their neighbors in Colchester. Some Indians being after geese in spring, shot one which fell on the ice. Seeing that it could not escape, they did not go for it at once, when some persons coming along in a sleigh picked it up. The Indians, however, came up and claimed it. The others refused to give it up, saying that they had shot it. "Where you from?" said an Indian. "From Truro," was the reply. "Sartin bleev so; Pictou man no shoot dead goose."

Another, usually known as Beetle John, is especially worthy of notice, as having been the owner of a shallop. It was built on the Big Island of Merigomish, in a small cove at the head of the French channel, and for some time he traded in her.

As to their numbers, we have no doubt that the views entertained by many and expressed by themselves, regarding the large numbers of the Aborigines in this and other parts of North America, are greatly exaggerated. On the other hand, the assertion made by some who have studied the subject, that the Indians in these quarters are now as numerous as ever they were, is not correct, as to this part of the country. In the year 1775, a return was made by Dr. Harris, by which they were estimated at 865. This may not have been an exact census, but could not be far from the truth. Yet in the report of the Indian branch of the Department of the Secretary of State for the Provinces, 1872, the number in the County of Pictou is estimated at 125, and in Antigonish at 93, while Colchester has only

31. This agrees with the recollection of the old people, who speak of seeing 40 or 50 of them travelling together, or of 60 camps in one place. Before the first of these periods, there is a tradition of a great mortality among them by small pox. During the war against the English, a number of them had killed and scalped a family sick with the disease, by which they caught the infection, which spread through the tribe. It was their practice, as soon as one felt himself unwell, to plunge into the water, and all who did so died.

Since the period referred to, the free use of fire water, the diminished supply of food to be obtained by fishing and hunting, and epidemics at different times, have diminished their numbers. But of late there has been an improvement among them. Their supplies of food from their old resources having failed, they have been engaged in industrial employments, such as cooperage, and supplying the markets with fresh fish, in which, from their activity and skill, they can earn a more regular and better living than formerly. We believe, too, that both from the laws against selling liquor to them, and their own sense of the evils which it has brought upon their race, there is now much less drinking among them than years ago, and that not only are they better off, but their numbers are beginning to recruit.

The Government, at various times, have projected measures with the view of inducing them to adopt more settled habits. With this view a series of queries was addressed, in the year 1800, by Judge Monk, the Superintendent of Indian affairs, to leading men in the different sections of the Province, seeking information regarding their willingness to adopt the employments of civilized life, or to have their children receive education or training in useful arts. Mr. Mortimer replies for Pictou, and to the enquiry: " Are there any who have shown a disposition to settle, or who have taken up trades?" He says,

"Joseph Purnall has made several attempts to settle by planting potatoes, Indian corn, beans, &c. Indeed, the greater part of the Indians who frequent this quarter have shown a disposition to settle, by planting a little, as above, in several parts of this district. An Indian from 'Mathews Vineyard,' named Samuel Oakum, who has married into this tribe, is a tolerable mechanic in several branches, particularly coopering and rigging vessels, and is also a pretty good sailor." Proposals were also made to teach the women knitting and spinning.

It is but just to add that the benevolent of this county, from Dr. McGregor downward, have been interested in the improvement of their social and spiritual condition. On various occasions attempts were made to educate young Indians, but these failed, partly from their own repugnance to the restraints of civilized life, and partly from the opposition of their spiritual guides. In the year 1828, a society was formed in Pictou, called the Indian Civilization Society. But all these efforts produced no permanent result.

The year 1799 is notable in the history of the county as that in which the first contested election was held within its bounds. From the first settlement of Halifax, society embraced churchmen and dissenters, and thus contained all the material for Whig and Tory parties. The American Revolutionary war, and the influx of Loyalists, the majority of whom were Tories and high churchmen, tended to strengthen the hands of power, and repress everything like popular influence. Still there had been a growing feeling of opposition to the irresponsible power of the Provincial rulers, and an increasing desire to bring the Government under the control of public opinion. This state of feeling, which afterward swelled to a flood, under the guidance of S. G. W. Archibald and Joseph Howe, first found expression in this election, and had for its exponent and apostle, W. Cottnam Tonge, said to have been a man of brilliant talents, an eloquent speaker, and having many qualities fitted to make him a popular leader. At all events, he was now the tribune of the people and everything that was bad in the eyes of Sir John Wentworth and his official clique. To their no small annoyance, he now offered [for the County of Halifax, (which then embraced what now forms the Counties of Halifax, Colchester and Pictou) against the old members, Michael Wallace, Lawrence Hartshorne, Charles Morris and James Stewart (afterward Judge Stewart), who were friends of the Government, and who had issued a card jointly, appealing to the electors, though afterward they denied that there was any combination among them.

Another element, however, had perhaps more influence in the country. The representation of the county had hitherto been in the hands of the town of Halifax, and, indeed, of the Government officials. A feeling, however, was growing up through the rural districts, that their views were little understood and imperfectly represented at the Capital; and through Colchester and Pictou, there was a strong desire to have local members. Accordingly, the people of these districts generally combined in favour of Edward Mortimer of Pictou, and James Fulton of Londonderry, and united with the friends of Tonge in Halifax, to oppose the Government candidates.

The poll opened at Halifax on the 13th November, and closed on the 23rd. It was then adjourned to Onslow, where it continued for two days. It was thence adjourned to the town plot of Walmsley, as it was called, at Fishers Grant, where it opened on the 5th December, in the barn of James McPherson. By consent of the candidates, the last day's polling was at the East River, where it finally closed on the 13th December. In Halifax, the country candidates received very little support, and Tonge, though receiving more, was still far behind the Government candidates, but Colchester and Pictou went almost unanimously for their local members, and at the same time gave large majorities for Tonge, so that the three were triumphantly elected, and with them, the highest on the poll of the Government candidates. The result of this election to the Province was a systematic attack upon the old irresponsible regime, which, however, produced little fruit.

The election however had a special importance for the County of Pictou, as it was the origin of those party feuds, for which it has since been noted. The opposition to Mortimer formed the nucleus of a party, formed partly on political and partly on personal grounds. The division thus formed was fostered in after years by various circumstances, and unfortunately became mixed with ecclesiastical, we can scarcely call them religious, differences, which gave intensity to the feelings excited.

Mr. Wallace, who was defeated, was at that time treasurer of the Province, had been for some time a member of the House and became so again, by the unseating of Tonge, for want of a freehold in the county. He was afterward a member of Council, and several times administered the Government in the absence of the Lieut. Governor. He was a native of Scotland, but had emigrated to the Southern States, where he had been doing business as a merchant, but, on the American Revolutionary war breaking out, he espoused the side of the British Government, and removed to Halifax, where he was engaged in business, till he became Provincial Treasurer. He was thus described in one of the Halifax papers at a later period. "He is one of those who think the King can do no wrong, that the British constitution is the most perfect fabric the world ever saw. He hates a radical as he hates Satan himself. He would, if he had the power, shake all the liberals in the world over the crater of Vesuvius, but his heart would be too kind, to let them 'fa' in.' When he was a member of the

Assembly many years ago, his opponents used to take advantage of the irritability of his disposition, and generally put him in such a passion, as to deprive him of the power of speech. He was once sitting on the bench of the Inferior Court, and was engaged in some calculation of damages, when one of the counsel for the parties uttered something, which grated on the old man's ears, and forgetting for a moment the dignity of his office, he abruptly asked, 'What's that you say, you———rascal?'"

He was distinguished even in those days of irresponsible power, for his adherence to arbitrary principles, and his hatred of everything like popular rights. On one occasion the Speaker of the Assembly having presented an Appropriation Bill, which had passed both branches of the Legislature, to Dr. Croke administering the Government, the latter said, "I do not assent to this Bill," and three days after summoned the House, and addressed them in a speech in which he told them, that the Government would appropriate the revenue of the Province more beneficially and economically, than the Assembly had provided for by their bill, "after which", as the journals of the House say, "Mr. Speaker offered to address his Honour the President, but was prevented in a turbulent and violent manner, by the Hon. Michael Wallace, acting President of His Majesty's Council, who declared the House prorogued." And on the President consulting the Council whether he should not draw warrants on the Treasurer, without an Appropriation Bill of the Assembly, Wallace alone voted in favour of the proposal.* At a later date, the House of Assembly having made some enquiries regarding the revenue received from the coal mines, Wallace replied that the disposal of it was none of their business.

But we have here to do with him as his influence

---

*Murdoch's History, III, 288, 293.

affected Pictou. He took his defeat with a keenness that
we can now scarcely understand, and Mortimer having
given free expression to feelings of triumph, natural under
the circumstances, he publicly vowed revenge. From
that time his course was one of unrelenting hostility not
only to Mortimer, but to the leading men in Pictou, both,
in church and state. Those being the days of irresponsible Executive power, and from his offices being always
influential, and at times administering the Government,
he had the machinery of Government very much under
his control, and was ever ready to exercise it for their
annoyance, and to nurture the personal and party feelings
that had begun at this election. Years after even the
descendants of those who opposed him, might be driven
from his office with passionate execrations, while for a
man to quarrel with his minister was sufficient to entitle
him to official favour.

He is said to have kept a book containing black and
white lists, of every man in Pictou at the time of the
election. When any application was made to Government
from parties in Pictou, his first care was to examine these
lists, to see what had been the conduct of the parties at
that time, and treat them accordingly. On one occasion,
a road having been laid out in such a way as to do a great
deal of damage to a man's intervale, a petition was forwarded to Government, to have its course altered. On
applying to Mr. Wallace, he asked if the course in which
they proposed taking the road would suit the public as
well. The parties said that they thought it would suit
better. He began making a favourable reply, but on
glancing over the petition, and observing the name of the
party interested, he stopped and exclaimed, "John D—."
Reaching down his black book and finding the name, he
said, "Take it where it was laid out, if it should go
through his house." He, however, afterward relented.
The following incident, which I received from my father,

however, will show that he could remember favours as well as injuries. Among those in Pictou who voted for Wallace was his father, old John Patterson. Some years after, my father and his brothers applied to Government for a grant of Crown lands. At that time there was an unwillingness to grant land in quantities, in consequence of parties taking it up on speculation. On applying to Mr. Wallace, he asked who they were. They said they were the sons of John Patterson, deacon. "Sons of Old John Patterson! Oh, yes, you'll get your land," was the immediate rejoinder. Other parties having applied about the same time, they were told to "go to Mortimer, and let him get them land."

An incident may be mentioned in connection with the close of the poll, as illustrative of the progress of the country. On the last day of the election, Dr. McGregor entertained the candidates and some strangers at dinner, and made for them a fire of coal. This was considered quite a novelty, and an important event for the Province. It was only the year previous that coal had been discovered on a brook, passing in rear of his and William McKay's lots. William Fraser (surveyor) in that year carried a sample to Halifax to the Governor, Sir John Wentworth, who sent him with it to Admiral Sawyer, who ordered a small cargo to be sent to Halifax, which was done, but it did not prove of good quality. Soon after the Dr. and some of his neighbours took out licenses from Government to dig coal, but undoubtedly he was the first to use it as fuel. He first opened a pit on what is still known as the McGregor seam, discovered on his own land, and used the coal in his house. This would be as early as 1801 or '2. From that time he regularly, in the fall of the year, got out his winter's supply, and sometimes sold some. Previously the blacksmiths had used charcoal, but now John McKay, of Pictou, commenced sending lighters up the river, and took the coal to Pictou for use in his smithy, and the other blacksmiths soon followed the same course.

CHAPTER XI.

COUNTY AND COURT BUSINESS.

As we have mentioned in a previous chapter, Pictou was, in the year 1792, set apart from Colchester as a separate district. One of the first steps taken in consequence, was the erection of a jail. It stood on the lot now occupied by the establishment of James D. B. Fraser & Sons. The lower part of the building, forming the cellar storey, was built of stone, with grated windows. The upper was of logs, and clap-boarded on the outside. This contained at the one end rooms for the jailor; at the other, a lock-up for prisoners. Below were cells, in which criminals were confined, the more desperate in irons. It was built by John Patterson, and the following account at this time, including this and other items, may be given as a curiosity:

The Magistrates of Pictou Dr. to John Patterson, Sr.

Brought forward £8 19s. 3d; To interest 10s. 6d.......... £9 17 9
1793—Captain Allardice, broak hinges and other damages........ 0 10 0
James Carmichael, 30s. for goin with the jury laying out roads 1 10 0
Thomas Harris, Senr., order for laying out roads........... 2 0 0
James Dun, for the use of his house some years for M. Mingo 13 10 0
To help to build the town brige........................... 10 0 0
Building the gaill.. 87 0 0
To Mr. Mortimer for provisions for a sailor............... 0 17 0
To goods by Mr. Dawson to the poor, by the overseers...... 1 15 7
To James Dun for stocks, Handcufs, lock, stove, &c., jury box... 1 15 6
To John Patterson, Senr., for three terms store rent...... 2 0 0

To Robert Lowden, for Clerk of the Sessions 1793..........	2	10	0
To Do Do 1792..........	6	0	0
To Taking of and Reparing and Putting on the Gaill lock, broke by Mirian................................	0	3	6
To 1 Double Paddlock Replaced for 1 Destroyed by Mr. Mirian..	0	4	0
To David McCoull, for cutting Cobequit Road..............	15	10	0
To colecting, Storage, wastage in and out of 94£, in Grain, Butter & Shouger at Ten per cent....................	9	0	0
To colecting £20 at Five per cent.......................	1	0	0
	£183	11	4

Contra,–Credit.

Brought forward..	2	4	7
1794—From James Dun, collector............................	15	6	8
From Alexr Robertson, do..............................	12	9	6
John Brownfield, do..............................	6	12	7
John McKenzie, do..............................	13	17	3
James Briden, do..............................	10	1	10
From clerk, Augt. 20...................................	1	13	6
From Do Decr. 29...................................	1	10	0
From Duncan McKenzie.................................	20	7	5
From Robert Lowden....................................	2	10	0
From Mr. Mortimer, Previous Tax for 1792................	5	5	6
From Mr. Scot, in grain for Road........................	3	16	9
From John McDonald, Road Tax in orders................	4	5	3
To Thomas Copland's Noat.............................	1	10	0
To William Fraser Noat................................	2	7	0
To Joseph Scot's Noat.................................	2	16	6
From James Dun, License..............................	1	15	6
From George Roy, Colector............................	4	2	0
Constables Colin McKay £4 5s. 7d., Duncan Cameron £2 14s. 1½d..	6	19	8½
	£121	7	1½

The stocks stood in front of the jail, and for many years were used for the punishment of offenders. In our younger days a pair stood on the east side of George street, but were not then used.

Dr. McGregor says that on his arrival, "As for lawyers, there was such good neighborhood, that we never expected to need a lawyer or a court house." But the above account shows that these days had passed; that they had use for a jail, and that on one occasion its occupant had been so dissatisfied with the accommodation, that he had taken forcible measures to be relieved from it.

By the act erecting Pictou into a separate district, a Court of General Sessions of the Peace and Inferior Court of Common Pleas, was appointed to be held at Walmsley, on the 3rd Tuesday of January and 3rd Tuesday of July.

The Records of the Sessions for the first few years have been lost. The first book commences with January term 1797, held before, Hugh Dunoon, Robert Pagan, John Dawson, Nicholas P. Olding and Edward Mortimer, Esquires. These, we may observe, were the leading justices of the county for many years. An account of the business done, will afford some interesting information regarding the state of the county.

First comes a large number of regulations, regarding animals of the male kind going at large, which we need not particularly specify. But we may remark, that the establishment of pounds in various parts of the district, and regulations for impounding cattle, occupy a large space in their proceedings in subsequent years.*

Then we find regulations regarding the preservation of the salmon fisheries, manifesting a care and wisdom, which we would not have expected at that early period.

" That no person shall sett a salmon net, seine or wear more than two-thirds across the channel at low water, in any of the rivers within this dis-

---

* Thus we find, in 1802, the various pounds located as follows :
" Harbour........................Where it now is.
Above the Town Gutt..............At a brook above John Patterson, Junr's.
Fishers Grant....................At James McPhersons.
West River.......................At Hugh Fraser's Brook.
Middle River.....................At Joseph Crocketts.
East River, west side............At Donald McNaughton's Brook.
East River, east side............At Alex. McLeans.
McLellan's Brook.................At William Frasers.
East Branch East River...........At Alex. Grants.
West Branch   " ....At James Camerons.
Little Harbour...................At Alex. McQueens."

trict, nor nearer to that of his neighbour's than the distance of fifty yards, to be measured in the direction of the tide or stream.

"And that no person or persons shall fish salmon in any manner of way whatever within this district after the 19th day of October, on transgression the net to be forfeited, and a fine of twenty shillings to be levied of the same, or from the goods and chattels of the offender, the same to be disposed of as in the case of boars (*i. e.* one-half to the informer, and one-half to the overseers of the poor,) no prosecution to be without the complaint is made within ten days after the offence is committed, the seine or net to be seized while in the water, and carried immediately to a Justice of the Peace, who is to judge of the offence, and if the seine or net is returned by the owner or any persons, the fine is to be doubled, any proprietor or householder may prosecute said offence.

" Also, that no person or persons within this district shall fish salmon with a spear or by sweeping with a net or seine, under the penalty of £5 for every offence, said fine to be disposed of as in the regulations respecting boars.

" Also presented and ordered that no person or persons shall fish salmon with net, spear or otherwise in that part of the Middle River between Archibald and Taylor's Mill-dam and Alexander Fraser, Sr.'s., under the penalty of £10 for each and every offence, one moiety to the prosecutor, and the other moiety to the overseers of the poor, for the use of the poor. And that no person or persons shall catch any salmon fish, in either of the two pools at the foot of the falls, on the West Branch East River, and also in the pool at the foot of the falls on the West River of Pictou, about two miles above William McKenzie's, under a penalty of 40s., to be disposed of as in the last case."

For the killing of a bear a premium of twenty shillings is allowed on presentation of the muzzle. This was afterward reduced to ten shillings.

Then comes an order to "tax every poll from 16 upward three shillings, every horse one year old and oxen and cows four years old three pence, every sheep one year old one half-penny each, every 100 acres of land, three pence per hundred, for the making and repairing highways within this District, the respective overseers and committees to lay out the work as formerly, in the most needfull places on the publick highways in the different settlements within this District."

The roads then existing, for which overseers were appointed are named as follows, viz., " Fishers Grant, West River, Middle River, West Side East River, East Side East River, East Branch from Colin McKenzie to

Peter Finners, Upper Settlement, West Branch, Little Harbour, Above the Town Gutt, Merigomish, Gulph." But at this meeting steps were taken for the opening of the roads from "John Blaikies on the hill (at what is now the Cross Roads Green Hill) unto Charles Blaikies on the West River," and "from Charles Blaikies to the mouth of the Middle River," and "from Alex. McLean's upper line on the East River to the head of the tide," and a Special Sessions was appointed to "lay out a road from Robert Marshalls to Hugh Frasers."

In regard to the collecting of taxes, we find the following :

"Presented and ordered that John Patterson, senr., be allowed ten per cent. for being Treasurer, and collecting the tax assessed in the year one thousand seven hundred and ninety-five, and likewise charge no leakage or weastage on any of the produce that he receives, but to be accountable to the District for the whole as he receives it."

The taxes were generally paid in grain or maple sugar, and were received by local collectors, who brought it to him, but we find in July the same year, that it was

"Presented and ordered that John Brownfield be allowed the sum of twelve shillings and six pence for bringing wheat and oats from Merigomish to Pictou to satisfy the jail tax."

The following are all the other items of expenditure :

"Also ordered that Mr. Patrick McKay be allowed the sum of five shillings for the use of his house for a Grand Jury room this present term, by an order upon the Town Treasurer.

"Also presented and ordered that the sum of twenty shillings be allowed to Edward Mortimer, Esquire, for the use of his house as a Court House, cuttin and Hauling fire-wood, putting on fires, &c., &c., by an order on the Treasurer."

Similar bills were allowed in subsequent years, the places of meeting being John Patterson's store, William Lyndsay's tavern, or other places.

The business was concluded by a criminal trial, the full record of which we must present to our readers.

"DISTRICT OF PICTOU, SS.,
"GENERAL SESSIONS OF THE PEACE,
Jan'y Term, 179 .
"THE KING
vs.

"For petty larceny. { Peter Tarbett, Sophia Tarbett, And Hannah, *alias* Rose. } Negroes.

"The prisoners having been brought forward and *arrainged* at the bar, witnesses sworn and interrogated, The Court having considered the evidence, and the parties being found guilty, do adjudge the whole of them be stripped naked from the middle upwards and receive as follows, viz: Peter Tarbett thirty-nine lashes, Sophia Tarbett, thirty-nine lashes, and Hannah *alias* Rose, thirty lashes on the naked Body by the hands of the proper officers, and be thence committed to Prison, untill the Court think it practicable to banish them out of this District, it being now an intense season.

"Pictou, January 18th, 1797.

"Jan'y 18th, 1797. Issued a warrant unto James Crocket, William Robertson and William Fraser, constables, to take the three negroes and strip them from the middle upwards, and whip them as specified in the sentence, &c. which warrant was returned Executed to-day.

"William Fraser, one of the constables, who being appointed to assist in whipping the negroes, refused to assist, was fined by the Court in the sum of forty shillings.

"His Majesty's General Quarter Sessions of the Peace for the District of Pictou stands adjourned until next term.

THOMAS HARRIS, JUNR.,
D'y Clerk of the Peace."

"*Vivant Rex et Regina.*"

We may mention that the punishment of flogging continued to be inflicted for years after. The last case took place in the year 1822. A person, residing near the Town Gut, having lost his wife, bought a full mourning suit. A darky stole the same, and having arrayed himself in the whole, even to hat and crape, started for Truro, but was arrested in this genteel rig, and was flogged, tied to a cannon, which still stood in our younger days at Yorston's corner, though on other occasions parties were tied to a cart.

At the July term we find the name of Robert Lowden

as an additional Justice, and the following added to the regulations regarding salmon fishing:

"No settler to buy or barter salmon fish, salt or fresh, from any Indian or Indians within said District, from nineteenth day of October, 1797, to the nineteenth day of May, under a penalty of ten pounds, one-half to the prosecutor, the other half to the overseers of the poor."

The jail seems to have engaged attention, for we find the following order:

"That the Lower story of the jail (order to be floored with two inch plank in the year 1794 be done away) be now floored with four inch pitch pine plank, with pitch pine sleepers laid upon stone, the sleepers to be four inches thick and six inches deep, five in number, the plank to be spiked to the sleepers, with spikes eight inches long—likewise that the jail be clapboarded with sawed clapboards, the first stroke to be 1¼ inch plank round the foundations; corners, doors and windows to be cased with weather boards on each end, the whole to be finished by the 30th October in a workmanlike manner—that John Patterson, senr., and Wm. Lowden, senr., be inspectors to see the work carried on—also that Daniel McKay, Wm. Monsieur and Thomas Fraser, carpenter, be a committee to see that it be finished in a workmanlike manner and that the same be vandued to the lowest bidder at Mr. John Patterson's store, on Tuesday, the 8th day of August next at 12 o'clock, meridian, and that David Lowden, John Patterson, junr., and Duncan Cameron be a committee to vandue the same—also ordered, that the Clerk of the Peace advertise the above in the most public places on the Harbour of Pictou, West, Middle and East Rivers, Merigomish and Gulph."

The following additional items of expenditure passed:

"Also presented and ordered, that Thomas Harris, junr., be allowed the sum of three pounds currency for his services as Clerk of the Peace for the year one thousand seven hundred and ninety-seven."

"Likewise presented and ordered that Thomas Harris, senr., be allowed the sum of two pounds currency by an order on the Treasurer for receiving, victualling and attending persons in jail in the year 1797."

As to the assessment for the year, it was now ordered

"That £140 currency be raised in the District, the same to be appropriated to the making and repairing a certain road leading from Mr. George McConnell's unto Truro, as far as the line of the District of Pictou, and £60 for defraying District charges."

Considering the state of the country at this time, we regard the above vote for the Truro road as exceedingly creditable. The raising of the amount, however, was afterwards deferred till the year 1799, at the January term in which year it was resolved, that

"All the license money now on hand, due or that may become due before the 1st day of July next, and not already appropriated, be added to the £140, the overseers to straighten the crooks as they may see most beneficial for the benefit of the public. Also that the £19 9s 6d in the Treasury, being the sum allowed the District for the express purpose of making roads, out of the sum raised here as Provincial taxes, be also laid out on such road. And further provided that in case any part of the £140 to be levied be not paid before July the amount be borrowed, so that the work may be done before haying."

David Archibald was selected as commissioner, to have, with John Archibald, of Truro, the whole management. He was to be paid 7s. 6d. per day, and to give his obligation to the Court to see the money laid out for repairing said road in the most advantageous manner. It was furthermore ordered that it " be made 12 feet wide, clear of every incumbrance whatever and to be thrown up from each side where it may be necessary."

The collecting this tax however was a work of time. For years we find in John Patterson's ledger, charges to the settlers of "Cobyquid Road Tax."

In the year following, the business was so similar that we need not repeat, but a few items of a different nature may be given. Thus in regard to salmon fishing we have the following in the year 1798 :

" That no person or persons within this district shall set or leave a net or seine in any of the rivers within this district, from Saturday at 12 o'clock, noon, until Monday following at 12 o'clock noon, and for each and every offence, upon the oath of one credible witness, shall pay a fine of 40s. And also that no person or persons shall chase, follow or drive fish into nets, or seines in any manner of way whatever in any of the rivers within this District, nor be seen carrying a salmon spear near any of the rivers within this District during the fishing season, and for every offence on conviction by the oath of one credible witness, shall pay a fine of £2 currency."

We also find in the year 1805, orders for a fish-gate at Archibald's mill-dam, Middle River, to be only one foot high from the bottom of the river in front of the dam and three and a-half feet on the back, four feet wide in passage—according to the frame given by David Archibald so as to allow passage at all seasons.

In the same year we find the following:

" It is ordered that Samuel Copeland, James McPherson, (Fishers Grant) Hugh Fraser, Sr., East River, and Alexander Chisholm (Gulph) be taken cognizance of for not doing their duty as overseers of thistles for the year 1797."

" That no person or persons within this District shall carry any unbroken flax into his, her or their dwelling house within the District, nor suffer the same to be done, and on conviction by the oath of one credible witness, shall pay a fine of £5, the same to be applied as in the case of boars."

" That Act 21 George III. be put in execution, to call out the inhabitants after deep falls of snow, with their horses, oxen and sleds, in order that the road be rendered passable."

" Also that all proprietors, agents or present possessors of lands in this District shall cause to be cut down all thistles growing on such lands, on or before the twenty-eighth day of July instant, and likewise all inspectors neglecting their duty, as also every person or persons refusing obedience to said regulation, shall be dealt with as the law directs."

" Also ordered that this regulation shall be published in the most public places within this District."

Criminals still required attention for the same year we find:

" Ordered that Mr. William Lowden, Senr., be empowered to get a good and sufficient pair of stocks made for this District in the cheapest manner possible, on or before the fifteenth day of January, 1799, and to be delivered to the sheriff or jailor, to be by him deposited in the jail, unless otherwise directed by the magistrates, the expense to be defrayed by an order on the Treasurer."

These did not seem to last long, for at the July term, 1804, we find an order, that two pairs of stocks be made, one for Merigomish, and one for the Harbour of Pictou.

Again in 1805, we find it ordered:

" That a board fence be built at the front of the District Jail, sufficient to prevent any communiation from persons on the outside to them within said jail, the fence to be about eight feet high and twelve feet from the front of the jail."

Then as to expenditure we find the following

District of Pictou Dr. to John Patterson.

1797—To premiums paid the Indians for the killing 4 bears........ £4 0 0
       To premiums paid the English for killing 3 bears at 20s. each   3 0 0

To setting up and taking down the Justices benches July term 1797.....................................	0	5	0
Court House, Benches and Firewood for the Court and G. Jury Jany. 1798......................................	1	0	0
Premium paid Jas Cameron for killing a bear...............	1	0	0
	£9	5	0

True Bill,

DAVID McLEAN,

Foreman, G. J.

" Presented and ordered that Edward Mortimer be allowed out of the Treasury the sum of 17s. on account of goods delivered a poor sailor, and also the sum of 3s. and 4d., on account of bread delivered the negroes, while in jail, amounting in the whole to one pound and fourpence.

"Also 18s. to John McKay for making three pairs handcuffs and repairing a lock in the jail."

The tendency of municipal expense is to increase, and hence we find, that the salary of the Clerk of the Peace, which was £3 in 1797 was in 1800 raised to £4, and his successor received £5. But there was great delay in collecting taxes, and salaries were long in arrears. Thus we find in 1803, the Court ordering " the £70 which was voted by the grand jury in 1801, for defraying public charges be raised immediately, the late emigrants, who have been only two years in this settlement to be exempted." And in the same year in July, £20 10s. was voted to Thomas Harris, for his services as clerk since 1799.

Of regulations of a general nature we may insert the following.

1799. "Presented and ordered that cognizance be taken of all persons within the District, guilty of not clearing away all putrid fish, meat or other nuisance from about their houses, stores or wharfs, which may be the means of causing an offensive air, and be prejudicial to the health of the inhabitants of the District."

1804. "That the law of this Province respecting shipping throwing their ballast overboard below high water mark, be put in force."

The following however we do not regard as so commendable. In the year 1801, it was resolved to " memorialize Government and the Legislature for authority that there might be three public fairs established within this District, viz., on the Harbour of Pictou, the week following the Court (Wednesday) ; the second on the East River, handy to the public road of the Lower Settlement on the West Side, the last week of September ; the third on the large beach at the east end of Merigomish, on Wednesday preceding that of East River." These fairs continued to a late period, the principal being held at David Marshalls, now Horns, place, Middle River. But they were little more than scenes of drinking.

The first notice of a ferry is in 1807, when we find "John Foster licensed to keep ferry at Fishers Grant, that he shall not charge more than 15d. for a single person, 9d. apiece for four, 6d. apiece over four, and for swimming across oxen or horses 1s. 6d. per head."

Again we have various orders against parties violating sound morality, such as the following :

1799—" That Mr. M. be taken cognizance of for selling with a half bushel said to be small, and neither branded or sealed."

1800—" Also presented, that inquiry be made into the conduct of Mrs. G. for keeping a house of bad fame, as reported."

1801—" Also presented and ordered, that the laws of this Province concerning vagrants, be duly put in execution, and that no person of suspicious character be allowed to come into the District, without producing a proper certificate, nor any person from places infected with contagious diseases be allowed to come into this District."

" That A. M. be fined in the sum of two shillings for swearing, the same being for the use of the poor, and also that J. H. be fined in the sum of two shillings for most notorious swearing."

1804—" That G. P. and A. G., young men in this place, who lead immoral and scandalous lives, such as getting drunk, cursing, blaspheming the name of God, fighting and insulting sober people, be bound over to keep the peace, from July term 1804."

" On the presentment of the grand jury, it is ordered, that A. C. and D. L be fined the sum of five shillings each for being intoxicated with liquor and swearing."

Sometimes a jury was empannelled and criminals tried under a regular indictment. Thus in 1805, we find an indictment against S. L. N. for assault on Mr. Mortimer, on which the jury found him not guilty.

The annual action on the laying out of roads shows something of the progress of the country. Thus at the July term 1797, we find an order for "inspecting the road from West to East Branch passing James Grants mill;" at the January term 1798, we find a committee appointed to lay out a road from the harbour to Scotch Hill, Angus Campbell, Wm. Fraser, Hector McQuarrie and Duncan Cameron being then mentioned as residents at the latter place, one to lay out road up McLellans Brook to Peter Frasers, and another to lay out a road from Fishers Grant to Little Harbour; and the road along the West River was confirmed with the following luminous description. "Beginning at a gate on the publick highway, at the foot of Anthony McLellans intervale, from thence crossing the river and leading up the river, on the road now occupied, and continuing its course as far as a pair of bars at Robert Stewarts, from thence to go to the Southward of the road now occupied, until it strikes the old road on the bank of the river, at a small piece of Intervale, and from thence to continue along the old road, until it joins the road already established." At the same meeting, a return was made of "the road on the East River from Donald Frasers to the Governors Road." In the following January term, we find the road from the West to the East River confirmed, but it is not till 1805, that we find a committee "to lay out a road from Hugh Frasers East River to Fishers Grant."

For the better protection of the roads we find the following orders :

"1799. That all gates on the King's highways be removed, and the highways be cleared of every incumbrance whatever."

"Also that a committee be appointed to inspect the road leading from Archibald's Mills to John Blaikies, as there are many obstructions on said road, by which the public is much injured."

"1802. Presented by John McKenzie, overseer of roads for the West River, that there are a number of windfalls on the road between the Sawmill Brook and Ed. McLean's. It is ordered, that the said John McKenzie may clear the windfalls out of said road, or employ who he may think proper, and the said person or persons be exempted so much of their statute labor for the present year, as they may be employed in performing the same."

The granting of licenses for the sale of intoxicating liquors occupied the attention of the Court from the first. All the merchants received license to sell liquor by retail. At the very first meeting, we find license granted to " Ed. Mortimer to retail spirituous liquors for the term of six months, he yielding obedience unto the laws of the Province." And there were always one or more "taverns or places of public entertainment" in town and several in the country. Thus in 1799 the list is—

Pictou Harbour.................Wm. Lyndsay* and Jas. Dun.
West River......................George McConnell.
Merigomish......................Robert Smith.
Gulph............................Wm. McGregor.

It is evident that all the travelling of those days would not require so many places of entertainment, and that they must have drawn largely for their support on the inhabitants. Still we find them increasing, for in the year 1801, it was resolved, " That not exceeding two tavern licenses be granted to innkeepers on the Harbour of Pictou, and only two within Merigomish, also that one be granted to some person residing on each of the three rivers in Pictou, and also one on the Gulph." The year previous there had been four licences granted at the Harbour.

Troubles however would arise from breaches of the law. Thus in 1798 we find parties summoned to a special sessions in John Patterson's store, for retailing spirituous liquors, and in the year previous, the following presentment of the Grand Jury :—

"On the presentment of Ye Grand Jury, it is ordered, that J. D. be summoned, to attend the General Sessions of the Peace, to give an account of his conduct, for the retailing spirituous liquors, contrary to the intent and

---

* He was from Scotland and built the house long known as Mrs. O'Neill's tavern, where he followed the same employment. His house was frequently the place of meeting of the Sessions.

meaning of the laws of the Province, as appears by the evidence, on the trial of three negroes on the 18th January, 1797, for petty larceny, one of said negroes having carried a pawn or pawns to the house of the said J. D. on the Lord's Day, and in' exchange thereof, received spirituous liquors to a small amount in proportion to the article lodged."

In the year 1800, the first bridge was built across the Town Gut, by John Patterson. Our readers may have an idea of the inconvenience of travelling, if they just reflect on the fact, that there was not previously a bridge of any size in the whole county. Arrangements were made for crossing small streams by felling trees across them. Sometimes a single tree was used, on which a traveller might cross on foot, but sometimes two or three might be placed together, which formed a rude bridge, not only more convenient for the foot traveller, but over which a horse might pass. But commonly the streams were crossed by fording where shallow, and often this involved a considerable circuit, while the deeper required canoes.

The bridge now constructed was built on wooden pillars, with stringers from one to another the whole distance from shore to shore, the present embankment not having been made till the next bridge was built, about the years 1818-20, and the channel in the centre having since that time been cut out by the waters thus confined. It was a creditable structure for the time. We give the Deacon's account, from which it appears that it was built partly by subscription on the part of those in town, and partly by the Sessions, without any Government aid.

The Magistrates of Pictou Dr. to John Patterson, Senr.

for Builden a Brige over the Town Gut, July 15, 1800	£60	0	0
1803—July To 3 years Intrest on Thirty pounds	5	8	0
1805—To 2 years Intrest on Twenty pounds	2	8	0
1808—To 3 years Intrest on Ten pounds	1	16	0
	£69	12	0

Contra credit July 15

1800—by the Treasurer	£10	10	0
by Robert Patterson Esq	2	0	0

by John Patterson Senr....................................	2	0	0
by Thomas Harris Junr...................................	2	10	0
by Edward Mortimer......................................	2	0	0
by Robert & Thomas Pagans............................	2	0	0
by Donald McKenzie.....................................	0	10	0
by John Dawsson......................................,	2	0	0
by David Patterson.......................................	1	0	0
by James Patterson......................................		10	0
by John Patterson Junr..................................	2	10	0
by James McDonald......................................		10	0
by Wm Campbell...		6	3
1801—April 17 by John Clark by Mr Dausson................		5	0
1801—by Edward McLean...................................		10	0
May 1802 James Patterson by Mr Dausson.....................		10	0
1803—March to one order on the clerk Lichens money..........	10	0	0
Augt to pay by William Murdock.......................		5	0
1804—By John McKenzie...................................	1	0	0
Augt 1805 To one order of Thomas Harris Junr for..............	10	0	0
	£50	11	3

It will thus be seen that at the time of the Deacon's death in 1808, eight years after the work was done, there was still a considerable part of the amount remaining unpaid.

About the same time other bridges were engaging attention. At the July term, 1801, the License money, amounting to £17 4s., was ordered "to be expended, £10 on the bridge over James McKay's Gut, East River, £3 to assist building a bridge over the East River of Merigomish, and £4 4 for a bridge over French River."

In the year 1803 the first bridge at New Glasgow was built, about a hundred yards above the site of the present one, but it was all carried away the following winter. The next bridge there was built by Robert Grant, miller, on piers, the portions of which below water have formed the foundations of all the bridges that have since been built there.

In the same year we find it ordered that " all the statute labour for this present year, from Anthony Cultons on the East River of Pictou on both sides, downwards, including

Fishers Grant, be laid out in erecting a bridge across McKays Gut, and also the deficient labour on Fishers Grant for last year be laid out on the same, and to be begun on the first of July next."

In July term of that year, £20 of license money was voted "to assist in erecting a bridge over the Middle River, but in case the above bridge will be built by Government money, the above to be expended otherwise."

In the year 1804, we find it voted "that the bridge on the Saw-mill Brook, on the road leading from the Harbour to the West River, be railed, the same to be paid out of the public Treasury." It was also agreed to expend so much of the public money as may be necessary for repairing the bridge on McCulloch's Brook, Middle River. These votes show that these had been previously erected.

As showing the progress of expenditure, we give the amounts voted at February term, 1808,—

Gut Bridge	£ 60	0	0
James McKays Gut Bridge	25	0	0
Public wharf	10	0	0
Saw Mill Bridge	50	0	0
Purchasing a lot for court house	30	0	0
Extra contingencies	125	0	0
Overseers of poor for the use of the poor	50	0	0
	£350	0	0

Altogether, an examination of these records impresses one very favorably, regarding the business habits and capacity, of those who first administered our county affairs. One who remembers them, says of those we have named. "They were all men of education and refinement, and their gentlemanly deportment and dignified manner induced a high respect for the bench, and gave a tone of order and refinement to society." We may add, that considering all the circumstances, the records were kept by the clerk, Thomas Harris, in a very creditable manner.

The Inferior Court, as it was commonly called, or Court of Common Pleas, though issuing process to any part of the Province, trying titles to land, and indeed doing the same work as the Supreme Court, with the exception of the higher criminal business, was not presided over by a legal mind. It was not till the year 1824, that the act was passed, by which a lawyer was to be appointed the presiding judge and also President of the Court of Sessions, an act which at the time was very unpopular. Even then he was to be aided by two lay justices. At this time however all the judges were laymen. The first book of Records of the Court now existing commences with the year 1804. The first Judges of the Court were Hugh Dunoon, Robert Pagan, and either then or a little later, John Dawson and Edward Mortimer. One who remembered them on the bench, says, " They all commanded respect. The former were in education superior to the last, but from natural gifts he after all exerted the most influence." *

Besides the Inferior Court, there was established in each county by an act of the Legislature, a court known as the Commissioners Court, consisting at first of five, and afterward of three, commissioners, which met monthly for the trial of cases of debt, up to a certain amount. Of this court in Pictou, Mortimer was the head and almost the body. The system did not continue long.

The first meeting of the Supreme Court of which we find a record, took place on the second Tuesday of June, 1806, George Henry Monk being the presiding judge. It

---

* The following is a list, so far as we have been able to make it up, of all who occupied the office till the abolition of the Court:

Hugh Dunoon, John Dawson, Robert Pagan, Edward Mortimer, A. McDonald, George Smith, Robert Lowden, Andrew McCara, James Skinner, William Mortimer, Abraham Patterson. Under the act of 1824, Jared I. Chipman was appointed first justice for the eastern part of the Province. He died on 2nd June, 1832, and was succeeded by William Q. Sawers, who filled the office till the abolition of the Court.

was usual at that time for two judges to attend, but, he being the only judge of the Supreme Court present, the judges of the Inferior Court, Hugh Dunoon, John Dawson, and Robert Pagan sat with him as associates. In the following years, Judge Monk usually presided, with either Brenton Halliburton or Foster Hutchinson as his associate. Thomas Harris (clerk) was the first Deputy Prothonotary, and his cousin, Thomas Harris, the surveyor, was sheriff at this time. The court sat once a year, till the year 1816, from which time it met twice a year, in June and September.

There being no court house, the first sittings of the Supreme Court were held in a building on the west side of George Street, a little below Church Street, now, we believe, Dr. Kirkwood's barn, but then a carpenter's shop. To the lower end, it was said, was sometimes a pig pen, which even extended under the building. The late D. Fraser used to say, that he was foreman of the first jury. On his describing the place of meeting, one asked him, "Where was your jury room?" He replied, "When allowed to retire to make up our verdict, we went to a grove in McGeorge's pasture," which was near the site of the Episcopal church. In this or places not much better, the Court continued to meet, till the erection of the court house in the year 1813.

The first movement for the erection of a court house was at the January Sessions, in 1801, when it was resolved, "That a proper piece of ground to build a proper court house on near to the Blacksmiths shop, on the north side of Pictou Harbour, be purchased from the owner. And as Walmsley is a very inconvenient place to hold a court at, Government ought be immediately memorialized to confirm the above spot, where the court house is to be built, and Sir John Wentworth to give a name to the township."

At a meeting of Sessions, presided over by judge Monk,

at the first sitting of the Supreme Court, it was "ordered that £200 be assessed for erecting court house," but this was not done that year, and accordingly in the following year, (1807) the Court resolved, that, "the Grand Jury not having assessed the amount necessary, the district be amerced £150, which appears necessary for defraying the District debts and charges, with £200 formerly for court house." In the estimates for 1808 appears £30 for "purchasing a lot for court house." It was not however till the year 1813, that the building was erected, being that now known as the old court house.

The only lawyer resident in town at this time was John Fraser, a son of Capt. Fraser, already referred to. He was also the first collector of customs and was hence usually known as Collector Fraser. A number of eminent lawyers began to attend, among whom S. G. W. Archibald was especially noticed, but there were also such men as Chandler, S. B. Robie and W. H. O. Haliburton, who interested the people by their eloquence or amused them by their flashes of wit.

We may at this place give the names of the officers of the Court till the present time. The offices of clerk of the Peace and Prothonotary have in this county always been held by the same person. Thomas Harris died in 1809, but we find George Smith holding these offices from 1806 till 1809. Walter Patterson succeeded, and continued to hold both positions till his death in 1821. At his death, his brother Archibald was in office for a few months, when Dr. James Skinner * was appointed. He died in 1836, and was succeeded by his son, James Skinner, Jr., who died in 1861, and was succeeded by David Matheson.

* Dr. Skinner was a son of the Rev. Donald Skinner, parish minister of Ardnamurchan, and grandson of Hugh McLean of Kingarlock, Argyleshire. He was for years active not only as a physician, but in the public business of the county.

Thomas Harris, Sr., was deputy sheriff till 1811, when he was succeeded by the late John W. Harris, who continued in office till the county was divided in 1836, when the Government appointed J. J. Sawyer, who had formerly been High Sheriff of the united county of Halifax, to be High Sheriff of the three counties of Halifax, Colchester and Pictou, and Mr. Harris was appointed his Deputy. But when the Legislature met, the House of Assembly set their faces against this plurality system, and in the following year, he was appointed High Sheriff, which office he continued to hold till 1857, when his son, Wm. H. Harris, succeeded him.

In connexion with this, we may here notice the first trial for murder in this county. The crime was committed on the 26th May, 1811, by a man named McIntosh. He had been originally a tradesman, but took up the idea of going into business. He went to Halifax and obtained a supply of goods, which he put on board a schooner to bring around to Pictou, but was detained all winter in Guysborough. Giving himself out as a person of some importance, he succeeded in marrying there a lady, renowned for her beauty. Arriving in Pictou in spring, he commenced merchandizing, and flourished while the goods lasted; but when they were done he found himself in debt. His creditors had him arrested, when a friend, named Dougald McDonald, obtained his release by becoming security for his appearance at court. When the time arrived, however, he failed to appear. The judge told his bailsman to take him wherever he could find him. The latter accordingly went with the sheriff to try to seize him. McIntosh shut himself in his house, which stood on the east side of Yorstons wharf, a little below where Hamilton's bakery now stands. McDonald took a crowbar and commenced prying open the door. As soon as he had it partially open, McIntosh fired a blunderbuss at him, the contents of which lodged in his body, so that he bled to death in an hour or two.

The magistrates met and immediately issued a warrant for the arrest of McIntosh. But he armed himself and defied any one to arrest him, threatening death to every person who should touch him. Leaving his house, he crept under the wharf, and perching upon some of the logs, it was no easy matter to dislodge him, and almost every person was afraid to venture near. At length, John Sylvester, of Middle River, a fearless old man-of-war man, undertook, with another, to make him prisoner. Taking a pistol, he went under the wharf, and, immediately presenting it, ordered McIntosh to come down, threatening to fire if he did not do so at once. Seeing his determination, the latter surrendered.

McIntosh was arraigned for murder at the Supreme Court, on the 3rd of August, 1811, and his trial came on in due course on the 5th, before Judge Monk. There being then no proper court house, the trial took place in the old Presbyterian church. As this was the first case of the kind in Pictou, great interest was excited, and the house was crowded. Trials were not conducted in so tedious a manner as they are now, but this was prolonged well into the night, so that the closing address was delivered by candlelight. R. J. Uniacke, the Attorney-General, conducted the case for the Crown, and the prisoner was defended by Halliburton and Chipman, who set up as a defence that an Englishman's house was his castle, and that he had a right to defend it against any person breaking in. The Attorney-General closed the case in a manner that excited general admiration. He had taken no notes either of the evidence or the addresses of the opposing counsel. But with his marvellous memory, he omitted no fact bearing on the case, and no point in the objections of the defense, disposing of all opposition with consummate ability. McIntosh was accordingly condemned, and on the 7th sentenced to be executed. But while he lay in prison, the jubilee of George III. was proclaimed, and he was pardoned. During his imprisonment, he seemed affected by his situation, and every Sabbath sent a request for the prayers of the church. But on obtaining his freedom all his concern vanished. He afterward went to St. John, where he was drowned. His widow married in the United States a very wealthy man, and was living till recently.

We may add here that in subsequent years, the county suffered much from litigation, especially regarding boundary lines, owing in a great measure to the manner in which the surveys for the early grants were conducted. Sometimes the surveyors were incompetent, but more frequently the system was to blame. The one rule adopted was to give more land than was named. Some of this was put down, as "allowance for roads, &c.," while such excuses as slack chainage or hilly land formed pretexts for farther additions. But besides this, surveyors exercised a sort of princely liberality, as it was regarded, in giving as if the land were their own, a considerable surplus. This was deemed kindness to the settlers, but from the disputes which these extra quantities produced, it would have been a real kindness to the county if each man's quantity had been exactly measured. Then sometimes the lines were not run round the whole lot, but merely corners set, and the courses marked, and thus the settler was often left to his own conscience how much he would appropriate. Again grants were given nominally for a certain amount, but to a certain boundary or some other grant, without the distance being measured, though it might include an additional quantity, as large as was originally intended. Then the possessor of the next grant might consider himself equally entitled to the land between them, and perhaps would get a surveyor to run his lines, so as to interfere with the other's. Or sometimes a second surveyor coming on the same lot, instead of endeavouring to trace the line made by his predecessor, would make a new one of his own. Such proceedings led to endless disputes, with the worst results. Two neighbours have gone to law about a piece of land till both lost their farms. We have known a litigious man ruin himself, and two neighbours in succession, upon the same adjoining lot. Then family connexions and friends would take sides in the quarrel, and the strife thus extend through a whole community.

Trials were not then the tedious affairs they are now. The judge took brief notes of the evidence, without delaying the examination, there was little squabbling over the putting a question, by which so much time is now spent, and lawyers as well as judges acted as gentlemen. Hence the trial of a cause, was often considered by many "as good as a play."

Of all the lawyers who attended the court, Archibald stood preeminent. At a later date, Johnston was his equal in knowledge of law, but Archibald was always unrivalled in his tact and skill in managing a jury. When all other means failed, he would laugh them out of a verdict. In private he was celebrated for a constant flow of genial humour, but sometimes he had the laugh turned against him. An old Highlander on the West River, who had been much at law, but was still in comfortable circumstances, had frequently invited him to stay at his house, when travelling that way. Mr. A. at length accepted the invitation. His horse was fed and himself welcomed to the house, and soon a savoury dinner was prepared, the principal part of which was a little pig, known as a roaster. Hungry from travelling, he relished it heartily, and when he had eaten, was loud in his praises of the delicacy, and of his host's hospitality, but at length made the unfortunate remark, "But you don't kill all your pigs so young?" "Ah!" said the old man, "no kill her at all; she pe drooned i' the brook!"

CHAPTER XII.

IMMIGRATION AT THE BEGINNING OF THIS CENTURY—
1801-1805.

The first years of this century brought large accessions to this country by immigration, principally from the Highlands of Scotland. Every year at least two or three vessels arrived with passengers, who gradually filled up the interior of the county, and spread to the neighbouring districts. It was at this time that the Highland proprietors were clearing their estates of the small tenants, with the view of turning their property into large sheep farms or deer forests, a policy involving suffering and hardship to many an humble family, but which has given to these and other colonies, some of their most deserving population, and ultimately proved to the advantage of the ejected themselves. The largest accession, which Pictou received in this way, was in the years from 1801 to 1805, as many as 1,300 souls landing in a single season, and at this time several new settlements were formed.

When they arrived in Pictou, they were taken into the houses of the previous settlers, who were sometimes relatives or old acquaintances; but, whether or not, the new comers found a truly Highland welcome, or, what was even better, a Christian exemplification of the precept, "Be not forgetful to entertain strangers," till they could select their own location. Indeed, the notice of the arrival of an emigrant vessel brought people from all quarters, to enquire for relatives on board, whom they took to their homes, or to find acquaintances or persons from their native districts, or even strangers, to whom they would freely extend the same hospitalities.

The new comers received freely, but made the best return in their power by their labour, till they could obtain a lot for themselves. Sometimes, indeed, they hired for some time, before they settled upon their own land, the young especially often remaining for years in the houses of their kind entertainers. As soon as they had obtained an allotment of a piece of Government land, the old settlers near combined, in helping them to erect an humble habitation and to make their first clearing. The house was generally built of round logs, 15 to 20 feet long, undressed, the seams between which were closed with moss or clay. When their circumstances improved, larger houses, perhaps framed, but oftener of squared logs, were constructed; but in the meantime they had a home, which the hand of power had not allowed them in their native land, and which, though poor enough, was better than was possessed by the poorer peasantry of many parts of Britain and Ireland.

The same assistance was readily given in making their first clearing. An axe and a hoe were considered the only implements, necessary to commence a farm in the woods, and even these were often supplied in charity. A path was blazed or partially cleared from the residence of the nearest settler, and the goods of the new comer transported on the backs of men or horses. Neighbours gathered to cut down a portion of the forest round their dwelling. The trees were felled, lopped and cut into lengths, then set fire to, and thus the branches and small wood was consumed. The logs were then piled in heaps, and burnt, or rolled away for fencing, while the stumps were left to decay. This was very disagreeable and fatiguing work, but it was performed in the joy of having a home for themselves and their children, which no lordling could touch, and in the gladsome anticipation of future independence. Women and children aided in gathering and burning rubbish, or other work suited to their strength.

When the ground was sufficiently cleared, wheat was sown and covered with the hoe. Potatoes were planted in round hollows four or five inches deep, in which from three to five sets were placed. Thus the first season, which might be the year after their arrival, or perhaps the second or third, they might have from two to six acres under crop. The vegetable mould, formed by the leaves of successive years, and the ashes left from the burnt wood, rendered the soil very fruitful, and the new settler never failed to reap a bountiful return for the amount sown. Potatoes, it was supposed, would never fail. Such was the commencement made by hundreds at this time in this and other places in the Maritime Provinces. They often felt discouragement enough, particularly as in many cases they had come out under highly colored representations of the country.* But many of these, who thus commenced in the woods in destitution, afterward became independent, and left their families in comfortable circumstances, and had reason to bless the selfishness of Lairds and Dukes, who had turned them out of the little holdings, possessed by their fathers for generations, and pulled the roof tree from off their humble homes.

We may mention, that in the same manner, the sons and daughters of the old settlers in many instances commenced life. When the youth reached manhood, he either received a portion of his father's land, or took up crown land for himself, and erecting his log house, readily found some rustic maid, not afraid of labor, or of spoiling her complexion by exposure to the sun, ready to share

---

* One of the settlers on the Four Mile Brook having been engaged one day hacking at the big trees, which grew on his lot, with all the awkwardness of a Scotchman, becoming tired, sat down, and losing heart altogether, began to cry. His wife coming out, asked what was the matter. He told her his feelings. She immediately returned to the house, put on an old coat of his, and coming back seized the axe and commenced an attack upon a tree. He burst out laughing, took heart again, was never so discouraged afterwards, and ultimately became independent.

his joys and sorrows, his trials and successes. Duly yoked to bear the burdens of life together, they went to their humble log house perhaps on foot, or at best "riding double." Such was the style in which sixty or seventy years ago, the majority of brides were brought home. Commencing life however with stout hearts and in the fear of God, they enjoyed their full share of domestic bliss, and reared a race, who for vigor and worth, may shame their degenerate successors.

We must now, however, give some account of the commencement of the settlements formed at this time, either by young men brought up in the country, or by these immigrants.

The first clearing on Mill Brook was made by Thomas and John Fraser, sons of Kenneth Fraser, Middle River, either in the year 1800 or 1801. They went up the bed of the brook, from where Kerr's mill now stands, carrying their supplies and implements, and erected a camp on what is now Wallace Monroe's farm. They chopped and cleared on that same place, and having put in some seed, they left for the summer and returned in the fall to gather the proceeds. The bears were so numerous, that they did not venture out of their camp after night, even to the brook for water. Being in the habit of returning on Saturday to their father's homestead, on Middle River, they used on leaving to set a bear trap, baited with the remains of their week's provisions, and very commonly found one secured on Monday morning. Hence the place was long known as Bear Brook. Having got their land surveyed and divided, Thomas built a small house on his side of the lot, on the lower part of what is now his son Richard's farm. He had been married, and while they were at work his wife used to come up to cook for them, and perhaps help otherwise, but remaining most of the time at Green Hill. But now, probably in the year 1802, he came with his family to reside here. They came up the brook on foot, carrying their eldest child, between one and two years of age, and their articles of household gear. In some parts of the brook there were small patches of intervale, over which they passed; but where the banks were steep and close together, they were obliged to walk along the rocky bed of the brook, until they came near their home. Here there is a pretty fall, perhaps forty feet in height, where the water dashes over a steep declivity into a narrow gorge. The banks just below the fall rise some fifteen or twenty feet higher than the fall, and so steep that it is impossible to ascend them. They were thus obliged to climb the face of the fall, which they were able to do, as the water inclines to one side. Up this they carried their child and all their utensils.

Soon after a path was blazed to the Middle River, at what is now William Munroe's place. There was not a settler above them, nor for a considerable distance on either side. They were followed shortly after by Alexander Ross, who came to Pictou in the year 1802, and settled where his son Kenneth now resides, Alexander McDonald, who arrived in the following year, and Robert Gordon who came about the same time.

Seven or eight years after settling there, Thomas Fraser put up a mill in the gorge below the fall. He built no dam, but introduced the water into the mill directly from the fall, by a short race. He made a sort of road along the bank on one side, to the mill, but it was still difficult getting up and down from it, and a few years after a freshet carried the whole away. After this he built a mill above the fall, where he had not only stones for grinding wheat, but had the second oatmill in the county, about the year 1817 or 1818.

In the year 1801 came out two vessels, full of passengers, brought out by Hugh Dunoon, Esq. He made representations similar to those, by which interested parties have often deluded people across the Atlantic, such as that the same tree would yield them soap, sugar and fuel, or that they might get sugar from the tree, and gather tea at its roots. One vessel, chartered by him, called the Sarah, brought out 700 souls, though two children being counted as one, and infants in arms going free, they were reckoned as 500 passengers. They were crowded together, and their rations were scanty in quantity, as well as inferior in quality. Small-pox and whooping-cough broke out among them, so that the ship might be said to have realized the horrors of the Middle Passage. They were thirteen weeks on the voyage, having sailed in June, and not having reached Pictou till September, and in that time forty-seven died. During the passage they were boarded by a man-of-war, which pressed 25 of the able bodied passengers, but on Dunoon going on board, and representing himself as a Government agent, they were released. When the vessel arrived in Pictou, sickness still prevailed, so that she was kept in quarantine at the Beaches for some time.

The other vessel, called the Pigeon, sailed later, but arrived befor her. She was a small vessel, and had only a small number of passengers.

Of those on board these vessels a number were Roman Catholics, most of whom removed to Antigonish or places further East. The others took up land in various places, forming new settlements or filling up the older ones. Some of these formed the first settlement on Mount Thom. Among these were Alexander Stewart (afterward known as Post), John McLean, Kenneth McLeod, John Urquhart, Wm. McDonald, Alex. Chisholm, John Fraser, Hugh Cameron, Alexander Cameron and James Fraser. The land had previously been laid out in lots, and each selected his, but there was no settler further up than Dalgliesh, at what is now Robertsons place, on the Eight Mile Brook. Alexander Stewart kindled the first fire on Mount Thom, on the evening of 31st December

(New Year's eve) of that year. His house was on the old Halifax Road, where he afterward kept a public house, and where his son Murdoch has since resided. He had, of course, only a rude hut in the woods. His wife, as she gazed through the partially open roof at the waving tree-tops overshadowing them, and within, at her shivering little ones clinging round her, and thought of the comforts she had left behind in the old land, declared her wish to be back in Scotland, if it were even to be in a jail.

Soon after he became mail-courier to Halifax. George McConnell, at the Ten Mile House, owned a horse, and so did David Archibald, at Salmon River, but between these two places there was not another, and for years he made his trips on foot, carrying the mail on his back, or sometimes in his vest pocket, and, at the proper season, carrying a gun to shoot any partridges which might cross his path. His trips were made regularly, though not so frequently as, in these days of railroads, would be deemed satisfactory, being in fact only once a month, his remuneration being at the rate of ten shillings for each trip. Some time after he purchased a little black pony, on which he made his trips fortnightly, and this continued to be the mode of conveyance for several years.

Of those brought out by Dunoon, another body occupied McLennans Mountain. This was so named from the brook, which runs by it, which received its name from the first settler at its mouth. There were upon the brook at this time the following settlers: Thomas Turnbull, John Fraser (Squire), William Fraser, Elder, son of Simon the first Elder, Alexander and Peter Fraser, sons of McAndrew, John Fraser, son of Simon (Basin), who was suffocated by the fumes of charcoal in his own cellar, and John and James Cassidy.

On the banks of this stream, the limestone is cavernous and contains numerous deep interstices. One of these on the farm owned by Peter Fraser, forms an entrance to what is known as "the cave." It is at the foot of a hill, and by stooping the visitor may enter this "dark retreat." There he finds himself in an apartment about one hundred feet in length, and on an average six feet wide. A small stream of pure sweet water flows along its floor, beautiful stalactites hung from the roof, which have now generally been removed by visitors, and the rude masonry of the walls is only equalled by the projecting masses, which seem ready to fall from above. From this chamber narrow passages lead to other chambers, which have never been explored. At one time the owner spent his summers here, having laid a floor at the entrance, and having fitted it up with a door and window. Here the visitor was welcome to his scanty accommodation, but he has long since been removed to a still narrower house.

Orders were now issued by Government to William Fraser to survey the land on the mountain, and to divide it into lots for the new comers. This being done, a band of twenty-three of them occupied the whole block, each selecting his lot. This would probably be the year after arrival. Most if not all of them were from Lord Lovats country, near Inverness. A list of them will be found below.* With them there settled one person brought up in the country, viz, Simon Fraser, deacon Thomas' son.

Although the country had improved much since the first settlers came, and trade was now brisk, yet we may well suppose that a number of persons from the old country, who had never handled an axe, settling down in the midst of an unbroken forest, without roads or other

* List of first settlers on McLennan's Mountain :—Don. McDonald (tailor), Donald McDonald, James Fraser, ———— Grant, William McLean, Finlay McDonald (piper), Donald Fraser, Finlay McDonald (carpenter), James Cameron, Thomas Cameron, John Fraser (Buie), Alexander Cameron, Finlay McIntosh, Alexander Fraser (weaver), James Fraser (Bann), Hugh Cameron, Alexander McDonald, John Fraser, Peter Stewart, James Fraser, John McRae, Donald McPherson, Angus Fraser (Deacon).

conveniences, had before them a task requiring stout hearts, and for years involving toil and sacrifice. Some of the tales of their ignorance of the country are rather amusing. The following is too good to be lost. They were much afraid of the bear. On one occasion, one of them being in the woods saw a porcupine on a tree, and at once concluded that it was the dreaded foe. He therefore at once gave the alarm to his neighbours. All the men near and some of the women gathered without delay. One of them had a gun, which was put in requisition. Thus armed they advanced boldly, but with due caution to meet the monster. Nine shots were fired at him, by which he was at length laid low. Inspired by curiosity, and in the proud consciousness of their victory, they proceeded to examine their vanquished foe, but found matter of still greater astonishment, in the manner in which the quills stuck in their hands.

The stories they had been told about getting sugar off the trees led to some amusing mistakes, with them and others of the new comers. On one occasion, after the landing of a company of emigrants, a number of them were put to sleep in a barn near town. Early in the morning, when the children awoke, they were heard saying to each other, "Come, let us go out and see if we'll get some sugar on the trees." One woman asked to be shown the trees from which it was obtained. When this was done, she picked off some of the bark with her fingers and commenced chewing it, expecting to enjoy the saccharine juice. After they had learned how it was made, one man, as the season for sugar making advanced, finding the supply of sap beginning to fail, fastened a strong withe round a tree, under which he drove a wedge tightly, determined to squeeze out of it the last drop of juice.

Yet these men, in this and other places, surmounted the difficulties of settlement, and became independent in their circumstances.

Others of the Dunoon passengers settled in various parts of the county. Archibald McKay and Donald Cameron, from near Inverness, settled on Frasers Mountain. Three other settlers, however, were there before them, Donald McKay (Squire's son), who was the first, William Fraser (surveyor) and Charles Brown.

A number of those who came this year, occupied the upper part of the East Branch East River. Among these were Donald Kennedy, Robert McIntosh, James Chisholm (blacksmith), John McDonald, Duncan McDonald, Archibald Campbell, John McDonald, John Thomson Alexander Thomson, and John Grant, the most of them from Glen Urquhart, and in the years immediately following, they were joined by others. Some years after, Duncan McDonald, then an old man, was lost in the woods, with his grandson. The latter followed the county line, and, after three days travelling without food, came out upon the settlement. The old man took a different course, and after a search by the inhabitants, was found dead, after five days' absence.

In the year 1802, came William Cummings, from near Inverness in the following year settled on what is called the Blanchard Road, and commenced the Blanchard settlement. This road was originally cut out by Colonel Blanchard of Truro, to reach a large grant of his at Lochaber, in the County of Antigonish.

In the year 1801, and in others about the same period, Captain Lowden also brought out a number of persons from the south of Scotland. Some of these came to work at his vessels, but others as settlers. Among these may be mentioned Robert Bone, George Reid, afterward of Green Hill, James Gordon and Samuel Wilson.

James Gordon was a cartwright, who is worthy of notice as having made the first fanners ever in use in Pictou. They were built in the year 1803 for Captain Lowden, who had brought out the irons from

Scotland. As they were the only set in the place, they were carried about over a circuit of ten miles, but they have continued to do duty to the present day, and may be seen in the possession of the Captain's grandsons at the Beaches. We tried them in November, 1876, and found them likely to do good work for years, if not to wear out another generation of their degenerate successors.

The year 1802 witnessed the arrival of a large number of emigrants. In the month of August, 370 landed, natives of the Island of Barra, and all Roman Catholics. As they had been accustomed to the fisheries, Governor Wentworth located them for a time on Pictou Island, and the shores adjacent, but they all moved away eastward to Antigonish or Cape Breton. A number of Protestants also arrived, who settled in various places, but we are not informed of any settlement formed by them.

One who arrived in this year thus describes the state of the town at that period. North of Front street and east of Coleraine street, down to Alpin Grants, was covered with good hardwood, and people cut their firewood there. The farthest east house in the town was Mr. Pagan's, already mentioned. There was Lowden's salt house on the east side of Coleraine street. On the latter Dunn had his tavern, back of where the Royal Oak stood. Thomas Fraser, carpenter, had a house where St. Andrews Church stands, and a ship carpenter named Young had a small log house opposite. Following Water street westwardly, Joseph Begg had a log house on the site of his stone building, lately taken down. Then John Dawson had his house, store and wharf, in rear of what is now the Taylor House. On the site of Messrs. Yorston's store, Captain Lowden had his dwelling and store in one building. On George Street was McGeorge's tavern, on the site of the property lately owned by John Proudfoot. On Yorston's Wharf was John Patterson's store, still standing, and John McKay's blacksmith shop. On the site of the drug store of James D. B. Fraser & Son, was the jail. William Lyndsay kept a tavern on the site of what has since been Mrs. Cameron's Inn. In front of this was the open shore, the tide coming up to the opposite side of the road, and sometimes over it.* To the end of this to the east, J. Connell had a small log house. Thomas Harris (sheriff) had a small house on the lane back of the establishment of the late Peter Brown. Then Helier Houkuard had a red house, near where the late H. Hatton, Esq., resided. He was a Guernsey man, who went out fishing in summer in his schooner, which was then the only vessel owned in Pictou. He had a wharf near the site of where the post office now is, with a small fish house upon it. Here sometimes small vessels were built. Westward Copeland had a barn where the market house now is, in one end of which S. L. Newcomb was teaching school. Beyond was a fine hayfield. The only building to the west was Copeland's house and store on the same site, and indeed partly the same building as occupied by John Crerar, Esq. He had a wharf, where Dr. Johnston's now is, and a small one about where the property of the late James Dawson was. Farther back John Patterson had erected his house and made a small clearing on the top of his hill, near where his grandson, A. J. Patterson, resides. The hill was then so steep, that on certain parts of it, he was obliged to make steps. On the north side of Church Street, then called Queen street, at the corner of George Street, he had erected what was usually known as the Old Barracks. It consisted of a range of small dwellings united. It had three doors with a tenement on each side, making six in all.

In the year 1803, it was stated that there were 5000 inhabitants in Pictou, and that 1000 more were expected

---

* As late as the year 1820, there were stones placed along Water street near Meagher's Slip, to enable passengers to pass dry shod at high water. When the tide was in, it formed a pond to the north of the street.

that season. On the 6th August, the Lieut. Governor wrote that 845 had arrived. Of the immigration of this year the voyage of one vessel was long remembered. She was called the Favourite, of Kirkcaldy, and was commanded by a Capt. Ballantyne. She sailed from Ullapooll, without a clearance, and arrived in Pictou on the 3rd August, having made the passage in five weeks and three days, being regarded then and for some time after as the quickest ever made. She had 500 passengers on board, and landed one more than she took on board, one birth and no death having taken place on the voyage. But almost immediately after the passengers' goods had been landed, she sank in the harbour. Such a strange occurrence might well excite enquiry as to its cause, and as we have received, from a most reliable and worthy old man, who when young was a passenger on board, a veritable account of the whole particulars, we shall give them as we have received them. It appears that shortly before the vessel left, one man who came in her was out one evening looking after his cows, when he saw a little creature like a rabbit going round to them, and sucking the milk from them. He immediately took his gun, and tried to shoot it, but found it impossible to do so. Suspecting the cause, he put a silver six pence into the gun, and again fired, when the creature limped of, leaving traces of blood in its track. The next day he made enquiries, if there were any person in the parish hurt, and sure enough found, that one old woman was confined to the house, by some injury she had received. He called at her residence, but could not see her. On his engaging his passage in the Favourite, she was heard to declare, that with him on board the vessel would never reach America. In consequence of this, the passengers applied to the authorities to have her confined, until the vessel should arrive. As we have seen, she had a remarkably quick passage, and when on the banks of Newfoundland, they spoke a vessel homeward-bound. On the arrival of the latter, the captain said that they might let her go, as the Favourite was doubtless safe in Pictou by that time. They did so, but my readers may judge, just soon enough.

But this is a sceptical age. The tendency now is to attribute all such events to natural causes. Hence on conversing with an elderly lady in my congregation, who had been a passenger on board, and asking her how the vessel happened to sink, she said, "Oh, *they took the ballast out of her,*" as if that would account for such an event. We wish that every reader who thinks it would, had seen the indignation, with which our first informant reproved the incredulity of one, who doubted the possession of such supernatural powers. "What, don't you believe your Bible?"*

There was not then a settler on the Four or Six mile Brook, except James Barrie, a native of Perthshire, who had settled there only the year before, where the mills are. The most of that section of the county was now occupied by these immigrants. In the following year they commenced operations. In that year Alexander McKenzie made the first clearing on the Four Mile Brook. There was then no settler above John Rogers' place, now Alex. McLellan's. He was joined by Donald McKenzie, Murdoch Innis and others. On the Six Mile Brook, McBeath, who afterward removed to New Brunswick, Murdoch Sutherland, William Gunn, Donald Sutherland and George Sutherland settled about the same time, and on the Eight Mile Brook, Hugh Sutherland, Murdoch Munroe and Alexander Graham, besides others whose names we have not received. About the same time arrived William Munroe and Hugh McPherson, who had served in Lord Reay's Fencibles. †

---

* He was kind enough to say for our comfort, that there were no witches in America. This, however, is by no means admitted by others.

† As this regiment yielded so many settlers to Pictou, we may mention,

A number of those who arrived this year settled at Rogers Hill. There was not till that time, a settler between McCaras place and River John. But we have failed to obtain any particulars of interest.

Of those who came this year, however, a number were from the parish of Lairg, in Sutherlandshire, who took up land farther up the Middle River, and formed a new settlement which they called New Lairg, after the name of their native parish. Among these were Angus McLeod and John McLeod, and perhaps some others, who settled there soon after arrival. The same year, or about that time, arrived Donald Murray, Hugh Murray, John Murray, John McKay and John McKenzie, who had served in Lord Reay's Fencibles in the suppression of the insurrection in Ireland, who settled around them. In subsequent years, others took up land till they got so far on the way to Stewiacke that the soil became poor, and a number of them abandoned it.

On the 4th of July the same year arrived the brig Alexander, of Stornoway, owned by a Mr. McIvor of that place, with passengers mostly from the Lewis. The captain died on the passage, and the owner, who was on board, took sick, when the vessel was taken charge of by

---

that it was raised in that portion of Sutherlandshire, known as Lord Reays country, and embodied in the year 1795. They were soon after sent to Ireland, where they saw some hard scenes during the rebellion. Stewart, in his History of the Highland Regiments, says: "Such was their good conduct, that Lord Lake had his own guard formed of them, to whom he became so much attached, that he seldom passed any guard or post, without alighting from his horse, going among and holding conversation with them. At the defeat of Castlebar, he frequently exclaimed, 'If I had my brave and honest Reays here, this would not have happened.' At Tara Hill on the 26th May, 1798, three companies of the Reays, under a spirited and judicious veteran, Captain Hector McLean, supported by two troops of yeomanry, drove back and scattered a body of rebels, who were in great force on this strong and elevated position. So conciliatory was their conduct, that where they were quartered, the inhabitants were quiet and apparently less disaffected than elsewhere. They were disbanded in 1802."

Mr. David McGregor, father of John McGregor, afterward M. P. for Glasgow, and Secretary to the Board of Trade, but then a child. The vessel returned the following year with another cargo of passengers from the same place. They were encamped for a time in the woods to the north of Front St., but the majority of them moved to the Gulf shore of Wallace, where they commenced a settlement, but a number settled in different parts of this county.

The first settlement on the back shore was made about the year 1803, between Toney River and Cape John, by George McIvor and Allan Munroe, Highlanders from the Island of Lewis, the former of whom afterward removed to Cape Breton. About the same year, Norman McLeod settled on Toney River, where he was the first settler, who afterward moved further along the shore, and Donald McLeod, both of them from the same island. About the same time Roderick McDonald and Alexander McDonald settled on the shore, the former of whom, however, afterward removed to Wallace. In the year 1810, John Stromberg, a Swede, settled farther toward the cape, and a man named Smith, on what is now Skinner's farm, about the same time.

This section of the country was distinguished by its splendid pine. One of the first settlers on Carriboo River loaded three vessels from his own land with white pine timber. About the year 1810, James Mills, a gentleman from England, erected large mills on Toney River, and vessels loaded at its entrance for Great Britain. But no part of the county exhibited such an extent of superior pitch pine. In some places, nothing could be seen but its peculiar foliage. Trees rose clear of limbs to a considerable height, and, though never equalling in size some other wood, yet were large compared with anything now to be seen. The writer's father has told of getting into a grove of this kind, where every tree squared fourteen inches clear of sap, which, however, in no case exceeded an inch in thickness.

From Rogers Hill settlement was gradually creeping westward. James Fitzpatrick, a native of the North of Ireland, settled on the hill, which has since received the name of Fitzpatricks Mountain, which presents one of the finest prospects in the Province, embracing the whole country between it and the shore, and the coast from Pictou to River John, with the Straits of Northumberland and Prince Edward Island. Andrew McCara, Esq., settled further out on the farm now occupied by Duncan McLeod, as early as the year 1800. He was a Lowland Scotchman, who had received a collegiate education, being a fellow student of Dr. McGregor. He had emigrated to Philadelphia, whence he was driven out by one of those terrible visitations of yellow fever, which then sometimes desolated American cities as far north as New York. Many persons wondered that a man of his education, should have contented himself with his situation in the woods at Rogers Hill. On one occasion he was visited by some old friends from Philadelphia, who used all their influence to induce him to return. On their representing the advantages enjoyed there, he replied, "Yes, but you've got the yellow fever there." They went on to state this and the other point of superiority of Pennsylvania, and this and the other disadvantage of Nova Scotia, but to each argument of the kind, the old man had but the one reply, "Yes, but you've got the yellow fever there."

At length land was taken up on the West Branch River John, the first settler being Rod. McKenzie, who made the first smoke there in the year 1805. His son Murdoch erected the first mill there. When the Philadelphia Company's grant was escheated, Dr. Harris having died previously, Government agreed to give each of his children a certain amount of crown land wherever they might select. One daughter, married to John Moore of Truro, received her portion on the West Branch River John, and settled there in the year 1812. A year later they were joined by Thomas McKay from Rogart in Sutherlandshire, and two years later by Donald and William Murray from the same parish, and Henry Marshall, originally from Germany. Of this settlement we may say here, that the first schoolhouse was erected in the year 1825, that the first preaching was by the Rev. Hugh McLeod, of Saltsprings, but the first minister, who supplied them regularly, was the Rev. William Sutherland of Earltown. The first church was built in the year 1837, being the same occupied at present.

A number of those who immigrated at this time, settled in Carriboo, on the Cochrane grant. Previous to this, John and Thomas Harris, sons of Matthew, had erected a saw mill on Little Carriboo River, and about this time James erected another on the Big Branch, but, not having secured his title to the land, another party came in before him and obtained a grant of it, so that he abandoned it, when he was about ready to commence work. A short time before, Thomas Patterson, son of the Squire, and one of the Rogers, made the first settlement on Carriboo Island, the former on the place afterward purchased by Donald McKenzie, and now occupied by his son Roderick, and the latter on the place since occupied by Hector and John McKenzie. Patterson was drowned in the year 1806, under melancholy circumstances, as thus described by his son, the Rev. R. S. Patterson.

"I remember yet my father's death. I was then between five and six years old. We had been to Pictou, and were returning home to Carriboo Island. My mother had a frightful dream the night before, and refused to go with my father in the boat. He and a sailor went in her. They had a couple of cannons for some vessels, of which there were a number in Carriboo harbour at that time. It was war time, and merchant vessels took some guns to defend themselves against privateers. My mother and I, with a servant girl, who assisted to carry my youngest brother, David, who was then an infant, walked through the woods over the peninsula, and crossed to the island in a flat. On arriving at home, we saw the boat in which my father and the sailor were, coming up the harbour. A few moments after we looked, and no boat was to be seen. Search was made, but she was not found for some time. The body of the sailor was found about nine days after, and it was ascertained, that the boat had upset and sunk. Her masts were seen at low water."

The new comers occupied the Cochrane Grant, without title, and after they had surmounted the first difficulties, Cochrane made an attempt to dispossess them. He employed several of the ablest lawyers in the Province. Finding the title defective, he, doubtless under legal advice, went round among the settlers with a lawyer, kindly offering them leases, which, through the "Oily Gammon" powers of persuasion of the latter, some were induced to accept. The causes came on for trial at the Supreme Court, when the late Judge Wilkins, who presided, scouted Governor Patterson's title, ridiculed the horse and saddle transfer, and denounced the lawyer's conduct, in deceiving ignorant people into acknowledging Cochrane's title by taking leases from him. This led to a furious altercation between the judge and the plaintiff's attorney, the late J.W. Johnston. So angry did each become, and so violent was their language, that the audience looked on in amazement, some almost in terror, and that night it was fully expected, that the affair would end in a duel between the lawyer and a friend of the judge on his behalf. The course of the judge produced somewhat of a sensation, and excited the indignation of parties in Halifax, who threatened to take measures for his dismissal, but the lawyer was obliged to apologize. The result of the case, however, was that a number of the settlers compromised by paying Cochrane a small sum, but others firmly resisted all his claims, and their heirs or assigns hold the land undisturbed to this day.

We may mention here that the usual place of burial for the people of this settlement is at a point inside the Beaches, known as Burying Ground Point. Some suppose it to have been a French cemetery, but others connect the commencement of it with a solitary man, usually known as Martin Day, who lived there. No person knew whence he came or anything about him. He had but little intercourse with any person, and few desired intercourse with him. Indeed, he was generally supposed to have been an old pirate. Finally he was found dead in his house, and his body was buried near.

In the year 1805, a vessel arrived with passengers from Gairloch in Ross-shire. Three of them, Philip McDonald, Alex. McKenzie and Donald McPherson, took up land on what they called Gairloch Brook, after their native parish, and commenced the settlement of Gairloch. About the same time David Ferguson settled there.

We may mention that among the immigrants of these years were some then young, who have since occupied a prominent place in the affairs of the county. Among these may be named the Hon. John Holmes, who came, a lad of thirteen, in 1803, the Hon. James Fraser, who came, a child, in 1804, and John McKay, Esq., stipendiary magistrate of New Glasgow, who came, a boy of twelve, in 1805.

For some years later Pictou continued to be the *Point D'appui* for vessels with Scottish emigrants to the shores of the Gulf, but now the most desirable localities in the county being occupied, and the rich lands of other counties, particularly of Cape Breton, attracting attention, a large proportion of those who landed here found their way thither, or to Prince Edward Island, or even New Brunswick.

We may here observe that the business of carrying emigrants was at this time often conducted in a very reprehensible manner. McGregor thus describes it:
"Men of broken fortune or unprincipled adventurers, were generally the persons who have been engaged in the traffic, long known by the emphatic cognomen of the "White slave trade," of transporting emigrants to America. They travelled over the country among the labouring classes, allured them by flattering, and commonly false accounts of the New World, to decide on emigrating, and to pay half of the passage money in advance. A ship of the worst class, ill found with materials, and most uncomfortably accommodated, was chartered to proceed to a certain port, where the passengers embarked. Crowded closely in the hold, the provisions and water indifferent, and often unwholesome and scanty, inhaling the foul air generated by filth and dirt, typhus fever was almost inevitably produced, and as is too well-known many of the passengers usually became its victims."*

The results was that the British Parliament was obliged to interfere and passed stringent regulations on the subject. These, however, were often evaded, and some years later, one of the worst cases of the kind occurred in connection with the emigration to Pictou. An individual engaged in the business, induced a number of persons in the Highlands to sell off their cattle and other goods, and give him the money. But when they reached the port, whence they were to sail, no vessel was provided. Their condition was described as heart-rending, and the heartless deceiver was brought before the Sheriff and was sent for a time to taste the sweets of prison life. But a case perhaps even worse than this followed almost immediately after. A number of passengers were shipped in a small vessel from the North of Scotland. Soon after

* Hist. B. N. America I. 457.

sailing, she met with a storm, in consequence of which she put back to Stromness. By this time they had partially examined their supply of provisions, and now a complete examination took place, with results to fill all honest minds with astonishment and indignation. Casks, labelled bread, were found to have two layers on top, while the centre was filled with rotten potatoes, stones, straw and earth, and casks labelled pork had one layer on top and rotten fish below. In fact, had the vessel not put back in time, those on board must have perished. The result was that the owner of the vessel, who, however, was innocent in the matter, having only chartered her, was subjected to a penalty of £800. When the vessel arrived in Pictou, some of the passengers revealed these facts, when the man had the effrontery to prosecute them before the Supreme Court for libel. The facts were clearly proved, and the jury did not take a long time to give a verdict for the defendants. He also prosecuted Robert Patterson, Esq., for taking their deposition, but the jury without leaving the box, threw out the case.

CHAPTER XIII.

FROM THE BEGINNING OF THE CENTURY TO THE CLOSE OF THE FRENCH REVOLUTIONARY WAR.
1800—1815.

The period which we are now to consider is noted, as that in which the timber trade from Pictou was at its height, though it had begun some years previously, and continued on a diminished scale for some years later.

When the first settlers arrived, the whole of the county was covered with timber of the finest quality. Of this the white pine was particularly prominent, but oak and the various kinds of hardwood, were found of large size and in great abundance, alike down to the very margin of the sea, and up to the very summits of the highest hills. From the first settlement, this had proved one of the most valuable resources of the inhabitants. From the year 1774, when the first cargo of squared timber was shipped to Britain, the trade in that article had continued to increase, and after the closing of the Baltic against British commerce, the price rose to an unprecedented height, and the trade from Pictou increased proportionally. In the year 1803, about fifty vessels were loaded here with timber for Britain, and in the period from 1800 to 1820, it was calculated that the exports from Pictou, of which timber was the principal, amounted on an average to £100,000 sterling per annum. It is to be observed, however, that this included trade from the outports as well, Pictou being the only port of entry for the North Shore of this Province. Still it was the centre of the whole trade, and the larger portion was from the harbour itself.

And now the cutting, hewing, hauling, rafting, and shipping of ton timber, became for some years almost the one business of the people of Pictou. The farmer not only spent his time in winter in cutting and preparing it, but also much of the spring and summer in rafting and shipping it. As to his fields he thought only of hastily committing his seed to the ground in spring, and gathering at harvest time what crop had chosen to grow, and paid no attention to manuring, rotation or other improvements in agriculture, in some instances the dung being allowed to accumulate round the stables, till the sills rotted, and it became a question whether it were easier to remove the mass or the barn, unless where an individual with more foresight, had erected his building by a running stream, which served to carry away the filth.

While however lumbering was *the* business of Pictou at this time, yet even the partial attention, which people gave to their farms, brought plentiful returns. The soil was so rich, that in many places people took crop after crop of wheat, it might be to the number of a dozen, and in one case of which I have been told, of seventeen, in succession. The abundant crops of potatoes enabled the farmers to feed large numbers of swine, and the high prices of all kinds of produce, especially of cattle, in consequence of the war, rendered it a time of unbounded prosperity to the agricultural population.

The lumbering business proved most injurious to the social habits and moral condition of the community. It brought a large influx of population of a very loose character, and it had its usual demoralizing effects upon the residents. Most of the farmers had wood on their farms, so near their dwellings that they could make timber without removing from their homes, but many adopted the system of living in the woods in winter, as still practised in the great lumbering districts of Canada. In the autumn a number of men uniting would go to the woods with a supply of provisions, erect a rude camp in which they spent the winter, with the exception of visits to the settlements for the supply of necessaries, of which rum was deemed the most important, and was commonly the first exhausted. They then proceeded to cut down timber, to square and haul it to the neighboring streams. In the spring, when the melting of the snow caused a large rising of the rivers, the lumber was floated down to the tide, where it was formed into rafts, and transported to the place of shipment. This mode of living, separated for a time from the humanizing influence of civilized society, tends to brutalize men; while the exposure to cold and wet, particularly in river driving, forms a strong temptation to hard drinking, and tends to break down the strongest constitution.

Another evil soon appeared. The first settlers had had a hard struggle to obtain the necessaries of life, but now in the life time of those, who made the first inroads upon the forests, and endured such hardships as we have formerly described, money became so plenty, that people lost all moderation or economy in the use of it, and an era of extravagance trod closely upon the heels of an era of privation. " The farmer," says Dr. McGregor, " neglected his farm and went to square timber. The consequence was, that he had to go to the merchant to buy provisions, and the merchant persuaded him that he needed many other things beside provisions. If the farmer scrupled to buy more superfluities, he would ask him, ' Why do you hesitate? You know that a stick of timber will pay it.' Thus a taste for vanity and expensive living was introduced among us." " We have suffered from emigrants settling among us from different parts of the Highlands, but more from merchants and traders from England and the North of Scotland. The ignorance and superstition of the former have not done us so much evil, as the avarice, the luxury, the show, and the glittering toys of the latter."

But the great evil, we might almost say the great characteristic of the times, was the great extent to which rum was consumed. The first settlers used very little, in fact had not the means of getting it in any quantities. But the extent to which it was now used seems absolutely incredible.

The habit of drinking was most prevalent among the lumberers. We have heard for example, of a man being employed through the winter at five shillings a day, with an allowance of two glasses of liquor, but yet being in debt in spring, although the money had gone for little else but rum. When a lumbering party went to the woods, it was customary to initiate their proceedings with a carouse, which might make such an inroad upon their supply of liquor, as to render an early visit to the settlement necessary to have it replenished. When they did get to work, they daily consumed quantities, which, if they had been using some modern liquors, would have quickly prepared them for the undertaker, while at intervals their labours were arrested for the enjoyment of a carouse, which might last two or three days. Thus in spring they still found themselves in debt to the merchants, from whom they had received their supplies in autumn, and the only course that seemed open, was to go through the same process the next season, with a fresh supply of articles from the merchant, which he was very ready to afford, with the view of obtaining their timber. In this way many farms were mortgaged and never redeemed.

But though drinking was specially prevalent among the lumberers, yet all classes were affected by it. In the most moral settlements, every third or fourth family would have a puncheon of rum, for the supply of themselves and neighbours. In some instances, where there were a number of sons in a family approaching manhood, the whole might be consumed with very little assistance from others. In one large settlement, it was calculated on one occasion, that there had been introduced in the fall at the rate of half a puncheon for each family, and before spring the supply of some was exhausted. I have heard of a tradesman at his bench taking his glass regularly every hour. A person who worked in a shipyard told me, that the allowance to each workman from the employer was three glasses a day, while he was confident that on an average each man drank as many more. A member of my congregation told me, of himself and others working at a job for ten days or a fortnight in the heat of summer drinking each their quart bottle of rum a day, and not at the time feeling the worse of it, though at the close of that period, they felt unfit for work of any kind for the next week or two. Men, not content with a glass, would sometimes drink a half pint at a time, or even a pint, and I knew a man who at one time undertook to drink a whole quart at once, and did so, but it nearly cost him his life. He was in such a state that his friends were summoned to him as dying, but he recovered and lived for years, drinking to the end, though he never attempted such a feat as that again.

If these be regarded as extreme cases, yet the habitual use of liquor was common among all classes. The minister took his dram as regularly as parishioners. The elder sold liquor. No respectable person thought of sitting down to dinner without the decanter on one corner of the table. The poorest would have felt hurt, if a friend called and he had no liquor to give him. No workman was employed without his daily allowance, and that commonly not less than two glasses. As likely it was three, and even that quantity was often supplemented by an additional allowance on private account. No bargain was consummated without a dram. On all occasions of public concourse, liquor flowed freely, and scenes of family interest, births, burials and bridals were consecrated in a similar manner,*

---

* A well-to-do farmer having died, his nephew was seen going home from

while the visits to the shore of the sailors from the shipping in port made the streets frequently scenes of drunkenness and riot. But how the same habits prevailed among the genteel, may appear from the fact of a lady boasting, that the liquor bill for her house amounted to £400 per annum.

It must be said that the pure West India rum then drunk, did not produce such injurious consequences as the liquor now in use. It had not the same maddening effect at the moment, nor did it produce such evil results afterward. Hence men lived to old age, after the consumption of liquors to an amount that now seems incredible. Still, this drinking was a tremendous evil, and the period we are describing was such, that those who can remember it regard it as the worst morally that Pictou has seen before or since. Well might Dr. McGregor say, "Once in a day I could not have believed that all the vices in the world would have done so much damage in Pictou, as I have seen drunkenness alone do within these few years." It may be observed that a similar state of things widely prevailed at the time throughout America.

It might have been expected, that the prosperity of this period, would at least have had an important influence upon the improvement of the country. But it would be difficult to find in any land, an example of such prosperity, leaving so few permanent results for good even upon its material progress. The land was depreciated in value,

---

his funeral under the influence of liquor. On being remonstrated with, he replied, "Ah, its not every day I have an uncle John buried." It having been the regular practice even among the most sober, that at a funeral every man should take two glasses, one on his arrival and one on the procession starting, Dr. McGregor, on one occasion, addressing those assembled, urged that henceforth they should be content with only one. Scarcely had he finished, when an old elder, whose conservative notions had been hurt by the proposal, stepped up to the table, filled a glass, and as he raised it to his lips said, "Here's to the man that will take his two glasses," and then drank it off.

by having the valuable timber removed from it, without its being cleared or rendered fit for the plough, while a ruinous system of farming impoverished the land already under cultivation. Farms, in which the soil was originally excellent, became thoroughly exhausted, and the evils of this state of things have to some extent continued to the present day. Merchants fared no better. Partly owing to the credit system, and partly to the changes in the price of timber, most of them were unsuccessful in the end.

Of the trade carried on in this county at this period, by far the largest portion was in the hands of Edward Mortimer, and this is the proper place for a more extended notice of him. He was a native of Keith, Banffshire, Scotland. He arrived in this country, as many a Scottish youth has gone abroad, with only his own energy and steady habits for his fortune. I have heard of his once speaking in depreciating terms of this country, in presence of old John Patterson, who immediately replied, "Ye needna talk when ye came to it, I dinna ken whether ye had twa shirts, but I ken ye hadna two jackets." It is of course all the more creditable, that by his energy and skill, he in a short time became the foremost man in Pictou, or in the eastern part of the Province. He first visited this place about the year 1788, employed by the firm of Liddells, in Halifax, in a schooner trading round the coast. Soon after he commenced business here, at first in partnership with them, but soon after on his own account. He first located himself a little above the point, which has so long gone by his name, on the front of Squire Patterson's lot, whose daughter he married. Here he put up a small building, intended for both house and store, of which the cellar can still be traced, and also built a wharf, of which portions of the foundation are still visible. Afterward he removed to the point, near the stone house, where he had his dwelling house and stores close by the shore, and where he built two long wharves out to deep water, the remains of which can still be seen.*

He is said to have been a man of commanding presence, tall, broad-shouldered and portly—as one from Britain described him, with "the appearance of a great man, and the address of a great man." Indeed, he was manifestly a born leader of men, and one that would have exercised a commanding influence, in any society into which he might enter. But he must have been a man of first-rate business capacity, for he now had nearly the whole trade of the place in his hands, and by his influence the trade of the Gulf was concentrated at Pictou. Persons still living can recollect, when the point above the town, where he did business, presented every season a forest of masts. He is said to have loaded 80 vessels in one year, not, however, all in Pictou harbour, but many in surrounding ports, his business extending to Bay Verte and Prince Edward Island. His book-keeper stated, that in one season, in seven successive weeks, he shipped timber to the value of £35,000, or at the rate of £5,000 per week.

Though the timber trade was his principal business, yet he did also a large business in the fisheries. The Arichat and other fishermen came here for their supplies, and traded their fish. At that time seals were still taken in considerable numbers in the Gulf, and the oil was manufactured in James Patterson's Cove, so that at the proper season, when the wind blew upon the town, the inhabitants were regaled with what fishermen would regard as a savory odour.

To so large an extent was the business of the place in his hands, that he regarded any person commencing a

---

* After his removal his old house was regarded as haunted, it was said persons who attempted to stay in it being frightened out of it by a noise as of the rolling of barrels, and persons who approached from the water seeing it lighted up, but when they landed finding all in darkness. The stone house was erected only a short time before his death.

general trade, as an intruder upon his legitimate domain, and he did not hesitate to use measures to crush him, which now would not be considered fair between man and man. For example, after men had agreed to give their timber or produce to other parties, he would have no hesitation in persuading them, or concussing them into giving it to him.

By the system of credit which prevailed, he had almost every inhabitant of the county in his books, and thus, in a measure, under his control, and business was then conducted, so as if possible to keep them in that position. Not only were goods pressed upon them, but they were kept in ignorance of the state of their accounts, as a means of securing a continuance of their custom. For a debtor to demand a settlement, seemed to indicate an intention of dealing with some other party, and to prevent this, the policy—not of Mortimer alone, however—was to keep his name in the ledger.

His influence, however, especially with the country people, was largely owing to his frank manner and real kindness of heart. He celebrated many of their marriages, as the dissenting ministers were not allowed to marry by license, and on such occasions he and Mrs. M. danced with the common people, and mingled freely with all ranks, in a manner that gained their good will. Besides, he was a man ever ready to do a favor. The poor and the friendless were freely helped, and ever after retained a grateful recollection of such services. We have conversed frequently with country people, who recollected that period, and their general testimony was, that in any difficulty, they had only to apply to Mortimer to receive ready help. Though he wished to have people in his books, and loved the power that this gave him, yet he was never disposed to deal harshly with them. On the contrary, his inclination was rather to act the Lord Bountiful. And it was only after his death, when his estate came to be settled up, that the people felt the evils of the credit system, under which they had become so deeply involved.

From the time of his election in 1799 till his death in 1819, he continued to represent the County of Halifax. His natural gifts gave him great weight in the Legislature, at a time when personal influence was [more potent] than it is now. This power he used earnestly for the interests of Pictou, and the liberal grants which he obtained for local improvements, caused him to be regarded as a public benefactor, people looking to him almost as if the money came from his own pocket. In other districts, parties used to apply to him, when wanting Legislative assistance, and were accustomed to say, that he was better for them than their own members.

Though opposed by Wallace, he also gained such influence over successive Governors, that generally all local patronage was entirely at his disposal. The Earl of Dalhousie, after his death, said, "I found in the late Mr. Mortimer a country gentleman, whose liberal mind and patriotic principles were an honor and a blessing to his neighborhood. To him I gave my confidence, with authority to use the power vested in me to the fullest extent, except as being subject to my confirmation. With his zealous assistance and influence, I know that astonishing progress has been made in opening the forest land." It is not surprising, under all these circumstances, that he should be entitled King of Pictou.* It must be said of him, that he was a sincere and earnest worker for the good of the county and of the Province. He was liberal in giving and hearty in promoting measures for the public weal. His fault was his love of power, but if ambition

---

* A wag once wrote a humourous production, entitled, "Chronicles of Pictou," which began somewhat in the following terms, "There was a King in the East and his name was Edwardus, and he was the chief of the tribe of the Pattersonians, and he ruleth the Pictonians with a rod of iron."

be the last weakness of noble minds, we may excuse the manifestation of it in one naturally so fitted to rule over men, and who was by circumstances placed in such a commanding position. And we may be thankful that such power was in the hands of one who, on the whole, used it so well.

The result of his business was that he rapidly accumulated a large fortune. In a few years, he counted himself worth £100,000, we doubt not the largest fortune acquired in the same time in Nova Scotia. But, alas! scarcely could a case occur more strikingly indicating the instability of earthly greatness. He was cut down in the prime of his days in the year 1819, after two or three days illness, when about 50 years of age, and his estate actually proved insolvent. Legacies for religious and charitable purposes were never paid, and a portion reserved of his real estate as dower, afforded a moderate competence to his widow. We have never fully ascertained the causes of this, but know that one was the disastrous failure of the firm of Liddells, and perhaps another was the want of his master mind in settling his affairs. He never had any children. He died 10th October, 1819, and his tombstone has the following inscription :

SACRED TO THE MEMORY
of
EDWARD MORTIMER, Esq.,
Who departed this life, 10th October, MDCCCXIX,
In the fifty-second year of his age.
He was
A native of Keith, in the shire of Banff,
North Britain.
In early life he removed to this Province,
where
Occupying himself in mercantile pursuits,
He acquired a reputation honourable to himself,
And
Concentrated in the Port of Pictou,
The greater part of the trade of the adjacent coasts,.

For twenty-years
He represented the County of Halifax
In the General Assembly of the Province,
And during this long period,
His public conduct
Founded upon enlightened and liberal principles,
Gained him the confidence
Not only
Of his constituents but of the Province at large.
He was also
A Judge of the Court of Common Pleas,
And for many years
Chief Magistrate of Pictou,
And
In the discharge of the duties of these offices,
As well as in his private capacity,
A strenuous promoter of the good order and peace
of society.
To his public exertions,
The Pictou Academy is deeply indebted.
These and his private munificence
Have rendered him its principal founder.
In Pictou
He is remembered as the poor man's friend,
And
The inhabitants of the Province at large,
Retain a grateful recollection
of his valuable services.

Of those in the same line of business with Mortimer at this time, the principal were John Dawson and Thomas Davison. The former was a native of the parish of Irongray, in the County of Dumfries. He was a man of education and mind, and filled several public situations creditably. He bought the lot on Water street, to the east of the road leading to Yorstons Wharf. There he erected a large two-storey house, nearly on the site of the Taylor House, with two wings to the north. In front of this property he built a wharf, which has disappeared. John Patterson built an extension from the end of his wharf at right angles to the eastward, leaving a narrow passage between it and Dawson's wharf. Inside of this, there was thus formed what was called the dock, in which boats and even schooners were safely moored. At that time all the trade along shore was by boats, the settlers bringing their produce and carrying away goods from the merchants in the same way. In this way for some years a larger portion of the country trade was concentrated at this point. Dawson's health having failed, he sold out his business to his son-in-law, William Kidston, afterward of Halifax, and removed to a farm two miles out of town on the West River road, where he died on the 2nd January, 1815, aged 54.

Thomas Davison was originally from Londonderry, N. S. He erected the house on the north-west corner of George and Water street. He was for a time an active man in church and state.

At this time William Matheson began business. He at first started peddling, we have been told, on a loan of £20. He afterwards sold his farm at Rogers Hill and removed to West River, where he did a country trade, exchanging goods for timber and country produce, but taking care to risk nothing in ships, ship-building or shipping timber, so that he could say that all he had ever lost by sea was one hat. Cautiously proceeding thus, he accumulated money, and was the only man of that period who came out of it with anything like a fortune. In his later years, he was distinguished by his gifts for religious purposes, and at his death he devoted the larger portion of his property to the British and Foreign Bible Society, and to the Seminary of the Presbyterian Church.

At this period a number of others were attracted to Pictou, and did business for a time. Among these may be mentioned Hector McLean, who had married a daughter of Captain Fraser, of the 82nd. He was heir to the estate of Kingarloch, in Scotland, which he sold, and, investing the proceeds in goods, he commenced on the Deacons Wharf, in company with his brother-in-law,

Simon Fraser. He failed, however, some time after. He built the house in which John R. Noonan now resides. As to many others who attempted business, then and afterward, we cannot do better than give the picture, drawn by the author of the letters of Mephibosheth Stepsure, of the career of Solomon Gosling.*

"About thirty years ago, his father David left him very well to do; and Solomon, who at that time was a brisk young man, had the prospect, by using a little industry, of living as comfortably as any in the town. Soon after the death of old David, he was married and a likelier couple were not often to be seen. But unluckily for them both, when Solomon went to Halifax in the winter, Polly went along with him to sell her turkeys and see the fashions; and from that day the Goslings had never a day to do well. Solomon was never very fond of hard work. At the same time he could not be accused of idleness. He was always a very good neighbour; and at every burial or barn raising, Solomon was set down as one who would be sure to be there. By these means he gradually contracted the habit of running about; which left his own premises in an unpromising plight. Polly, too, by seeing the fashions, had learnt to be genteel; and for the sake of a little show, both lessened the thrift of the family, and added to the outlay; so that, between one thing and another, Solomon began to be hampered, and had more calls than comforters.

"Though Goose Hill farm, from want of industry, had not been productive, it was still a property of considerable value: and it occurred to Solomon, that, converted into goods, it would yield more prompt and lucrative returns, than by any mode of agriculture. Full of the idea, accordingly, my neighbour went to town; and, by mortgaging his property to Calibogus, the West India merchant, he returned with a general assortment, suited to the wants of the town.

When a merchant lays in his goods, he naturally consults the taste of his customers. Solomon's, accordingly, consisted chiefly of West India produce, gin, brandy, tobacco, and a few chests of tea. For the youngsters, he had provided an assortment of superfine broadcloths, and fancy muslins, ready made boots, whips, spurs, and a great variety of gum flowers and other articles which come under the general denomination of notions.

"When all these things were arranged, they had a very pretty appearance. For a number of weeks, little was talked of, but Mr Gosling's store; for such he had now become by becoming a merchant; little was to be seen, but my

---

* These letters were from the pen of Dr. Thomas McCulloch, and were originally published in the *Acadian Recorder* for 1821 and '22. As may be judged from the above specimen, they are light satiric sketches of rural life at the time, and in regard to its follies, so held the mirror up to nature, that we know no work from which we can obtain a better idea of the state of society in Nova Scotia at that period.

neighbours riding thither to buy, and returning with bargains; and during the course of the day, long lines of horses, fastened to every accessible post of the fences, rendered an entrance to his house almost impracticable. By these means, the general appearance of the town soon underwent a complete revolution. Homespun and homely fare were to be found only with a few hard-fisted old folks, whose ideas could never rise above labor and saving. The rest appeared so neat and genteel upon Sundays, that even the Rev. Mr. Drone, though I did not see that his flock had enabled him to exchange his own habiliments for Mr. Gosling's superfine, expressed his satisfaction by his complacent looks.

" Mr. Gosling, too, had in reality, considerably improved his circumstances. The greater part of my neighbours being already in debt to old Ledger and other traders about; and considering that if they took their money to these, it would only go to their credit, carried it to Mr. Gosling's store; so that by these means he was soon able to clear off a number of his old encumbrances, and to carry to market as much cash as established his credit.

" Among traders punctuality of payment begets confidence in the seller; and the credit which this affords to the purchaser, is generally followed by an enlargement of orders. My neighbour returned with a much greater supply; and here his reverses commenced. Credit could not be refused to good customers who had brought their money to the store. Those, also, who formerly showed their good will by bringing their cash, proved their present cordiality by taking large credits. But when the time for returning to the market for supplies arrived, Mr. Gosling had nothing to take thither but his books. These, it is true, had an imposing appearance. They contained debts to a large amount; and my neighbour assured his creditors, that, when they were collected, he would be able to pay them all honourably, and have a large reversion to himself. But, when his accounts were made out, many young men who owed him large sums, had gone to Passamaquoddy; and of those who remained, the greater part had mortgaged their farms to Mr. Ledger and the other old traders; and now carried their ready money to Jerry Gawpus, who had just commenced trader by selling his farm. In short, nothing remained for Mr. Gosling but the bodies or labours of his debtors; and these last they all declared themselves very willing to give.

" About this time it happened that vessels were giving a great price; and it naturally occurred to my neighbour, that, by the labour which he could command, he might build a couple. These, accordingly, were put upon the stocks. But labour in payment of debt, goes on heavily; and besides, when vessels were giving two prices, nobody would work without double wages; so that the vessels, like the ark, saw many summers and winters. In the meantime peace came, and those who owned vessels were glad to get rid of them at any price. By dint of perseverance, however, Mr. Gosling's were finished; but they had scarcely touched the water, when they were attached by Mr. Hemp, who at the same time declared, that, when they were sold, he would lose fifty per cent upon his account for the rigging. Such was my neighbour's case; when, happening, as I have already mentioned, to step into Parson Drone's, I found that Mr. Gosling had been telling his ailments, and was receiving the reverend old gentleman's ordinary, clerical consolation. ' What can't be cured, must be endured; let us have patience.'

"'I'll tell you what it is, parson,' replied my neighbour, 'patience may do well enough for those who have plenty; but it won't do for me. Callibogus has foreclosed the mortgage; my vessels are attached; and my books are of no more value than a rotten pumpkin. After struggling hard to supply the country with goods, and to bring up a family so as to be a credit to the town, the country has brought us to ruin. I won't submit to it. I won't see my son Rehoboam, poor fellow, working like a slave upon the roads, with his coat turned into a jacket, and the elbows clouted with the tails. My girls were not sent to Mrs. M'Cackle's boarding school to learn to scrub floors. The truth is, parson, the country does not deserve to be lived in. There is neither trade nor money in it, and produce gives nothing.—It is only fit for Indians and emigrants from Scotland, who were starving at home. It is time for me to go elsewhere, and carry my family to a place that presents better prospects to young folks.'

"In reply, the parson was beginning to exhort Mr. Gosling to beware of the murmurings of the wicked; when Jack Catchpole, the constable, stepped in to say that the sheriff would be glad to speak with Mr. Gosling at the door. Our sheriff is a very hospitable gentleman; and when any of his neighbours are in hardship, he will call upon them, and even insist upon their making his house their home. Nor did I ever know any shy folks getting off with an excuse. As it occurred to me, therefore, that Mr. Gosling might not come back for the parson's admonition, I returned home; and soon learned that my neighbour had really gone elsewhere, and made a settlement in the very place where Sampson turned miller."

The large number of vessels loading every summer at this port during this period, rendered it a favorite hunting ground for the press gang. For several years, scarce a summer passed without a visit of this kind, and many exciting scenes were the result. No sooner did a man-of-war cutter appear in sight, than it proved a signal for the boats to put off from the ships, carrying their crews to the land, who hastily betook themselves to the bushes, which were then close upon the town. "I have witnessed," says one, " the desperate race between the pursuer and pursued, and observed both parties land, the former somewhat behind; and it was a thrilling moment, when the press gang, with their cutlasses flashing in the sun, and firearms discharging, followed close upon the flying sailors. The firearms did little harm, and were, perhaps, not intended to be deadly. The sailors generally escaped to the woods. On one occasion a large force of seamen came from their retreat, each armed with a cudgel, and drove the man-of-war boat from the wharf. The commander threatened to bring the armed cutter up to the town and take revenge. He did not, however, execute his threat, perhaps considering that discretion was the better part of valor."

On another occasion, some sailors taking refuge in a store of Mortimer's, their pursuers fired two shots through the door after them. Mortimer complained to the Admiral, and such was his influence, that the officer in command was reprimanded and ordered to apologize to him. He however only replied, that he could accept no apology for the act, as it was not an affront to him, but human life that had been endangered.

Among the visitors, was a Capt. Elm, a regular old sea dog, who more than once beat off a press gang, and it is said on one occasion, knocked a hole in the bottom of a man-of-war's boat, coming along side his ship.

On one occasion an embargo had been declared, and a convoy promised to protect the fleet to their destination. About fifty vessels were assembled, and were delayed a good part of the summer. One of the Captains died, and the funeral was a very pretty sight. The boats of all the ships were formed in line and with colours at half mast, followed in regular procession the remains to the shore. They were landed at the Deacons wharf, and the sailors, all dressed in their best, followed them to the grave yard, where till recently a painted board marked the last resting-place of Capt. Sturm of the ship Symmetry.

The event however, connected with the press gang, which made most noise, was the pressing of two landsmen, Edward Crae and Matthew Allan, in the year 1808. They were two stout men, Allan being notorious as a bruiser.

They had made themselves obnoxious to persons at Carriboo, who occupied the Cochrane grant as squatters, by cutting timber on it; and hence their capture was not objected to by some, and it was even believed, that it had been instigated by these parties. It took place on the day of a general muster, when the whole adult population were in town, and produced great excitement. They were taken on board ship and carried to the West Indies. When the House of Assembly met, they voted the proceeding oppressive and illegal, and requested the Governor to interfere in their behalf, which he promised to do; but before the order arrived in the West Indies for their relief, they had effected their escape by swimming ashore at Antigua.

Except, however, in its effects on trade, little was seen or directly felt in Pictou of the war. Annually the militia were called out, first for company drill in the different sections of the county, then for general muster and battalion drill, usually at or near town, and commonly in Mr. Mortimer's field. Many thus attained considerable knowledge of military exercises, and they were even exercised in target firing with muskets. These occasions were to the youth scenes of amusement, and, alas, too often of drunkenness. In the year 1807, a portion of them were drafted to Halifax, and for some time were employed there, doing garrison duty, cutting fascines and erecting a palisade around the town. One who served with them at this time told me of another purpose, which their presence in Halifax served. A regiment was stationed there, composed of men who had been compromised in the Irish Rebellion, and who had enlisted to save their lives. Such was the desperate character of many of them, that until the militia arrived, their own officers were afraid to trust them in the town.

Besides the ordinary militia, however, there was formed at this time, in Pictou town, a company of volunteer artillery, which was put through a regular system of drill, and attained to a respectable measure of efficiency. This company continued for years after to turn out and fire salutes on public occasions, even to a period within our own recollection.

During the later period of the war, particularly after the Americans had joined the foes of England, vessels loading were obliged to wait for a man-of-war, as a convoy, until sometimes there would be as many as fifty of them in the harbour at one time. So long were they obliged to wait that in some instances an adventurous captain stole out of the harbor, went to England, discharged cargo and returned before the rest sailed. We have only heard of one Pictou vessel captured. It was a schooner or brigantine commanded by Captain David Fraser, whom we have already mentioned as the first child born to the Hector passengers, and as an illustration of war times, we may here give some account of his adventures.

When about twenty years of age he went to Halifax, and thence to sea. Afterward we find him sailing out of the United States, and having risen to be mate of an ocean-going vessel. While in this position, he was taken by the Algerines, and the whole crew kept in close confinement. Fever or plague broke out among them, of which one after another died. The survivors were obliged to bury their companions, only a sort of wooden shovel being allowed them for the purpose, with which they scooped out a shallow pit in the sand. At last he was the only survivor. He was then given or sold as a slave to an old woman, to whom he was compelled to do all her drudgery. His old clothes became worn to tatters, and his skin blistered by the sun, but she allowed him no new supply. In this condition he obtained from a vessel a piece of an old sail, with which he made a sort of garment, more expressive of the ingenuity of the maker than of fashionable elegance. Suspicions being allayed, so that he was not watched very closely, he one night swam out to a British vessel off the coast. He was taken on board, supplied with clothing and taken away in her. Falling in with an American vessel, in which the mate had died, the captain engaged him to supply his place. Thus he arrived back in Virginia, when the captain refused to pay him his wages. Fraser, however, having been previously in the American service, succeeded in compelling him to do so.

He next engaged in the secret trade carried on by the Americans with Europe, and for a time was successful, but finally his vessel was captured by one of Bonaparte's cruisers, when she had on board three barrels of dollars, one of which belonged to himself. He was deprived of all and made an appeal to the Emperor, pleading that the Americans and French were not at war, but received the reply, "When I pay the other bills of the Americans, I will pay that too!"

From France he made his way to Stockholm, and thence to England, where he married. Soon after he returned to Pictou, with his wife and one child, after he had been just twenty years absent. Here he received the command of the vessel referred to, which was owned by Mr. Mortimer. But his ill luck seemed to follow him, for she was captured by Commodore Rogers, of the American navy. After a few things were taken out of her, she was by his orders set on fire. The crew were taken prisoners to Salem, but he made his escape and travelling by land, reached British territory, and the crew obtained their liberty the next spring by the return of peace. The vessel had originally been an American prize, and fitted up more handsomely than was usual in colonial vessels at that time. When she was about to sail, one of his friends expressing sympathy for those who had lost her, he replied, "Oh, it's just the fortune of war." After his return home, he was describing, in the presence of the same friend, her capture by Commodore Rogers, and expressing his indignation at seeing her burning. "Oh," said his friend, "it's just the fortune of war." "—— the fortune of war," was his irreverent reply.

This war brought to the county a number of coloured people, who had been originally slaves in Virginia, but who had escaped to the British fleet in Chesapeake Bay. John Currie, who only died in 1876, used to tell of swimming out to a man-of-war in a shower of bullets. Several families settled in the neighbourhood of the Town Gut stream, but nearly all have since moved to other places.

At this period commenced the making of roads fit for carriages, the first being that towards Halifax over Mount Thom. The troubles of travellers previously are thus amusingly sketched by a writer in the *Acadian Recorder*, in the year 1826:

"Many a story have I heard from my father, of the disasters which befell travellers in his time, when there was only one road in the Province deserving the name, viz.: that from Halifax to Windsor and Annapolis. And with wonder I have heard him tell, that it cost as much as would pave it all over with dollars. The people of the best settlements found their way to this road or to one another by a blaze; that is a mark made on a trunk of a tree here and there, in the proper course, for the purpose of directing travellers; but, in the younger settlements, travellers had to provide pocket compasses, and guessing their course, find their way through the forest, much in the same way as sailors do along the sea.

"In going by the compass, the traveller sometimes, widely mistaking his course, missed entirely the intended settlement, and came in upon another, or missed all settlements, and travelled on, till he lost all hope of seeing a house, in which case he often believed the compass itself went wrong, and discrediting it, he would wander he knew not whither. Sometimes the traveller would be confounded desperately, for the compass needle would obstinately refuse to traverse, and he could not know east from west, north from south.

"Travelling by a blaze was little better. He told us strange things of losing the blaze, and the impossibility of finding it again, of striking out a straightforward course, independent of the blaze, and yet, by and by, coming upon their own track again,—of the snow being so driven against the trees as to hide the blaze, and causing frequent stops to rub it off,—of its being so deep as to cover the blaze, and causing frequent stops to dig away the snow in order to discover it,—of travellers being benighted by such stops, and lodging in the forest, where they had to kindle large fires on the top of the snow five or six feet deep, and there (dismal to be told !) one side next the fire was roasted and the other frozen. I have heard him tell of experienced travellers, who in such a case would kindle two fires, at a proper distance from one another, and lie down between them, and thus enjoy themselves luxuriously between two fires. In those days swamps were avoided as intolerable. The steep mountain sides were preferable, hence there are still many hills on our roads, which might now be easily avoided.

" I have heard him tell of great dangers and hair-breadth escapes from drowning, in crossing brooks and rivers swollen with unexpected rains; for in those days no journey would be undertaken immediately after a heavy rain. He had himself went different times for two or three days nearly fasting, until the subsiding of the water rendered the road passable. He told of horses swagging in swamps almost to the ears, and of the great difficulty of their riders. There were few taverns, but every man who had a hut was hospitable."

Any roads hitherto made were merely bridle paths. But now through Mr. Mortimer's influence in the Legislature, liberal grants were obtained, and by means of these, and sums voted from the county funds, the road over Mount Thom was cleared of roots, and somewhat cast up. It was still rough and soft enough, but carriages could pass over it. Similar operations had been going on from the Truro side, and the two parties met at Salmon River, and celebrated the completion of the work in the manner usual in those times. This we suppose would be about the year 1810. In the year previous, Sir George Prevost, the Governor, visited Pictou accompanied by Michael Wallace and L. H. Hood, Esqs., and his A. D. C. Capt. Prevost. Miller, in his " Record of the first settlers of Colchester," speaks of this as the first occasion of a four wheeled carriage, passing through Truro. But it did not come to Pictou, for the Governor and his party all arrived there on horseback. The first wheeled carriage that ever came to Pictou, is said to have been brought through, a year or two later, by Judge Monk, when attending the Supreme Court. Soon after, Mr. Mortimer imported a two wheeled carriage, which was the first owned in Pictou.

The late Mr. Matheson, however, used to boast, that he had taken the first loaded team over the road. It was on this wise : Mortimer's supply of tobacco had become exhausted, and the other merchants were supposed to be in a similar situation. It being winter, no supply could be got by water. He therefore started by land with a rude sled, and returned with a puncheon of leaf tobacco, which was the only kind used then. The runners were not shod with iron, and so rough was the road, that he wore out three wooden shoeings.

In making roads at first, the circumstances above described rendered it absolutely necessary that they should be made on the high grounds. The lower lands in the driest seasons were troublesome, and in wet seasons, impracticable. Hence the first settlers, in choosing a line of road, simply selected the highest hills they could find, and made their way from one to another, where they could find the least low ground to traverse. When improvements began to be made, it was natural to make them upon the lines in use, especially as settlers had located themselves beside them. But now when large sums were being expended in making roads, it was certainly short-sighted policy, not to have sought something like level lines. By not doing so, much money was spent upon roads, which, as far as general travel was concerned, had to be abandoned, and from the interests of so many persons involved, it was difficult afterward to change for a better. Every miller, every blacksmith and every tavern keeper would fight hard against its removal. Hence for many years the road to Halifax went over the summit of Mount Thom in this county, and beyond it others as steep, such as Half-Moon Hill, Black Rock, &c., while the other roads were constructed on the same principle. Thus the road to the east climbed Green Hill at its steepest part, and went over the whole length of Frasers Mountain, besides a number of hills not so lofty.

Turning now to the ecclesiastical affairs of the period under review, we notice that in the year 1803, Pictou received an accession of one who had afterward a more than Provincial reputation, and one to whom, in some respects, Pictou is more indebted than to any other individual. We allude to the Rev. (afterward Dr.) Thomas McCulloch. He was a native of the parish of Neilston, Renfrewshire, Scotland. He received his philosophical education at the University of Glasgow, studied theology at Whitburn, under Professor Bruce, and was ordained as minister of a congregation in Stewarton, Ayrshire, but did not long remain there. He arrived in Pictou in the month of November, 1803, with his wife and family, on his way to Prince Edward Island. Owing to the lateness of the season, he was unable to obtain a passage thither that fall, and was engaged for the winter to supply the congregation of the "Harbour," as it was called. In the following spring, he was called to be their pastor, and inducted on the 6th June, the very day that parties came from Prince Edward Island to take him over.

At that time the town, as it was beginning to be called, consisted as we have seen of sixteen or eighteen buildings, including barns, a blacksmith shop and the jail, closely environed by the woods. There was no church, but a place was fitted up in a shed of Captain Lowden's, on Windmill Hill, where service was held in summer, but in winter time it was in private houses, most frequently in the "big room" of McGeorge's tavern, which stood on the west side of George street, on the site of the long building erected by the late John Proudfoot, and which, we may here remark, was long one of the institutions of the place. That fall (1804) the frame of the church was erected on part of the same lot on which Prince Street Church now stands, but fronting down Margaret Street. It had a small belfry and in it a small bell, which has had rather a curious history. It was originally a ship's bell, but was stolen from Mortimers Point and carried to Miramichi, where it was recognized by a gentleman who had been in Mr. Mortimer's employment, and restored. It was presented to the church, and continued to be used till about the year 1822, when it became cracked, and was sent to Scotland to be recast. It was at the same time enlarged, and afterward presented to the college, where it is still in use, while the congregation ordered a larger one, for which, in the year 1824, they put a spire upon their church.

The sphere of Dr. McCulloch's labours, as far as his congregation was concerned, was but limited. His was a mind, however, which in any place must have made its influence felt beyond the single spot where he might be located. As early as the year 1805, he projected an institution for the higher branches of education, especially for the benefit of dissenters, and particularly with the view of training a native ministry; but the scheme was not carried out, though it was not lost sight of. But of this we shall speak fully in another chapter.

From as early a period as 1807-8, we find him contributing to the public press; but circumstances soon brought him before the public in a discussion, which established his reputation as having no superior, perhaps no equal, in the Province, in learning, literary skill and controversial power. A controversy had arisen between the Church of England Bishop and the Rev. Edmund Burke, afterward Roman Catholic Bishop, a man of great learning, as well as an able writer. We have not seen the writings of either party, but it was generally agreed that the former was not a match for his antagonist, when Dr. McCulloch took up the cudgels, and, after some preliminary skirmishing, joined battle in a large 12mo. volume, published in Edinburgh in the year 1808, entitled "Popery condemned by Scripture and the Fathers; being a refutation of the principal Popish doctrines and assertions, maintained in the remarks on the Rev. Mr. Stanser's examination of the Rev. Mr. Burke's 'Letter of Instructions to the Catholic Missionaries of Nova Scotia,' and in the reply to the Rev. Mr. Cochrane's fifth and last letter to Mr. Burke."

This produced a rejoinder from Dr. Burke, who was in every way "a foeman worthy of his steel," and to this Dr. McCulloch again replied in a volume even larger than the first, entitled " Popery again Condemned by Scripture and the Fathers; being a reply to a part of the Popish doctrines and assertions contained in the remarks on the Refutation, and in the Review of Dr. Cochrane's letter by Edmund Burke, V.G., Que." To this, Dr. Burke attempted no reply. These two volumes, for learning and ability, excel anything we know of produced in the colonies. A portion of them has only a temporary interest, as connected with the controversy with Bishop Burke. But much the larger part is of permanent value, as a discussion of the great questions at issue between Protestants and Romanists.

Of his labors in connection with education we must reserve an account for another chapter.

The year 1808 witnessed another accession to the ministry of this county, in the Rev. John Mitchell, who settled in River John, taking the oversight of the people in that and the neighboring settlement of Tatamagouche. Mr. M. was a native of Newcastle-upon-Tyne, born in the year 1765. He was in early life a rope-maker, and had not received a classical education, but animated by an earnest desire to preach the gospel, he entered Hoxton Academy, when about thirty years of age.

In the year 1808 he was sent out to Quebec by the London Missionary Society. In the autumn of that year he removed to New Carlisle, on the Bay Chaleur, where he had his home for several years. Here he married Miss Shearer, a member of a Loyalist family, that had been obliged to escape from the United States during the war, with the loss of all their property, reaching British territory, under the guidance of two Indians, each of whom carried a child.

In the summer of 1803, he undertook a long missionary tour through New Brunswick and Nova Scotia, in the course of which he visited Pictou and most of the settlements along the coast. In autumn he removed to Amherst, where he continued to labor for some years.

In the year 1808, he removed to River John, where he labored for a year without connection with any ecclesiastical body in the Province. But in the following year, though originally a Congregationalist, he joined the Presbytery of Pictou.

At the time of Mr. Mitchell's settlement in Pictou, there were fifty families in River John, only three English, named West, Hines, and Gammon. Here he continued to labor with all diligence and faithfulness among his flock, pursuing the usual routine of a Presbyterian minister's duties. But he also extended his labors to Tatamagouche, which became part of his regular charge. There was no road worthy of the name between the two places, and consequently the travelling between them involved severe labor. Some time after, when New Annan was settled, he extended his labors to that settlement. In the work of the ministry over this field, much of which was in a wild, uncultivated state, he underwent much bodily fatigue, but he did it with the greatest cheerfulness.

In the year 1826 Tatamagouche and New Annan were formed into a distinct congregation, when his labours became much less severe. He enjoyed excellent health till near his end. A violent attack of gravel terminated in his death on the 8th May, 1841, when he was in the 76th year of his age.

" Mr. Mitchell was above the ordinary size, well formed

and sinewy, of a fair complexion and cheerful countenance. Although he made no pretensions to extent of learning, he was acute and possessed of a respectable share of general information. He was a good man, and his memory is much and justly revered."

In the year 1813, was formed the first Bible Society in Pictou. It was the second in the Province or in British America, that in Truro having been the first. But for several years previous, through the zeal of Dr. McGregor contributions had been forwarded to the Society. This had been done as early as the year 1808, so that it is admitted, that the first contribution to its funds from any British colony, came from Pictou. We find the Secretary in a letter of 4th June, 1809, acknowledging a sterling bill for £80, and referring to one previously sent for £64. These sums were probably in part for Bibles sold, but in part also were a free contribution. For the better promotion of the objects of the Institution, it was deemed advisable to organize an auxiliary society. A meeting was accordingly held for the purpose in the old West River church, on the 16th day of April, of this year. The Rev. Dr. McGregor preached from I. Tim., iii., 1., and a society was formed, embracing the whole county, with Ed. Mortimer, President, and a committee of directors, consisting of so many from each congregation. In the first year, they remitted £75 to the parent society, of which £50 was a free contribution, and £25 for the purchase of Bibles and Testaments. In the second year, £50 was sent as a free contribution, and in the third £75. In subsequent years the amounts diminished, but still something was done annually, and to this day the Institution has been supported more liberally in the County of Pictou, than in any other county in the Province.

In the year 1815, Pictou received its fifth minister, the Rev. William Patrick. He was a native of the parish of Kilsyth, County of Stirling, Scotland. In his younger years he was brought up in the Reformed Presbyterian Church, but connecting himself with the Secession Church, he studied theology under the Rev. Archibald Bruce, of Whitburn. He was for a number of years minister of a congregation in Lockerby, Scotland. On his arrival he was cordially called by the people of Merigomish, and inducted as their pastor on the fifth of November.

From that time he diligently performed all the duties of the pastoral office over that district, until increasing infirmity obliged him first to diminish his labours, and finally to relinquish them altogether. On the 7th May, 1844, the Rev. A. P. Miller was ordained as his colleague, after which he performed no public service. His weakness gradually increased, till suddenly, on being seized with a fit of sickness, which his exhausted constitution could not sustain, he calmly expired on the evening of 25th November, 1844, in the 73rd year of his age.

We may here give a few miscellaneous items connected with this period. The following weather notes from Dr. McGregor's " Memorabilia," may be of some interest :

" In 1802 the winter was remarkably mild, all along till March 22nd, and then it grew severer in proportion as it was expected to depart, so that the beginning of May was more wintry than January. Little snow fell and it continued short, because of frequent thaws till March. March 22nd was more stormy than any preceding day. On April 6th, I crossed the river on very good ice. I could not cross the harbour in a boat April 18th. On April 27th and 28th, was the greatest storm of snow that had come through the whole winter. There was also a considerable storm on May 4th and 5th, and the wind almost constantly from the north till May 31st. There was much snow in the woods on May 9th. Ploughing was begun on May 6th, 7th and 8th No wheat was sown till May 11th. Provender was so scarce that some could. not plough for want of food for the oxen. It was the sickliest season that I remember. The principal complaint was a kind of pleurisy, owing, I suppose to the uncommonly changeable state of the weather.

" In 1807, on the night of February 15th was a dreadful storm of wind and rain, which broke open the harbour, so that boats could pass and repass next morning, but on the next morning again, the harbour was frozen over. The bridge of the Middle River was carried off and the bridge of the East River injured by the storm. Boats were passing and repassing between Mortimers and Frasers Point the last week of February. On February 24th the river could not be crossed on the ice."

In the year 1807, the district was divided into three townships, Pictou, Egerton and Maxwelton, the boundaries of which have been already given.

On the 12th November 1813, took place what was long remembered as the big storm. Many buildings were unroofed, in some instances, the roofs being carried bodily to some distance. Forests over a large extent of country were levelled as completely as they would be in a chopping frolic. Its severity lasted only a little over two hours, when there was a complete calm. Of its power in Halifax, Haliburton says, " It commenced in Halifax at 5 o'clock P. M., from the south east, and blew with extraordinary violence till seven. Upwards of 70 vessels were driven on shore, sunk or materially injured, and many lives lost."

About this time an attempt was made to manufacture salt from the saline springs, which rise from the Lower Carboniferous rocks, at the foot of Mount Thom, and which give the name of Saltsprings to that settlement. The projectors were in England, and sent out an agent to superintend operations. He sank a shaft 200 feet deep, as if searching for the bed of salt. But from the position of the pit every body was satisfied, that it was in the wrong place. One man remonstrated with him, but he replied that he was getting £500 a year to find it, but he would receive so much, naming a larger sum, if he did not. A large quantity of iron was sent out to construct saltpans. It was hauled up at great expense, but the next spring it was all sold for old iron, the company having failed or ceased operations. About ten years later, parties commenced manufacturing salt from the brine of the spring. It proved of good quality, but they soon abandoned the work.

On the 26th of May, 1814, intelligence having arrived of the entry of the Allies into Paris and the abdication of

Napoleon, a salute of 21 guns was fired from the Battery, and in the evening the town was illuminated and bonfires kindled on the surrounding heights.

The close of this period was signalized by the commencement of New Glasgow. A lot containing 500 acres, extending in front from R. S. McCurdy's store to below the new burying ground, had been originally granted to John McKenzie (the captain's father), but was by him sold to Alex. McKay, the squire's son, for £20. He employed Wm. Fraser (surveyor) to lay off the front in acre and half-acre lots. He gave a lot at the bank, to the south of where the bridge is now, to a man named Chisholm, usually known as Daddy Chisholm, who built upon it a small log-house on the bank of the brook. Here he and his wife (he had no children) lived and for a time were the only inhabitants of New Glasgow. About the year 1809 the late James Carmichael bought from McKay the lot adjoining, to the east, and erected a log building on the site at present occupied by his son's stone building, and commenced business in partnership with a Scotchman named Argo. This house was burned down in the year 1811, after which Mr. C. built another on the same site and resumed business, but by himself. He first traded with the people for ton timber, but afterward took butter, pork, and other farm produce. In the next period he was one of the most active merchants of the county, but here we may say of him that he was a man distinguished by his kindness of heart, his public spirit, and his readiness
' for every good work

Soon after, Donald McKay bought what is now Bells corner, and erected a blacksmith shop, where the shop of James Fraser & Son now stands; and Hugh Fraser bought the lot between it and the bridge, on the same side, and commenced business there. Kenneth McAskill, a tailor, purchased the corner on the opposite side of Prevost street. The first inn was kept by Angus Chisholm, in the corner house now occupied by Henderson, where for long one of the old swinging signs invited the traveller to enter and be refreshed. The first two-story building in New Glasgow was James McGregor's, now the Sheffield house, but it was not built till several years later.

CHAPTER XIV.

IMMIGRATION AND NEW SETTLEMENTS AT THE CLOSE OF THE WAR.

The depression of business in Britain at the close of the war brought a large immigration to this Province; and in the years immediately following a large number arrived at Pictou, of whom a large proportion removed to other places, but a good number settled in various parts of the county, filling up the settlements already formed and forming a few new ones. The latter we shall here notice.

At this period Dalhousie Mountain was settled. A large grant had been taken up there some years previous by persons in different parts of the country, under the idea that the soil was of very superior quality. Two of them gave each fifty acres gratis to Peter Arthur, a native of the Orkneys, on condition of his settling there. He accepted the offer, and located himself in the woods five or six miles from any settler, the nearest being at what was recently occupied by Mr. Charles Rogers. There, for months, he would not see the face of a human being. He built a log barn without the assistance of a single individual.

At the conclusion of the French war, the prices of farm

stock in Britain fell to one-half their former rates, which led to a large emigration from the Lowlands of Scotland. A number of these (among whom may be mentioned five brothers Rae—John, George, William, James, and Robert—and John Adamson), all from Dumfriesshire, settled on Dalhousie Mountain about this time (1815-17). The land was covered with heavy hardwood timber, and they entertained high hopes, which were strengthened by the first few crops, which were good owing to the burning of the hardwood upon it. But the land proved rocky, the soil shallow and soon exhausted. The snow, too, in winter was deeper than in other parts of the county, and lay longer in spring. Their crops, too, suffered injury from frost. So that while, from their thorough Scotch industry, some of them did well, and all earned a subsistence, yet a number found it prudent to abandon their farms, so that places on which considerable labour had been expended, and comfortable buildings erected, are now unoccupied.

About the same time, a number of persons came from the Lowlands, particularly Dumfriesshire, and settled in various places. They were distinguished by steady industry and rigid economy, and they generally not only made a living but saved money. As an example of their sturdy energy, the following may be given. Three brothers Halliday settled between the Middle River and the West Branch East River. For five or six years all their cultivation was by the hoe. But at length one of them having a piece of land sufficiently cleared, was desirous of getting it ploughed. For this purpose, he brought a pair of oxen, plough and necessary gear, from Kerrs, on the Middle River, through the woods, over three miles, in the following fashion : He fastened the yoke to the horns of one ox and the chain on those of the other, and getting a boy to drive them, he put the plough on his own shoulders and carried it all that distance. Of their success we may give an example. The late Thomas Kerr, of Middle River, and James Roddick, having served their time together as millwrights, came out in the same vessel. Mr. Kerr described their position when they landed as follows: " I had just half a sovereign and Roddick had just aughteen pence, and he bought half a pun' o' tobacco wi' it." Yet they died worth some thousands of pounds in property and money.

It is proper here to give a short notice of the early settlement of Earltown, which commenced about the same time, for although it is beyond the bounds of the County, it is both as to its origin and population closely connected with this County. We may mention that the settlement embraces that portion of the County of Colchester lying between the east line of the township of Onslow and the Pictou County line. It was first surveyed in the year 1817, by Alex. Miller, who gave it its name, in compliment to the Earl of Dalhousie, then Governor of the Province.

The first settlers were Donald McIntosh and Angus Sutherland, who took up their residence in the unbroken forest in the year 1813. The next to join them was Alex. McKay (tailor). Others followed soon after, among whom may be mentioned George Ross, Robert Murray, John Sutherland (father of the Rev. Alex. Sutherland), who afterwards moved to Rogers Hill, Paul McDonald, John McKay, Peter Murray, John McKay (miller, father of Rev. Neil McKay), William Murray (father of Revs. William and Robert Murray), R. Murray (tailor), William McKay, &c.

Of the early sttlers, nearly all came from Sutherlandshire, chiefly from the parishes of Rogart, Lairg and Clyne. There were families from Inverness, two or three from Ross, and three or four from Caithness. All the original settlers spoke the Gaelic language, and it is still generally used by their descendants. Indeed, it is more generally spoken in Earltown than in any part of

Nova Scotia proper. Still it received some admixture of others, for while it had old soldiers who, in the Highland regiments, had gone through the Peninsular War, and at least one who had fought at Waterloo, it at the same time had a foreigner, who had been in the same battle under Napoleon, and the two, instead of being ready to embrace as brothers, were rather disposed to fight their battles over again.

Like all who take up their abode in the woods, the first settlers had many difficulties to encounter. They were for years without a grist mill. During that time they got their grain ground partly by the handmill, and partly at a grist mill at the West Branch River John. As there were no roads to the West Branch, and they had no horses, they were compelled to carry their grain on their backs to and from the mill, over a rough track. John McKay, known as the miller, put up the first grist mill, at a fall fifty feet high, resembling the Fall of Foyers in Scotland. The mill-stones that were used in it were taken from the West Branch, a distance of fourteen miles, on a drag hauled by 36 sturdy Highlanders. Mr. McKay, we may here observe, was proverbial for his kindness to the new settlers, and his hospitality, which was shared by many a stranger.

The early settlers were strong, industrious and economical. They were poor at first, but with great perseverance, they made themselves comfortable homes. There are men in Earltown to-day, who settled forty years ago in the woods without a guinea in their pockets, who have fine houses, large barns, excellent farms and considerable sums at interest. The inhabitants at that time were all connected with the Church of Scotland, but for several years they were without a minister. In consequence of this, persons sometimes carried their children to Pictou, a distance of twenty-five miles, to be baptized. They were occasionally visited by a Minister of the Church of Scotland, and on such occasions it was not uncommon to see him baptize twenty or thirty children at once. Rev. W. Sutherland was the first minister who settled at Earltown. He was never called or inducted into the congregation, but remained ministering to a few who adhered to him till his death. The Rev. Alexander Sutherland, of the Free Church of Scotland, was the first minister who was called by the people, and ordained in the place. He was settled in the year 1845. Though the people were for years without a minister, they did not forsake the assembling of themselves together. There were among them men eminent as Christians, intimately acquainted with the truths of religion, and able to express themselves in a manner fitted to edify others. "The Men," as they were called, held meetings regularly each Sabbath in the several parts of the settlement, and were the means of maintaining vital godliness among the people.

About the same time, the settlement of New Annan began. It lies about seven miles to the south of Tatamagouche, in the County of Colchester, and forms an oblong square about ten miles long by seven wide. The first settler there was Mr. John Bell, a native of Annandale, in Dumfriesshire, Scotland. He emigrated to Nova Scotia in 1806, but did not settle in New Annan till the year 1815. For some years previous to this, he had worked in Tatamagouche, but attracted by this seemingly fertile, and withal somewhat romantic district, with its well wooded hills, he selected as his future home a place on the banks of the French River, and about the centre of the present settlement—quite a pleasant, pretty spot, and occupied by his descendants to this day. Here he cut the first tree, and erected the first house, of course a log one, in New Annan, to which he removed his family. For six long and dreary years he dwelt alone in the wilderness. During all that time his nearest neighbor was six miles distant, but others followed. Speedily William Scott,

James McGeorge, Thomas Swan, and Mr. Byers, all from the same district in Scotland, and Mr. James Munro, took up positions near him.

Mr. Bell was a pious man, and so were the others, and they therefore soon felt the want of public worship. The nearest place of preaching was Tatamagouche, to which the Rev. Mr. Mitchell, of River John, gave part of his services. Many a day the more vigorous of them trod their seven miles on a Sabbath morning, over hills, through marshes, covered with fallen trees, and across the French River to hear the Gospel. But all could not travel such a distance, or surmount such difficulties. Mr. Bell therefore and a few others formed themselves into a prayer meeting, and held worship in a school house near Mr. Bell's, for such as were unable or unwilling to travel to Tatamagouche.

After a time Mr. Mitchell extended his labors to New Annan, giving them occasional supply. But his visits were valued all the more for their rarity. At that time there was not the semblance of a road about New Annan, or even Tatamagouche. He had to travel by the seashore or blazed paths through the woods. Several of the young men were in the habit of going to meet him on his journeys, and now grey-haired sires, tell of their exploits, as skating down French River and along Tatamagouche Bay, to attend on sacramental occasions at River John.

From their sturdy Scotch industry and frugality, these settlers soon attained to comparative comfort, and many of their descendants are in good circumstances. But though the district has a considerable amount of good soil, yet portions of the hills, which appeared to be rich, and which, when first cleared, gave good crops, were found in a few years to lose their fertility. The population is estimated at between twelve and fifteen hundred. As in many other places, the majority of the young of both sexes go abroad when they reach the age of eighteen or twenty.

To the same period belongs the settlement of Pictou Island. It is about five miles long and on an average about a mile and three quarters wide, and contains an area of 3263 acres. It lies off Pictou Harbour, a little to the northward, the east end being distant about ten miles from its entrance, and the west end five and three quarters miles from Big Carriboo Island light house. It has no harbour even for a boat. Toward the east end, the land rises to the height of about 150 feet, but in other parts it is occupied by swamps. The soil is fertile, being generally a sandy loam, and yields good crops of hay, grain and the vegetables of temperate climates, but there is no fruit raised upon it, partly we have no doubt, because the exposure to sea air is unfavourable to its culture, but partly, we believe, from want of attention on the part of the inhabitants. The only wild animals are the fox, the rabbit and the musk rat. There are no squirrels, rats or toads, and the frogs are but little larger than grasshoppers. As to serpents it is as free from them as old Ireland itself.

One or two incidents connected with the island previous to its settlement may here be given. On one occasion an Indian and his squaw in proceeding from Prince Edward Island to Pictou landed at Rogers Beach, and remained for the night. In the morning the unfaithful husband sent her into the woods for something to repair his canoe. But after she left, he started with it, and landed at Carriboo, where he asserted that his wife had died on P. E. Island, and that he had buried her on the beach. This was in the beginning of winter. The unfortunate woman erected for herself a rude hut in the woods, where she subsisted all winter on shellfish and rabbits, clothing herself and covering her hut with the skins of the latter. In the following spring, she was rescued by some Indians, who had disbelieved the deserter's story. It is said that he was burned by the other Indians.

The island containing originally some good wood, the late John Brown of Browns Point and another man, spent a winter on the island making staves. Their supply of provisions became exhausted before the ice broke up, and they resolved on making the perilous attempt of crossing on the rotten ice and open water to Carriboo. The inhabitants of that place seeing their situation came to their assistance, and they were rescued.

For two successive springs, two men named Campbell, and Patterson of Pictou were burning lime on the island, for which they brought coal from a small seam on Carriboo Island. On one of their trips to Pictou, they were nearly suffocated, by the sea breaking over their boat and slaking the lime.

The island was originally granted to Admiral Sir Alexander Cochrane. In the year 1814, he sent William Cumming as his agent to settle the island. He was accompanied or soon followed by three families, named Boyd, Hogan and Morris, all four being from Ireland.

In 1819 John McDonald, Donald McDonald and Charles Campbell arrived. In the following year Kenneth McKenzie, who had served in the 78th Highlanders, purchased the property of Cumming, and acted as agent. These were all Highlanders, and there soon arose a strife between them and the Irish, which became so bitter as to sometimes leave marks of violence on the persons of the contending parties. Probably in consequence of this, the Irish soon after left the Island. Just before they did so, a fire broke out, which consumed the greater part of the forest, the origin of which was attributed to the wife of one of them.

Shortly after, John McDonald (2nd) and Hugh McCallum arrived, and they were soon followed by several of their relatives. The population is now 129, 57 males and 72 females.

We now turn to notice the progress of settlement on the other side of the county. William McKenzie, a native of Sutherlandshire, who had emigrated in the year 1803, and had first settled at Lower Barneys River, removed to the Upper Settlement in the year 1807, where he was the first settler. He was the father of the surveyors, and located himself at Kenzieville, where his sons still live. Donald Robertson, who had emigrated from Perthshire in the year 1801, and at first settled at the foot of the river, removed in 1819 to the Upper Settlement, and took up his abode near McKenzie's, about a mile farther down the river. About this time (1819-21), Angus McKay, the elder, a native of the parish of Clyne, Sutherlandshire, and with him Simon Bannerman, Gordon Bannerman, old John Sutherland, of the kilt, a man who never wore trousers in his life, with his family, who were numerous, and several others, all from Sutherlandshire, settled in the upper woods of Barneys River. There were also a few Lowlanders, among whom may be mentioned William Irving, from Dumfriesshire, who settled at Barneys River in the year 1820, who has left a large number of descendants there.

It may be mentioned that in the years just previous, (1810-1816), James Haggart, from the parish of Kenmore, in Perthshire, with others from Blair Athole, settled in the valley of Piedmont. This name was given to it afterward by the Rev. Dugald McKeichan, the first minister of Barneys River, from its situation at the foot of a range of hills. James Mappel settled in Marshy Hope, a valley in the Antigonish Mountains, leading into Antigonish County, where Angus McDonald now lives. When his neighbours were in the habit of advising him to leave that marshy place, because the frost injured all his crops, his uniform reply was, " I hope it will improve."

Hence his neighbours made the remark, that his hope was a marshy hope, from which circumstance arose the name of the Valley. John McLean (the poet) settled at East Branch of Barneys River, and after a time removed to Antigonish County. He was a native of the Island of Coll.

Between the years 1830 and 1840, a number of other families came from the counties of Sutherland and Perth, and took up land in the same settlement, among whom were John McDonald, the weaver, and his brother Duncan, James Forbes, William Sutherland, Donald Bruce, James Leadbetter, John Bannerman, Donald McKay, Robert Ferguson, besides others.

William Urquhart, from Glen Urquhart in Scotland, settled at Blue Mountains soon after the year 1815, and was the first settler there. William Ross, the elder, came out from the same parish and joined him in the year 1818. He was the first who gave the name "Blue Mountain" to that district of country. Along with him came Roderick McDougald, Senr., John Austen, and others; and in the year 1820, came Donald Campbell, John Munroe, and others, who settled at Moose River. About the same time also other families, McLarens from Argyleshire, Kennedys and McDougalls from the Island of Mull, settled on the old St. Marys Road.

The people of the Blue Mountain are chiefly from Glen Urquhart, and the neighborhood of Beauly and Kirkhill, in Inverness, with a few from Ross-shire and a few from the Lowlands, among whom may be mentioned the Meikle family, descendants of James Meikle, Senr., who came from the South of Scotland near the English border.

Sometime after 1830, William McDonald from Caithness came to the Garden of Eden and took up his abode there. He was called the "Adam" of the garden, because he was the first settler and the oldest man there. Along with him came his sons, John, Alexander, and George, also his son-in-law William Miller, and he was followed by others from Caithness and Ross-shire.

At the beginning of this period, the district of St. Marys was attracting a good deal of attention, as a desirable place or settlement, and quite a number of persons moved thither from this county. It was specially noted for its magnificent timber, extensive intervales, rivers teeming with fish, and the abundance of game in the forest. The very first settlers were from Truro, who built the first house in Glenelg in 1801, but the great body of those who followed were from Pictou. In the year 1810, William Kirk, one of the old 82nd, removed from Green Hill and settled at Glenelg, and in the same year, John McLean and his son James came from West River and settled at Stillwater. In 1813, Alexander Hattie moved over and settled on the East River of St. Marys, about two miles beyond the county line, and in the year following, McKenzie from Green Hill settled at the head of Stillwater.

At the same period, Caledonia was settled, almost entirely by Pictonians. In the year 1810, Angus McDonald moved over and settled in Lower Caledonia. His posterity are now numerous along the river. About the same time, Simon Fraser, located himself near him, but he subsequently moved to Glenelg. In 1812, Donald Cameron came over from Green Hill and settled in Middle Caledonia, and in the same year, John McDonald from West Branch, East River, took up land in the same vicinity. In 1814, Angus, John and Neil McQuarrie, took up lands still farther up. They had come from Scotland in 1810, but had been living in Churchville. Others from Pictou followed, but these may be given as the pioneers.

A part of the East River of St. Marys belongs to Pictou County, and we must now more particularly notice its early settlement, which began at this period. The land had been previously granted by Government to David

Archibald, previously of Truro, then of Sherbrooke, and others, in blocks of 2000 acres each. The grantees however did not settle or make any improvements upon their lots, and the actual settlers had to buy from them.

The first man who crossed the water shed with his family, to settle on the East River of St. Marys, on the Pictou side of it, was Mr. Alexander McKay. He came over from Fish Pools in July, 1815, taking with him his wife and seven children. From Webster Mills, McLennans Mountain, to St. Marys, a distance of 22 miles, was an unbroken forest, with a bridle path through it for half the distance, and only a blaze for the rest. The only possible modes of travel were on horseback or on foot. Four horses were secured for the journey. The larger children were lashed on the backs of the horses, with the luggage, while the smaller were carried in the arms of their parents or of the drivers, who trudged along on foot. One of the horses belonged to Mr. McKay, but soon after the poor brute, being tired of the lonely life in the forest, set out to return to the haunts of civilization, but lost sight of the blaze, or in confidence in its own superior wisdom, took a straighter course and perished in a swamp. It was five or six years after before McKay could obtain another, not altogether however from the objections of the animal to live in such a solitude.

Later in the same summer (1815), three others followed in McKay's steps, and settled farther down the river, viz., John McBain, John Matheson and Hugh Fraser, and later still in the season, Angus Cameron arrived, adding the fifth family to the little community. McBain and Cameron settled on the west side of the river, the others on the east. By the last running of the county line, McBain's land on one side of the river, and Fraser's on the other, are thrown into Guysborough County. We may add that at this time there were but three settlers about Glenelg Lake, and only one farther up the river.

The traveller passing through this section of country at the present day, admires its broad, fertile, and well-cultivated intervales, hemmed in by ridges of forest-clad hills, and dotted by feathery elms. But a grander sight met the gaze of the early pioneers, from the brow of the mountain they had to cross in reaching it. On the west of the river down to the margin of the valley, was a pine forest, which stretched away without a break to the Musquodoboit River, while to the north and east a sea of rolling hills extended in the direction of Barneys River and Lochaber. Much of this forest remains, the glory of the hills, to this day. The timber of the valley is said to have been very large. Elms three and four feet through ran up without knot or limb, for 50 or 60 feet, and maples, oaks, and birches of equal or greater size, with here and there a giant pine, covered its surface, and seemed to smile defiance upon the puny efforts of the feeble band, now attempting to dispossess these monarchs of the forest, that had occupied the ground undisturbed for centuries.

All these pioneers had visited the locality several times before taking up their abode there, and had done some chopping and burning, had erected rude huts, and that spring had planted a few potatoes. That year, however, as we shall see presently, was the year of the mice, and thus their only crop became the prey of these creatures. The larger tubers they scooped out, eating or carrying away the contents, and the smaller they dragged to their holes in stumps and logs. The following spring, the settlers carried their seed from McLennans Mountain, except a little that they brought from Lochaber, nearly as far. But this again was the year of the frost, so that they were doomed to reap no harvest, except a little fodder for their cattle. We may add that for many years, until the forests were well cleared away, the frost continued to be very destructive along the valley. In consequence, for the first five or six years, they had little bread of any kind. Even after their clearings enlarged, and the frost and mildew became less destructive, they still had many hardships to endure. For years they had to dry their oats in pots, and shell them with their feet in barrels and boxes, and then carry the groats on their backs, or in winter drag them on handsleds, to Archibald's mill at Glenelg to be ground.

But otherwise they did not want for food. The river swarmed with fish, especially trout, salmon and gaspereaux. Eels also were abundant, but they were regarded rather as an enemy, for unless the nets were watched, they would eat the salmon in them as soon as they were caught. They would do this in a way that was always a marvel to those who were witnesses of the feat. They would eat the whole fish except the skin and the backbone, turning the skin inside out without breaking a single joint of the latter. Moose and Caribou were numerous, and partridges and rabbits were snared or looped, and often proved an acceptable addition to their larder. Even the porcupine sometimes found its way to their table. Bears were very numerous, but having abundance of food in the forest, very considerately abstained from disturbing the settlers for a number of years.

In subsequent years, settlers continued to arrive. In 1817, another Angus Cameron came over from McLennans Mount, and settled on the East side of the river between McKay and Fraser. In 1821, he was followed by Alexander Sutherland, who located himself farther up the river. In 1826, James Cameron removed from McLennans Mount, and occupied the place vacated by John Matheson, who had left. The same year John Hattie settled on the West side of the river. In 1831, three families of Gunns arrived from Scotland and occupied lands, between McKay and Sutherland.

Among the hardships to which these people were exposed, particular mention must be made of the want of a road. When they first settled here, they had above twenty miles of forest to traverse to reach a neighboring settlement, and one incident may illustrate the inconvenience of such a situation. Twenty years after their settlement, a young man, whose head is now white with the snows of age, had exchanged his solitary life for the social, and had erected his humble log home, but, not many weeks after his marriage, discovered that housekeeping was inconvenient without dishes. So he set out on his snowshoes for New Glasgow, and there purchased a set of cooking utensils, consisting of an oven, two pots, a kettle, two teapots, a half dozen cups and saucers, a half dozen plates of large size, with as many smaller ones, and a half dozen knives and forks. He succeeded in getting these conveyed on a sled, as far as there was any thing of a road, or to within twelve miles of his home. He then donned his snowshoes, fastened all the above mentioned articles about his person, and though heavily laden, he never came home with lighter heart, to meet the partner of his life.

Not until Captain McKenzie became member for Pictou in 1855, and in the year following obtained a grant for a road from the Garden of Eden to St. Marys, was it possible to ride in a wheeled vehicle, between these two places. Even in winter it was long before a sleigh road was opened. The fathers still tell of the way in which they used to fasten two poles to a horse after the fashion of shafts, with the lower ends trailing behind, and kept together by a cross piece, and with long wooden pins driven into the upper side, on which they laid their pork, which they dragged in this manner to the Garden. Even down the river toward Glenelg, there was no possible road to market or mill, for fifteen years after they settled here. In the winter the hand-sleigh, and in the summer the canoe, when the river was sufficiently swollen, were the only means of conveying burdens, excepting on shoulder or horseback.

Of the first settlers, Alex. McKay, the pioneer, deserves special mention, as probably possessing the greatest amount of strength and activity combined, of any man that was ever in the County. He was a son of Alex. McKay, whose name appears among the immigrants of 1784, and was born near Beauly, Inverness. He came with his father to this Province, when a boy of about 12 years of age. On the passage, he performed a feat which showed his daring and dexterity. He and another boy having climbed up the mast, two sailors followed with ropes, intending to tie them. One of them caught the other boy, but McKay seizing the top-mast stay, swung himself from it by his hands, and then passed hand over hand to the other mast, by which he descended to the deck, while the bystanders looked on in terror.

When a young man, he chased and caught a caribou calf. The Indians have a saying regarding the young of the animal, "one day old, Indian catch him, two days old, dog catch him, three days old Mundous himself no catch him." However McKay being in the woods with some others, they started a herd, and this calf became separated from the rest. He pursued it, endeavouring to drive it in a direction the opposite from that in which its companions had gone, at the same time watching lest the dam should turn upon him. After a short chase it tripped in crossing a tree, and fell, and in an instant he was upon it. He took the animal home where it became quite tame. He afterwards exchanged it for a heifer with Squire McKay, who sent it as a present to the Governor, Sir John Wentworth, receiving in return a present of 2000 acres of land in St. Marys. It was afterward sent to the Tower of London, where it continued for several years, being the first animal of the kind in that collection.

Many stories are told of his mingled physical power

and dexterity in his mature years. We give a specimen. A bull had become wild, and was shut up in a barn, where none dared to approach him. McKay was sent for. On arrival he gave orders to open the door, while he stood beside it. As the animal rushed out, he seized him by the horns, threw him on his back and held him as long as necessary. He was much engaged in lumbering, but neither river-driving nor the other concomitants of that mode of life, ever seemed to affect his constitution. When between 80 and 90 years old, he could mow his swathe with younger men, and he lived to be 97 years of age.

All the settlers were economical, industrious and religious. About the year 1826, they hired Hugh Cameron, subsequently of Wentworth Grant, as their first teacher, and about the same time, they started a sabbath school; and with slight interruptions, both have been continued since with the happiest results.

The first minister who preached to them was the Rev. Dr. McGregor, who about the year 1817, in one of his missionary excursions to Glenelg and Sherbrooke passed up the bed of the river on horseback. But with the exception of McKay, all the settlers adhered to the Church of Scotland. At the Disruption, the majority joined the Free church, and obtained a portion of the services of the Rev. Alex. Campbell of Lochaber. Two small churches were built, but now the people have all united with the Presbyterian church in Canada, and have erected a larger and more comfortable place of worship.

CHAPTER XV.

FROM THE PEACE TILL THE FINANCIAL CRISIS OF 1825-6
1815 - 1826.

Up to this period the history of the county had been one of continued and for some time rapid progress. Population and wealth had increased at a rate, which, compared with what has since been seen in the Western States, might even be considered slow, but which at that time was regarded as quite remarkable. From this time forward however for some years, its progress was very slow, and indeed in some respects it seemed for a time to have been stationary, or even to have retrograded. In the town of Pictou or its neighbourhood, property sold as high in 1815, as it did forty years later, and some farms brought larger sums at the former period, than they would now, and from that time there has been more or less emigration from Pictou, many especially of the young going abroad.

The first interruption to its continued prosperity was by the peace of 1815. The change largely affected the whole Province. The author of Agricola's letters thus describes it :—

" During the war, money here arising from the expenditure of the British Government, and from the sale of the rich cargoes and ships, which were daily brought in by our cruizers, was not only in brisk circulation, but in great abundance. The ships of war, which lay in the harbours, the various establishments of dockyard, ordnance and barracks, the strangers who resorted hither on commercial speculation, contributed to create an uncommon demand for all sorts of produce ; and as these were before inadequate to the ordinary wants of the community, they fell now infinitely short of the extraordinary consumption, to which the exigency of the times gave rise. During the whole of this period, the prices obtained by the occupier of lands for whatever he could bring to market, were prodigiously high and far outran the cost of production. Hay sold at from ten to twelve pounds per ton, and was frequently at fifteen ; beef and mutton varied from 8d. to 10d. per pound ; potatoes, turnips, and beets were oftener above than below 5s. per bushel, and all vegetables were exorbitant in like proportion. During this unprecedented prosperity, no exertion was needed by the farming body to earn a subsistence. The rewards of the most moderate labour were so ample, that they begat habits of indolence and luxury ; but excited not to new energy or a more spirited cultivation. Our landholders, satisfied with the enormous prices they obtained for beef and hay, and trusting that the springs of wealth, which flowed so copiously, would be perennial, discerned not the dark cloud at a distance, which was gathering round to overcast their horizon. Peace came and at once dried all the sources of this artificial prosperity. Real estate fell almost in an instant, trade declined, land produce was lowered by the effects of this general depression, and in about two years after the ratification of the treaty of Paris, an universal gloom had settled over the Province."

As the County of Pictou was less dependant on the war expenditure, than some other parts of the Province, and as the inhabitants had relied more on trade than on agriculture, the effects of the peace were not so disastrous or so immediate, as have been described, but still the effects were felt to a considerable extent. The rural population however were especially discouraged by two calamities which came upon the county at this period in successive years (1815-16), which we shall now notice.

The former year was long distinguished in this, as well as in the neighbouring counties of Colchester and Antigonish, as "the year of the mice." This was a most destructive visitation, from which this portion of the country suffered from these seemingly insignificant animals. During the previous season they did not appear in any unusual numbers. But at the end of Winter, they were so numerous as to trouble the sugar makers, by fouling their troughs for gathering sap, and before planting was over, the woods and fields alike swarmed with them. They were of the large species of field mouse, still sometimes seen in the country, but which has never since been very numerous.

They were very destructive and actually fierce. If pursued, when hard pressed, they would stand at bay rising upon their hind legs, setting their teeth and squealing fiercely. A farmer on whom I could rely told me, that having after planting, spread out some barley to dry in the sun before his door, in a little while he saw it covered with them. He let the cat out among them, but they actually turned upon her and fought her.

The late sown grain and the seed potatoes suffered from them ;* but it was when the grain began to ripen, that their destructiveness became especially manifest. They then attacked it in such numbers, that all means were unavailing to arrest their ravages. They have been known to cut down an acre in three days, so that whole fields were destroyed in a short time. One would nip a stalk off a little above the ground, and if instead of falling over, the end sank to the ground, leaving it still upright, he would bite it oft farther up, until it either fell over, or the ear came within his reach, when he would devour all the grain. Over acres on acres, they left not a stalk standing, nor a grain of wheat, to reward the labours of the farmer. They burrowed in the ground and consumed the potatoes. Cats, dogs, and martens gorged themselves to repletion upon them, but with little seeming diminution of their numbers. Trenches were dug and filled with water, but they formed but a slight barrier to their progress.

They passed away as rapidly as they came. In the Autumn, as the weather became colder, they became languid, scarcely able to crawl. One could trample them under his feet and finally they died in hundreds, so that they could be gathered in heaps, and their putrefying carcases might be found in some places in such numbers as to taint the air. At Cape George they went to the

---

* A man in Merigomish had made a clearing out at Piedmont in the woods. He carried out four bushels of oats to sow. On commencing, they came in swarms eating the grain as he sowed it. After continuing a while, he threw the whole to them in disgust, and returned home.

water, and there died, forming a ridge like seaweed along the edge of the sea, and codfish were caught off the coast with carcases in their maws.

Notwithstanding the unprecedented prosperity, which the country had enjoyed for about twenty years, such were the spendthrift habits engendered during that period, that the people were not prepared to meet such a calamity, and it was therefore felt very severely. But it was followed by what was long known as "the year of the frost," which left a large portion of the inhabitants in a suffering condition. The year 1816 was known throughout the northern parts of this Continent, and also in Europe, as "the year without a Summer." In the Northern States, frost, ice, and snow were common in June. Snow fell to the depth of ten inches in Vermont, seven in Maine, and three in Central New York. On the 5th July, ice was formed of the thickness of common window glass throughout New England, New York, and some parts of Pennsylvania. In August ice was formed half an inch thick. Indian corn was so frozen that the greater part was cut down for fodder. Indeed almost every green thing was destroyed. A similar state of things existed in England. During the whole season, the sun's rays seemed to be destitute of heat. All nature seemed to be clad in a sable hue. The average wholesale price of flour during that year in Philadelphia was $13 per barrel. The average price of wheat in England was 97s. per quarter.

Here the frost was hard in the woods in the month of June, provisions were high and from the destruction of crops the previous year by the mice, many were suffering, and nearly all the farmers were put to some inconvenience, for want of food for their families. Alexander Grant (miller), of the East River, went to Halifax to obtain a supply. He there bought 70 barrels flour, for which he had to pay £3 per barrel. On his way back on the 5th

June, he stayed all night at a tavern between Halifax and Truro, and in the morning the ground was frozen so hard, that it carried his horse. The flour came round by water, and he went down to town to bring it up the East River, which he did on a coal lighter. On his way up on the 16th, he saw a man trying to harrow his ground, where he had sowed some grain, and wearing a great coat in consequence of the cold. That night being Saturday, he put the flour into a barn owned by the late James Carmichael, Esq., who had shortly before begun to do business, where New Glasgow now stands. On Monday morning, before he reached the spot, there were as many assembled, as there were barrels of flour, and no sooner was the door opened, than a rush was made, and each man seized a barrel, asking no questions as to price, and it was with some difficulty that he saved one for his own family.

In the same year, Mr. Grant and his brother Robert, erected the first oatmill in Nova Scotia, probably the first in B. N. America, on the site still occupied in the same way, and known as Grant's Mills. Very little oatmeal had been used previously. Small quantities were sometimes brought out in vessels, and sometimes the country people manufactured a little in a coarse way, by roasting the grains in a large pot and afterwards separating and grinding the groats. But now Mr. Grant constructed a regular oatmill driven by water, of which the gear was made by a millwright, named Duff. It was still, however, somewhat rude in structure. Instead of iron over the kiln, the grain was supported on wooden slats, the edges of which were bevelled on the lower side, and there were no fanners. Indeed fanners were not then commonly used even by the farmers. Hence after the grain had been dried, they were obliged to carry it to the top of a hill near, and piously wait till Providence sent a wind sufficient to separate the shells from the kernels. But

the next year, he constructed fanners driven by the mill.
At this time agriculture seems to have engaged attention, and accordingly, the first agricultural society in the rural districts of the Province, was formed on the 1st January, 1817, at West River. This was before the publication of the letters of John Young, under the signature of Agricola. A meeting was held some time before, at which the resolution was adopted, to " form a society for the improvement of agriculture, to be called 'The West River Farming Society.' " Accordingly the Society was regularly organized at that date, 26 persons joining, and the following being elected office bearers :—Rev. Duncan Ross, President; Robert Stewart, Vice-President; Donald Fraser, Treasurer; John Bonnyman, Secretary; David McCoull, John Oliver, Anthony Smith, George McDonald, John McLean, Jonathan Blanchard, Committee. They seem to have had a poet among them also, for in the front of their minute book, the following appears as their motto :—

> Let this be held the farmers' creed,
> For stock seek out the choicest breed,
> In peace and plenty let them feed.
> Your lands sow with the best of seed,
> Let it not dung nor dressing want,
> And then provisions won't be scant.

By the rules then adopted, each member was to pay 5s. entry money and 1s. 3d. quarterly; no persons were to be admitted but farmers and freeholders of good moral character. And to insure continued good behaviour, it was enacted, that "if any member shall curse or swear or use any indecent language, or introduce any subject inconsistent with the business of the Society, he shall be fined by the President and a majority of the members present, in a sum not exceeding 5s."

The Society was to meet quarterly, and at each meeting a topic or topics, connected with rural economy, was to be discussed, " each member to come prepared either with a written essay, or to speak on the subject," the question selected for the first quarterly meeting in April, being, "What is the best method of preparing and increasing manure?" It served to elicit differences of opinion, for one man rose and said, that "instead of finding ways of making more, he wished they would find some way of getting quit o' it, for it was just a bother about his barn."

The Society continued to hold quarterly meetings, and to discuss agricultural topics. In the year 1818, they held a ploughing match in Mr. Mortimer's field, said to have been the first ever held in the Province. They imported seed grain, agricultural implements, and Ayrshire cattle. They also held some cattle shows, at which prizes were given for the best stock. They also gave prizes for the best acre of wheat and other crops, the greatest amount under summer fallow, and "to the person who should stump and plough fit for crop the greatest quantity of land never ploughed before, not more than three stumps per acre left on the land, and all stones that materially obstruct the operation of ploughing and harrowing to be removed, the quantity to be not less than two acres." In April, 1824, they offered £7 10s., in addition to the Legislative grant, for a flax mill. Anthony Smith, Esq., undertook to erect one. It was commenced that year, and in the following he received a prize for it, being the first of the kind erected in the Province. It did not however work long, as it did not receive sufficient employment to maintain it.

In the year 1819 the name was changed to the Pictou Agricultural Society, and Ed. Mortimer elected President, and some time after we find them presenting the Rev. Duncan Ross with a new plough, "to be one of Wilkies best, as an expression of their sense of his services to the cause of agriculture."

In the year 1820, we find a notice of a similar society on the East River, of which Dr. McGregor was Secretary.

Others were formed in other parts of the country, and continued for some time, aided by grants from the Central Board, and had considerable effect in improving the habits of our farmers.

In the year 1819, the whole community was shocked by the most dreadful murder probably ever committed in the Province,—viz., that by Donald Campbell of his father and stepmother. He was a simple ignorant man, but not previously regarded as violent or cruel. He was an only son, but his father had taken a second wife, and he was afraid, that in consequence he would lose his share of the paternal estate. This led him to form the design of destroying them both. Up till the time of committing the deed, he had given no such indications of hostility to them, as to excite any alarm. He lived at Earltown, but on the day before the commission of the deed, he was in town, and on his way back, called at his father's house, which stood on what is now Dinwoodies farm, and there obtained refreshments. He then started on his way homeward, calling at houses on his way as far as West Branch River John, with the design doubtless of producing the impression, that he had gone home. But when night came, he retraced his steps towards his father's house, which was a small one constructed of logs. Arriving there, he fastened the door by means of withes attached to the handle of the latch, and then set it on fire, while his father and stepmother were asleep. They were awakened by the fire, and succeeded in forcing the door open. They then commenced removing their things from the house, uttering at the same time loud cries for assistance. Donald was on the watch, and as his father was coming out with a large iron pot, he struck him with a heavy stick, and pushed him back into the house, where his bones were found next day.

His step mother succeeded in getting out. She was a stout strong woman and it was thought, that if she had

had fair play, she would have mastered him, but he struck her on the head with his dreadful bludgeon, and then drew her to the fire to cast her in. She was a heavy woman and either from her weight, or being alarmed before he accomplished his purpose, he only succeeded in putting her partially into the flames ; and in doing so, was somewhat scorched himself, a fact which afterward told against him on the trial.

In the meantime, their cries had brought to the scene a neighbour named McIntosh, who blew his horn to give notice of his coming. As he approached, Donald ran away. McIntosh saw his retreating figure, but did not suspect who it was, indeed supposed he had seen a ghost. He found Mrs. Campbell dead, and her body partly burned. He dragged it out of the fire, but was too late to save anything out of the house. He also found a little dog of Donald's at the spot, a circumstance which afterward excited suspicion.

Mrs. Campbell was buried without an inquest, although her brother, the late Angus Campbell, elder, Scotch Hill, at her funeral declared his belief that she had been murdered. A number of circumstances excited suspicion, and Donald was arrested. His stepmother's body was exhumed, and, on examination by the late Dr. Johnston, the marks upon it, left no doubt, that her death had been caused by violence. The stick was found with his father's blood and grey hairs upon it, and was afterwards produced in Court. A button was picked up, and on examination it was found to correspond with those on his coat, from which there was one missing. A gun flint was picked up on the spot, and a brother-in-law recognized it by a private mark, as one he had lent him just before the murder, and Campbell's gun was found without a flint. It was supposed that he had intended to shoot his parents, but that the flint had dropped out, and in the dark he could not find it, the great avenger having left it to cry from the ground against him. Parties testified to seeing him at various points on the road or taking to the woods before and after the fire. S. G. W. Archibald, who conducted the prosecution, took a large sheet of foolscap paper, and marking one point as the site of the house, and others at proper distances, to indicate the different points at which he had been seen, held it up before the jury, and showed how exactly the times of his being seen, coincided with the view of his going to and from his fathers house, though he denied having been near it that night. The whole circumstances left no doubt of his guilt, and accordingly at the September term of the Supreme Court, he was found guilty and sentenced to be executed.

The sentence was carried out at the scene of his crime, the gallows being erected on the site of his father's house. He was taken from the jail in Pictou in a cart, to Rogers Hill Church, which was as far as a wheeled carriage could go, guarded by a body of militia drafted from the several companies, and attended by several clergymen. There the fetters being removed from his legs, he walked the rest of the way under the same escort. Before his execution, he confessed his crime, but showed little appearance of contrition, although Dr. McGregor and the other ministers used all the means in their power, to bring him to a sense of his conduct and repentance for it. A large concourse assembled at the execution, and just before it took place, Dr. McGregor offered a prayer, which, from its earnestness and tenderness, lingered in the minds of most who heard it, while memory remained. But he was obliged to turn away in sadness, with the words " C Donald, I believe nothing will ever melt your heart."

The execution was superintended by the High Sheriff of Halifax, but was clumsily effected. When he gave the signal, the executioner attempted to draw the bolt, but it only slowly yielded to his efforts, and when the trap door fell with the unfortunate man, the knot slipped round to the back of his neck, which remained unbroken, so that he slowly choked to death. As the rope untwisted, he swung round with his face to the spectators on one side, and then as it recoiled, to those on the other, while his heavy breathing could be heard over the crowd, and, it was said, the pulsations of his heart, but perhaps rather, the heaving of his chest, could be seen by those near, presenting a spectacle, which led many present, never to see another execution.

We may here mention, that there have since been two executions in the county. The first was Neil McFadyan for the murder of James Kerr. In the fall of 1847, they had travelled together from Bay Chaleur to Pictou. Kerr's friends, not hearing from him, became anxious, and in spring enquiries were set on foot, when he was traced in company with McFadyan till near the house of the latter at Blue Mountain. The attention of parties in the neighbourhood was attracted by a stench from a neighboring wood, and on a search, part of a human body and clothes were found, which were identified as those of Kerr. Farther examination brought out a mass of circumstantial evidence, which left no doubt of McFadyan's guilt. And on trial before Judge Bliss, at the October term of the Supreme Court, he was condemned and sentenced to be executed. He was a bold, hardened villain, with no want of intelligence. The execution took place near the Beaches on the lot owned by the public, and used on occasions as a lazaretto. It was ordered to take place between ten and two o'clock. He was taken to the ground earlier than the hour intended. While they were waiting, it being a raw, cold day, late in the fall of the year, he said to the Sheriff, shrugging his shoulders, "It's cold here; you may as well put me through at once."

The other was the case of John McPhail for the murder of his wife. He was a poor, simple creature, who kept a low groggery back of the Albion Mines, on the road to the Middle River. His wife and he drank, and while both were under the influence of liquor, he beat her over the head and other parts of the body with a pick handle, even breaking her arm, so that she died. He was convicted principally on the evidence of his own child.

A criminal, who gave more trouble than either, and excited more alarm in the county, was a man named Jack Hines. He was an Englishman, who had come here and married in this county. He was a strong man and a great bully, so that he became a terror to the neighbourhood. He was at length arrested, tried and found guilty of burglary. The penalty at that time was death, but the jury recommended him to mercy. The judge however was bound to pronounce the sentence. The recommendation to mercy had to be forwarded to London, for the consideration of the Home Government, and such was the irregularity of communication, that though forwarded in February, an answer was not received till October, during which time he was kept in prison, his elbows kept close by his side, by means of a chain across the back attached to each. A pardon having arrived, there was considerable alarm at the prospect of his release, particularly among some of the magistrates, residing out of town. They therefore told him, that the condition of his pardon was, that he should leave the district. He was accordingly escorted to Mount Thom across the line, and left to pursue his way further as he pleased. Three nights after he robbed a store near Truro. He was arrested and on trial was sentenced to the workhouse in Halifax, but not long after made his escape.

We must now notice the trade of Pictou during the period we are reviewing. After the conclusion of the war, the timber trade still continued, though on a diminished scale, and we may here notice some of those engaged in it. Next to Mortimer, must be mentioned George Smith.

He was a native of Scotland, we believe of Banff, and was taken into partnership by Mr. Mortimer, their business being conducted under the name of Edward Mortimer & Co. After the death of the latter the business was conducted by Mr. Smith and William Mortimer, a nephew of Edward's, under the name of Smith, Mortimer & Co. Afterward, however, they dissolved partnership and each of them did business separately. For some years they were the most influential business men in Pictou. Mr. Smith represented the County of Halifax, from Mr. Mortimer's death in 1819 till the year 1836, when the County was divided, after which he represented the County of Pictou till the year 1838, when he was appointed to the Legislative Council. Mr. Smith was a man of fine presence and a gentleman of the old school. He was an able business man, and succeeded, as far as it was possible for one man in the advanced state of the country to do, to the place and influence of Mortimer. The misfortunes of his later years obscured all his former glory, and almost blotted from memory the services of his early career. It is therefore due to him to say, that at this period he was an active merchant, and did much for the business of the port—that he filled several public offices, judge of Probate, judge of the Superior Court, and Custos of the County in a most creditable manner, and that as a member of the Legislature, he commanded the highest respect, and was largely influential in promoting the interests of Pictou.

We may also notice John and Abraham Patterson, sons of old John Patterson. They began business in Pictou about the year 1815. Mortimer said when he heard of their commencing, that he was more afraid of those two young men, than of any persons that had ever begun business in Pictou. Though during the preceding years of prosperity, others had engaged in merchandizing, he still regarded any person attempting general trade, as an intruder upon his legitimate domain, and he employed his power to defeat their enterprise.

But in a short time, by their energy, and the confidence which they had inspired by their honorable dealings, they were doing a business, in its extent perhaps not surpassed by any in the eastern part of the Province. Their trade consisted principally in shipping timber to Britain, the fisheries, and the West India trade. In timber the article most in demand was squared pine, which was still obtained in considerable quantities in Pictou and the outports. They were not engaged largely in shipbuilding, their operations being confined principally to the building of small vessels for the fisheries or the West India trade. In fishing, the mode of doing business was to send to the various places to which the finny tribes chiefly resorted, small vessels, loaded with barrels, salt, and supplies of goods, such as fishermen required, in charge either of one of the firm or an agent, who exchanged these articles for fish. In this way they prosecuted the fishery the whole season, commencing with the Gaspereaux in Spring, then following successively the spring mackerel and the spring herring, the codfish and salmon, the fall herring and fall mackerel. In this way their business extended as far as Richibucto on the North, Rustico on the North coast of Prince Edward Island, Margarie and Cheticamp on the North coast of Cape Breton, and Canso on the South shore of Nova Scotia. At that time fish were taken in quantities which now seem almost incredible, five hundred barrels of mackerel at a single haul being considered a good, but not an extraordinary catch. Oftentimes they could not be cured, and heaps containing hundreds of barrels were left to rot upon the shore. The fishermen were generally a reckless set, depending on taking by a single haul enough to keep them for six months, and were dependant upon traders, for the supply of most of the necessaries of life. The fish brought back in these expeditions, was shipped, along with various kinds of lumber to the West Indies, and the vessels brought back return cargoes of West India produce. In this trade the brothers continued for a number of years. Doubtless there are many enterprising men in Pictou at the present day, but where is all this business now? For the decline of the timber trade, there is a good reason in the exhaustion of the supply, but why should not the fisheries be carried on from Pictou, as well as from ports, more distant from the places frequented by these finny tribes.

In their business dealings the two brothers were much esteemed. We have met in distant places in Cape Breton, and along the south shore of the Province, persons who looked back with the kindliest recollections to the days, when they traded to these quarters, and spoke with the warmest feelings of respect for themselves personally.

They retired from business in the year 1832. The older, so long known as "the deacon," many in Pictou will still remember. A man of but few words, he was ready for every good work. In the congregation he was *the* deacon and *the* manager, never putting himself forward, but always having work laid upon him, and doing it as naturally as if taking his meals,—a man of such entire negation of self, that he never seemed to feel that he was doing anything, yet the man to whom everybody looked when anything was to be done. Such was he in every society with which he was connected. He filled also public situations with honor to himself and advantage to others. He was a trustee of the Pictou Academy from its foundation, and for many years its treasurer. He was also treasurer to the Synod of the Presbyterian Church of Nova Scotia, when, however, the keeping of its accounts, and the disbursement of its funds, was not a matter of great labor. Guileless in character, lovable in nature and unassuming in all that he did, he passed away not only without an enemy, but amid universal expressions of profound respect. He died April, 1847.

Abraham, though in his later years living in a great measure retired from public life, was at this time for years one of the most prominent men in Pictou. In the year 1815 or '16, he was appointed a justice of the peace, which at that time involved something, having been recommended to the position by the unanimous voice of a public meeting of the inhabitants of the town of Pictou, and filled the office for more than fifty years. He was also a judge of the Inferior Court, at the time of the abolition of that tribunal. In the public movements of his time, he for many a day bore an honorable part. He died June, 1867.

The meeting for the election of magistrates referred to, took place at Taylor's Tavern, on the east side of the street leading to Yorstons Wharf, his biggest room being prepared for the purpose, when besides Mr. P., his brother, Walter, Robert Patterson, and, we believe, Robert McKay, were chosen, and in due course were appointed by Government. The meeting was harmonious, and not less so, when the nominations being over, one gentleman arose and said, " Mr. Chairman, I have another motion to propose." Attention being directed to him, he added, " I move that we now call for liquors all round." There is no record of the result, but we venture to say that the motion was carried, *nem: con.*, and, unlike many a better resolution, was immediately carried into execution.

Robert Patterson here mentioned, was usually known as Black Bob, to distinguish him from two cousins of the same name, his title being derived from the colour of his hair, all being grandsons of Squire Patterson. He lived above Dr. McCulloch's place on the old road, was now in business, and afterward an active magistrate.

Walter Patterson was the third son of John Patterson.

Few now remember him, but these few are always touched with tenderness, as they speak of him. By those who knew him, he is described as the ablest and finest of the first generation, that grew up in the town. He was a notary public, and filled such important offices as Clerk of the Peace, Prothonotary, and Clerk of the Commissioners' Court, besides more private ones, as Secretary of the Friendly Society. As has been said, wherever accuracy and good business habits were wanted, Walter Patterson was the man. He was specially beloved for his social habits. Though distinguished by a sobriety unusual for the times, yet a genial humor that never wounded, rendered him the joy of any circle he entered. He died in 1821, at Plymouth, England.

Among the other merchants of this period we may mention James Dawson. He was a native of Banff, and at first did business as a saddler, but afterward commenced trading, following the course we have already described in ship building, shipping timber, the fisheries and the West India trade. Finding trade prospering, he sent for his brother Robert, in partnership with whom he carried on business for some time, both saddling and merchandizing. But after a time they separated and each did business on his own account. The commercial changes of 1825-26, which we shall more particularly describe presently, involved him in pecuniary embarrassment. Being connected by marriage with Mr. Boyd, of the firm of Oliver & Boyd, Edinburgh, he was through them led to engage in the bookselling business, the first in the Province out of Halifax who did so. Encumbered with the debts of his previous business, which he had undertaken to pay with interest, he continued to prosecute it though without any very large profit to himself, maintaining a bookstore, which for many years surpassed those in Halifax, dealers there giving their attention mainly to stationery. In this way he was the means of circulating much healthful literature, and thus of promoting the progress of knowledge in the county and beyond. He was actively engaged in the religious societies of the day, and, though not having the knack of gaining popularity, he in these and other ways served his generation. He died at the residence of his son, Dr. J. W. Dawson of Montreal. His brother also was for many years a prominent man, especially in the religious movements of the day. We might also mention Robert McKay, Esqr., who, after the death of Mortimer, with whom he had been clerk, commenced business at River John, where he was successful for a time, but succumbed to the commercial storm just referred to. He was afterward shipping agent at South Pictou for the General Mining Association, and succeeded Mr. Smith as Custos of the county.

At this time commenced Henry Hatton. His father, Robert Hatton, was a lawyer, who emigrated from Ireland and settled in Pictou about the year 1813. His son first commenced business in one of the wings of John Dawson's building, but afterward had a large set of buildings, at what is now South Market St., but which was then not built up, and was known as Hattons wharf. He was afterward one of the largest shipbuilders in the county, and for several years represented the township of Pictou in the Assembly.

We may here mention a system of trade not uncommon at this time. Captains of vessels brought out supplies of goods, or sometimes solid specie, which they exchanged for timber. Of these the most deserving of notice is Stephen Lowrey, of Newcastle, Eng., who, afterward becoming a shipowner, traded with his vessels to Pictou to a considerably later period, and who, entering into partnership with the late James Purves, under the name of Stephen Lowrey & Co., did a large business in ship building at the head of Purves' wharf.

Trade being now thriving, a number of the merchants

combined to build a vessel, to be a regular trader between Pictou and the old country. She was a brig called the Enterprise, and was built by Thomas Lowden, who had a good reputation as a shipbuilder. She was launched in August, 1820, and the occasion was celebrated by a ball on board. She was a square box of a thing, carrying a large cargo for her tonnage, and always proving a safe and successful vessel, though a dull sailer. She continued to make her regular trips twice a year, bringing out British goods and taking back timber, till the crash of 1825-6, when she was sold. Thereafter she was employed in carrying coals from Newcastle to London, and was so engaged twenty years later, and may be yet.

The timber trade had declined both from the peace, and the diminution of the supply, but it was still of importance; and with the shipbuilding and other business still carried on, and the improvement in agriculture, still brought a fair measure of prosperity to the county, when there came the terrible financial crisis of 1825 and '26 in the mother country, which resulted in severe losses to all engaged in timber and shipping, whether there or here, and the utter bankruptcy of many. To show its operation, we may mention, that vessels which in the early part of 1825 brought £13 10s. per ton, in the following year would not bring £6, and men, who shipped cargoes of timber to England, were brought in debt for the freight. In St John, N. B., the day the intelligence of these changes arrived, was long known as Black Monday. Strong men wept, as in one day they found the work of a life-time swept away. A firm, that in the previous year, had shipped a hundred cargoes, now became bankrupt. In Pictou all the merchants suffered severely. Some had large stocks of timber on hand, which they had bought at high rates, but which were now unsaleable at any price. It lay in the rivers and outports for some time, till the outer portion was decayed or worm eaten, when it had to be hewn over again. Yet even after this expense, and with the quantity thus diminished, it sold for half the price per ton that it originally cost. Some became bankrupt, others never recovered from the blow, and for a time the trade of the port was laid prostrate.

This period is marked by the formation of societies of a religious or benevolent character. The most interesting of these is the Pictou Sabbath School Society. The first Sabbath School in the county, on the modern system, was commenced in town by the late James and Robert Dawson, according to the statement of the former, in the year 1814. They were joined soon after by John Geddie, Sr. But it was now determined to form a society, for the promotion of Sabbath Schools generally. Accordingly a meeting was held for the purpose, on the 25th March, 1822, in the old Court House, and a resolution was adopted, to form such a society " having for its object the encouragement, union and increase of Sabbath Schools." The rules were also agreed to, by which a payment of 2s and 6d annually, was to constitute membership for a year, and 20s for life. Meetings of the Society were to be held on the second Tuesdays of May, August, November and February, the last being the annual meeting. In town the teachers were to meet monthly, and in the country they were recommended to meet as often as possible. In all cases they were to bestow their labor gratis. In the year following, it was reported that there were 29 schools, with 1,000 pupils, in connection with the Society. The institution was for some years in vigorous operation, collecting funds to import books, sending agents through the country districts to establish new schools and to stimulate old ones, and in circulating religious literature, suited to the young. In the year 1827 we find reported as in connection with the Society, 77 schools, with 2,335 pupils and 198 teachers; also, that there had been imported books to the value of £104 6s 10d. sterling, and that the number circulated was 6,950, besides the libraries attached to many of the schools. Its last report that we have seen noticed, was in the year 1833, being the eleventh.

In the year 1823, the Bible Society was re-organized. From its first formation in 1813, it had scarcely ever met. The plan upon which it had been formed, of one Society for the county, with so many directors from each congregation, had not been found convenient in practice. Still these directors had collected money in their several quarters, which was yearly remitted to the parent institution, by Dr. McGregor, who also ordered and circulated Bibles, and generally did the business of the Society. But interest having fallen off, after some solicitation from him, the Society was now re-organized on its present basis, as the Pictou Auxiliary of the British & Foreign Bible Society, with Mr. James Dawson as secretary and depositary.

About the same time, the first subscription library in the county was formed in town. A public meeting was held for the purpose, presided over by Dr. McCulloch, who urged the importance of the proposal. The first importation of books was made in the spring of 1822. The institution continued for some thirty years, and during that time its books increased, until they formed a very respectable collection, the circulation of which did much for the promotion of intelligence and literary taste; but unfortunately interest in it declined, and it was finally dissolved and the books scattered.

Another institution of this period, which however we cannot commend, must be noticed. We allude to the Ballast Pier. From the number of vessels arriving in ballast, the discharge of which in lighters involved much labor and expense, a number of persons formed the idea, that it would be a profitable speculation to build a wharf on the edge of the channel, at which vessels might directly discharge. They also expected to fill in from it to the Deacons Wharf, and to make money by the lots to be reclaimed from the water. At the same time, the magistrates fearing injury to the harbour, by the manner in which ballast was being discharged, obtained in the year 1819 an Act of the Legislature, authorizing the Court of Sessions to make regulations for the good of commerce and the preservation of the harbour. In this act they were empowered to "fix such places in the harbour as shall be most convenient and proper for ships and vessels to discharge their ballast, and to make such agreement as may be needful and necessary with persons, for erecting and building wharves and other conveniences, for such ships and vessels to discharge their ballast upon," &c. Accordingly they contracted with this company to build such a wharf as mentioned, giving them the exclusive right to receive ballast on it for the next ten years, and empowering them to levy a remuneration of 3d. per ton register on every vessel so discharging. The wharf was accordingly constructed, and till the year 1824 vessels discharged there. But by this time the folly of the scheme began to appear. The wharf, from the wood of which it was built decaying, began to spread, and its contents to be discharged into the channel. The company, who had lost money by their speculation, wished to have their power extended to twenty years, but the magistrates refused, although they for a time permitted a practice not really any wiser, of vessels drawing up opposite the Battery Point and discharging their ballast there. The result is, that what was to fill the pockets of the projectors, not only proved a bad business for them, but remains an unsightly ruin, and an injury to the harbour. Sir James Kempt on visiting Pictou, when he came in sight of the harbour, seeing the ballast pier, asked what that was. On being told, he said, " You have spoiled your harbour," and to some extent this has been the case.

In an ecclesiastical point of view, the period we are

now reviewing, requires special notice, as that of the commencement of those religious divisions, for which the county has been since somewhat noted. The first ministers of the county were from what was then known as the Antiburgher branch of the Secession, but in teaching their people, they never introduced the peculiarities which divided Presbyterians in Scotland. The large majority of their original hearers were from the Established Church of Scotland, but they were glad to get the gospel, and, served as they were by men of superior powers, who cheerfully endured toil and privation for their spiritual good, raising no question as to Establishment or Secession, they not only fell in with their ministry, but became devotedly attached to them. There was thus entire harmony throughout the county, except as here and there opposition might be raised to a minister, by an individual of a litigious disposition. Afterward when every settlement was disturbed by strife, those who could remember this period, often looked back upon it with fond regret.

At the same time, there was no union among the Presbyterians throughout the Province. When Dr. McGregor arrived, like so many Scotchmen since, he thought that the difference which separated them in the old country, should be maintained here, and refused to unite with the Presbytery of Truro. " Taught by experience," he says, " that the peculiar rules of church communion observed in Scotland could not apply here, they offered to me the right hand of fellowship, which I, destitute of their teaching, did not accept." This want of union at first did little positive harm, as the congregations were separated, sometimes by wide tracts of wilderness, and there being little intercourse between them. In no case were two congregations maintained in the same place, or a congregation split in two, by any question which divided Presbyterians in the mother country. But as intercourse increased, the inconsistency of their position became apparent. The members of their congregations passed from one to the other, and were received without question and without scruple, and yet the ministers remained apart; and thus too, although the ministers were personally friendly, there was lost the benefit of united action.

They had sometimes met to consult on matters of common interest, and to some extent co-operated in promoting the Redeemer's Kingdom. But now the state of matters pressed itself upon their attention, so that after mutual intercourse and consultation, it was resolved to form a union, on the simple basis of the Westminster Confession of Faith, leaving all the questions, which divided Presbyterians in Scotland, as matters of forbearance. One measure, which at this time tended to bring this about was the Collegiate Institution, at this time' projected in Pictou. The greatness of the undertaking in their circumstances, and yet the pressing call for such an institution, in consequence of the deficient supply of ministers from abroad, rendered combined action necessary, to its successful establishment and maintenance.

The union was accordingly consummated in July 1817, the name adopted for the united body being, " the Presbyterian Church of Nova Scotia," and caused great joy. It embraced all the Presbyterian ministers in the Province, including two or three originally from the Church of Scotland, with the exception of Rev. A. Gray of Halifax, the constitution of whose congregation prevented his joining, but who remained on friendly terms with its ministers, and co-operated with them in their work as long as he lived.

This was the first of the Presbyterian unions, and was on a liberal basis. Its immediate results were happy. It was a leading cause of the union, which was accomplished in Scotland, between the two branches of the Secession in 1820, and at home, the Synod addressed itself energetically to its work. But looking upon it with the light, which time throws upon events, we can now see that the good men who accomplished it, were simply at least sixty years before their age, for in this 1877, we cannot enjoy the general union at which they aimed, and which they fondly hoped they had achieved. Already a cloud, seemingly no bigger than a man's hand, appeared on the horizon, and soon the commencement of strife proved as the letting out of waters.

We have already mentioned the commencement of party division in Pictou after the election of 1799. But from an early period an ecclesiastical element mingled with the personal and political feelings then excited. Mortimer was most friendly with the Secession ministers, while Wallace and the official party regarded a dissenter as a rebel, or worse, if such could be. Any man therefore who took offence at his minister became the friend of Wallace, and any one opposed to Mortimer was apt to quarrel with the church. Thus the two elements became mixed, and a party gradually sprang up opposed to the leading men both in Church and State.

In the years that followed, as we have seen, there was a large influx of settlers, from the Highlands and Islands of Scotland, very ignorant, only a minority able to read, yet like most people coming from the old country then and long after, looking with great contempt on ministers and every thing else in America, and this in proportion to their ignorance. In the Highlands, the Secession church was known only by report and that unfavourable ; and while, with few exceptions, the old settlers, who knew by experience the labours of their first ministers, and had a grateful recollection of the manner in which they had shared their privations, warmly adhered to them, the new comers began to decry them, as not preaching the gospel and to clamour for ministers of the Church of Scotland. The payment of stipend was to them a grievance previously unknown, and they regarded it as one of the glories of the Establishment to be free from it, and they expected by getting ministers of that body to enjoy the same immunities here. Those immigrants had now filled up the back settlements, so that the ministers here were unable properly to supply them with ministerial service. But knowing the natural prejudices of these people and being anxious to obtain for them ministers to their liking, and at the same time having learned to disregard the distinctions among the Presbyterians in Scotland, if they could obtain men of the right stamp, they applied to the leading ministers of the church of Scotland in the Highlands, such as Dr. Stewart, of Dingwall and McIntosh of Tain, to obtain ministers of that body to supply the wants of the settlers, still desiring and hoping to keep the Presbyterians here together as one body. These men fully approved of the union, and were anxious to meet the wishes of Dr. McGregor and his friends, but after a good deal of enquiry, they were obliged to write, that they could not get men to come, upon whom they could depend.

Just at this time others arrived, who adopted a different policy. The first minister of the Church of Scotland who remained in the county was the Rev. Donald A. Fraser, who arrived here in the year 1817. He was a native of the Island of Mull, of which his father was the parish minister. Being from the Church of Scotland, he was eagerly laid hold of by those who had been dissatisfied with the ministers here. Soon after he settled at McLennans Mountain, where there were at that time about forty families. There the next year a frame church, capable of seating 500 persons, was erected, and alongside of it, a log house for himself and his wife. This was the first church in the county, built in connection with the Church of Scotland, and we may say in the Province, for although there have been some others older, they were not originally built in that connection. A year later, another was built on Frasers Mountain, about six miles distant from that on McLennans Mountain, and two from New Glasgow, which then could scarcely be said to exist, and Mr. Fraser preached at these places alternately, giving also some supply to Blue Mountain, and preaching occasionally in other places, where parties were forming in connection with the Church of Scotland. There were at first only twenty-five families at Frasers Mountain, but they became the nucleus of the congregation of St. Andrews, New Glasgow. In the year 1828, the church was hauled down there, and placed on the lot on which their present place of worship stands.

But a person who at this time made more disturbance and excitement was Norman McLeod, who arrived in Pictou about the year 1818. He was not only not connected with any religious body, but denounced them all, even going so far as to say there was not a minister of Christ in the whole establishment. Those who have heard him at this time, describe his preaching as consisting of torrents of abuse against all religious bodies, and even against individuals, the like of which they had never heard, and which were perfectly indescribable. He had never been licensed or ordained, but regarded himself as under higher influences than the ministers of any church. "I am so full of the Holy Ghost, that my coat will not button on me," he said once in a sermon, as he made the attempt to bring the two sides together in front.*

But though so wildly fanatical, he was a man of great power, and gained an influence over a large portion of the Highlanders, such as no man in the county possessed. As Dr. McGregor said, "he will get three hearers to Mr.

---

* He did not seem to be always so favored. A gentleman told me that on one occasion he went to where he was preaching in a barn. As he passed the open barn door, McLeod stopped and said, "as soon as I saw that man, the Spirit refused me utterance."

Fraser's one, and the people will go much further to hear him, than any minister in Pictou." He took up his residence at Middle River, and the people of the upper part of the river, Lairg and neighborhood, who had hitherto been under the ministry of Mr. Ross, generally followed him, so that the latter relinquished to him his church at Middle River, which we may remark stood at Mr. Kerr' intervale. But his influence extended to many in almost every part of the county, and by his followers he was regarded with unbounded devotion.

After a time, however, a number became dissatisfied, when they found that he would not give them baptism for their children. Indeed during his lifetime, he found very few whom he considered qualified to receive the ordinance, and we are not certain if he found any to whom he would administer the Lord's Supper. He then induced a number of those over whom he retained his influence, to emigrate, and for this purpose to build a vessel at Middle River Point, which he called the Ark. In this they left, and afterwards formed the settlement of St. Anns, in Cape Breton.* Many in the county still remained his attached adherents, and were usually known as Normanites, and almost as long as he remained in the Province, when he visited Pictou they attended him wherever he went. It is but just to say, that these were regarded as among the most moral and religious of our Highland population.

In the year 1824, the Rev. Kenneth John McKenzie arrived in Pictou, and commenced his ministry among the

---

* At St. Anns he labored for many years, maintaining an unbounded sway over his adherents, which was used in favor of temperance and sound morality, but also we must say in nurturing a fanatical Pharisaism. He published a volume of some size, styled Normanism, besides minor publications. When an old man, he induced a number of his people again to emigrate, and for this purpose to build a vessel. In this they proceeded to Australia, and thence to New Zealand, where he died.

adherents of the Kirk in town. He was a native of Stornoway, and a man of superior talents. He at first preached in the court-house, but that summer, St. Andrews church was begun, and as soon as the outside was finished, service was held in it, the audience being seated on rude benches.

Up till this time, Mr. Fraser had been on friendly terms with Dr. McGregor, and though the spirit of contention had been rising, it was still hoped that permanent division would have been avoided. But from this time he broke off all association with him, refusing even the hospitalities of his house, at which he had been a frequent visitor. Thenceforward he and Mr. McKenzie devoted their energies to completing the work of division. As the people had been hitherto under the pastoral care of ministers of the Secession, this carried strife into every part of the county, and often into families. From the position of parties in the old country at that time, this division was probably unavoidable, but from the manner in which it originated here, and other circumstances, the feelings excited were very violent, and the results deplorable.

Other ministers of the body followed. The people of Gairloch and Saltsprings obtained the Rev. Hugh McLeod. Highlanders can stand a good deal in their minister, if he be of the true Church of Scotland, but he was more than they could stand, and after a few years they got rid of him, at the expense of a lawsuit. He went to Demerara, where he died, on the 10th May, 1832. In the year 1827, the Rev. John McRae, became minister of those who adhered to the Church of Scotland, at the Upper Settlement of the East River, where he continued till the disruption, when he returned to Scotland. The Rev. Dugald McKeichan, settled in Barneys River the same year, but only remained there three years. He returned in 1840, but also went to Scotland at the disruption. Rev.

Mr. McCaulay was the first minister of Rogers Hill. He is said to have been a relation of the historian. He removed to Prince Edward Island, where he relinquished the ministry, and was for years one of the ablest members of the Legislature

CHAPTER XVI.

DR. McCULLOCH AND THE PICTOU ACADEMY.

We must now turn back, to give an account of the efforts of Dr. McCulloch on behalf of education, and of the discussions, political and ecclesiastical, connected with the subject, which at one time occupied so prominent a place in the history of the county.

The want of ministers to occupy the numerous destitute settlements of the Province, had from an early period engaged the attention of those already in the field, and they sent urgent appeals to the bodies in Scotland, from which they had come, for additional laborers. The supplies thus received, however, were always irregular and inadequate, and hence was almost forced upon their attention, the question of the possibility of training young men for the ministry in this country. Under the influence of such considerations, Dr. McCulloch, as we have mentioned, as early as the year 1805, only two years after his arrival, projected an institution for the purpose of giving instruction in the higher branches of education—which would thus meet the object of the Presbytery, by giving young men desirous of entering the Gospel ministry, that literary and scientific culture, which the Presbyterian

Church has sought in its ministers, and at the same time afford the benefit of liberal studies to all who chose to avail themselves of them.

In that year, a society was formed for the establishment of such an institution, and subscriptions were taken throughout the district. The following is a copy of the heading of the one on the East River, the others, we presume, being in the same terms :—

"We, the subscribers, hereby declare our approbation of the Society formed in Pictou, for establishing a college of learning in this district. We are persuaded that such an institution would have a powerful influence to promote the interests of society, both by disseminating general knowledge, correcting the vices of youth, and instilling into their minds the principles of virtue. That this design may therefore be carried into effect, we bind and oblige ourselves, our heirs, executors, administrators and assigns, to pay the sums respectively subscribed by us, one-third part on the first Tuesday of May, 1806, another third part on the first Tuesday of May, 1807, and the remaining third part on the first Tuesday of May, 1808, to such person or persons as the society shall appoint for transacting their business."

This was headed by Dr. McGregor, with a subscription of £20, " provided the Harbour congregation pay me the sixteen pounds which they owe me." Others follow with subscriptions of £10, the whole amount in that settlement being £125, ($500). Writing at this time, Dr. McGregor says " The increasing demand for ministers seems to intimate the necessity of raising them in this country. The great expense of everything here renders this undertaking next to hopeless in our circumstances, yet Mr. McCulloch, who started the idea, has sanguine hopes. Pictou people have subscribed about £1,000, a more liberal subscription than they are well able to pay. We expect some money from the Province Treasury if we give our seminary a little name, as not rivalling the University, which Government has established. We expect great assistance from Britain and Ireland. We intend to send Mr. McCulloch home to beg."

The project was not carried into execution at that time. As far as obtaining Provincial aid, or even the legislation necessary for establishing such an institution, with Wentworth governor and Wallace at his back, we suspect that any expectations were found hopeless. Such a scheme would only appear to them as favoring a nest of pestilent disloyalty, which ought to be crushed as the serpent's brood. The country too was not in a state to support such a measure.

The idea however was not lost sight of, and with a view to its ultimate realization, the ministers took charge of promising young men, to whom they gave instruction, in the way of preparing them for entering such an institution, and at the same time raised funds to aid in supporting them. In the year 1814, we find Mr. Ross teaching five boys Latin and Greek, with a view to the ministry. Dr. McGregor also did something in the same way. In the meantime Dr. McCulloch, partly to improve his circumstances, for like most of the ministers of that period, he was imperfectly supported by his congregation, and partly with the view of raising the standard of education in the district, about this time opened a school of a higher class; and when in 1811, the Government passed an Act granting £100 per annum for a Grammar School in each county, and in the districts of Colchester, Pictou and Yarmouth, he obtained the grant for the one under his charge, and held it for a number of years. The building in which he taught, stood nearly opposite his gate on the old road out of town to the west. We may observe here that it continued to be used in the same way, till the winter of 1824, when it was hauled down to the lot, on which the engine house is now built. The ground was boggy, and it was placed on a foundation built up of squared logs, a few feet above the ground. After this it still continued to be used as a Grammar School, constantly till the year 1832, and again some years later.

The number and progress of the young men attending this institution, and studying in other quarters, revived the idea of a college ; and accordingly under the leadership of Dr. McCulloch, and the cordial approval of the Governor having been first obtained, a Society was formed for the establishment of such a seminary on a liberal basis. As Mortimer was then a power in the Legislature, and Sir John Coape Sherbrooke, the Governor, the most independent ruler that was ever at the head of our affairs, success was confidently anticipated. Accordingly on the petition of Dr. McGregor and others, both in Pictou and elsewhere, an act of incorporation was granted to the trustees in the year 1816.

At this time we should observe, that the only institution in the Province at that time for the higher education was the college at Windsor. It was established by an Act of the Legislature of Nova Scotia, about the year 1790, which at the same time provided £400 sterling a year (as currency then was—£444, or $1,776,) permanently for its support. The only restriction in the Act was that the president should be in holy orders in the Church of England. Subsequently a royal charter was obtained, by which the institution was designated "King's College," and the governors thereof authorized to pass statutes or by-laws for its government, which they did in reality. One of them ran thus:

"No member of the University shall frequent the Romish Mass, or the meeting-houses of Presbyterians, Baptists or Methodists, or the conventicles or places of worship of any other dissenters from the Church of England, or where divine service shall not be performed according to the liturgy of the Church of England, or shall be present at any seditious or rebellious meetings."

Independent of the bigotry of this, the conjunction of the first and last clauses is expressive. But another ran thus:

"No degree shall be conferred, till the candidate shall have taken the oaths of allegiance, supremacy and obedience to the statutes of the University; and shall have subscribed the thirty-nine articles of the Church of England, and the three articles contained in the thirty-ninth canon of the Synod of London, held in the year of our Lord, 1603."

The institution was modelled on the plan of the University of Oxford. The students were obliged, at a heavy expense, to reside within its walls, and its whole management was such, as would have excluded the great majority of the youth of the Province, even had its statutes been more liberal.*

It is *now* said that these statutes never were approved of by the Archbishop of Canterbury, who on the contrary expressed his disapproval of thus limiting to a small portion of the community, the benefits of an institution established by Government for the benefit of all, and was even determined to expunge the obnoxious laws; but that part of the trustees prevented the alteration. But these were the dominant party of the Institution, and these statements only present in a stronger light the bigotry of those, who in spite of representations from such a quarter, retained such regulations. Nothing was said at this time by the friends of the Institution about there being anything wrong about these statutes. For years after the Pictou Academy began, they were maintained in full force. Charles R. Fairbanks, one of the most brilliant public men that ever graced the Legislature of Nova Scotia, stated, in one of the debates on the Pictou Academy, that he had been educated at Kings College, but because he could not swallow the tests, he had been refused a degree. We may add here, that while thus restricted to about a fifth of the population, it had been receiving, besides the grant of £444 from the Legislature of Nova Scotia, £1000 sterling from the British Government annually since the year 1802.

With these arrangements the people were not satisfied, but scattered as they were, the majority struggling for the necessaries of life, and few of them thinking of

---

* T. C. Haliburton, in one of his speeches on the Pictou Academy, as reported, said that it cost a young man £120 per annum to live at Windsor College, and only £20 at Pictou.

collegiate education for their children, little had been said, and nothing had been done, to effect a change. But the proposal to establish an institution on a liberal basis, was generally hailed with satisfaction. The bill for the incorporation of the trustees, was introduced into the House by Mr. Chipman, a Baptist, and seconded by Mr. Wells, and passed unanimously. We think it worth while giving the names of the original trustees. They were, Edward Mortimer, Revs. Duncan Ross and Thomas McCulloch, Thomas Davidson, George Smith, Robert Lowden, Revs. William Patrick, James McGregor, Archibald Gray, and James Robson, S. G. W. Archibald, and James Foreman.

The intention was to found an institution specially for Dissenters, not indeed excluding Churchmen, but as Kings College was entirely under the control of the latter, it was expected that only the former would take advantage of the new institution, or combine in its support. Still they wished it equally free to all, and the act of incorporation was introduced into the Assembly and passed there without any tests whatever.

But the leaders of the Church of England, who were then dominant in the Council, took alarm at the idea of such an institution, which they judged would form a rallying point for Dissenters against the Church. They were willing, or at least the liberal minded among them were, to allow Presbyterians to have an institution, in which they might give their children such education as they could, but they feared the establishment of a college, which, combining Dissenters in its support, might become the successful rival of Kings. In consequence of this, when the bill was introduced into the Upper House, they introduced a series of tests of a very offensive and vexatious character. Every new trustee was to be either a member of the Church of England, or of the Presbyterian religion (not church, for that title was not conceded to such a body), and on his election, a majority of the trustees present must sign a formal certificate to that effect, and forward it to the Lieutenant Governor. He must also appear before the Supreme Court, and, if not of the Church of England, make the following declaration : " I, A. B., appointed one of the Trustees of the Pictou Academy, do declare that I do profess the Presbyterian religion, as the same is declared in the Westminster Confession of Faith." Until he did this, he could not legally act as trustee. Moreover, he was required to do this every three years, or his office became vacant. The same tests were to be applied to every person appointed a teacher, and he also was required to appear before the Supreme Court, and make a similar declaration. The trustees were also prevented from holding any property outside the District of Pictou.

The House of Assembly were obliged either to submit to these amendments, or lose the bill, and they reluctantly agreed to them. This act, to which a suspending clause had been appended, afterwards received the sanction of the Prince Regent, and become law. By this act the Trustees were empowered to pass by-laws and fill up vacancies in the board, subject to the approval of the Governor for the time being. Sometime afterward a charter of incorporation, in pursuance of the act, and under the great seal of the Province, passed to the Trustees.

It should have been mentioned, that to avoid exciting the jealousy of the friends of Kings College, who were really all powerful in the Government, it was resolved not to seek the right of conferring degrees or the other privileges of a college. Hence the name Pictou Academy, though from the first it was intended to impart the education usual in colleges.

To establish such an institution under the circumstances, was a task simply herculean. A large portion of the population in the rural districts were still struggling with the difficulties of a first settlement, and as to education, few thought of seeking for their children more than the ordinary training of a common school. Even that in many places was difficult to obtain, and when obtained very inferior. A large proportion of the inhabitants, did not feel the necessity of any thing better, and many did not value education at all. The population was sparse, and the several portions had but little communication with one another or with the capital. The tests introduced by the council threw the institution into the hands of the Presbyterians, and as they then consisted only of about twenty congregations, most of these in thinly settled districts and the members in humble circumstances, it will be perceived, that the Dr. had entered upon an undertaking, requiring a large amount of that faith, which can remove mountains.

Nevertheless the trustees addressed themselves to their work with great energy. They immediately proceeded to raise money by subscription, beginning with about £400 among themselves, for the purchase of land and the erection of a suitable building. In this way they collected about £1,000 ($4,000) a large sum under the circumstances. The following is the heading of the list :

"We, the subscribers, desirous of affording our concurrence and assistance to the society formed in Pictou, for providing the means of instruction in the branches of a liberal education, which are not taught in the Provincial Grammar Schools, hereby bind and oblige ourselves, our heirs and assigns to pay to the Treasurer of the society for the time being, the sums annexed to our respective signatures, the same to be paid when the society shall judge, that a sum has been subscribed sufficient to enter upon the execution of the said plan."

This is commenced by Mortimer, with a subscription of £100, who is followed by the three ministers, Messrs. McGregor, Ross and McCulloch, for sums of £50 each. Altogether in the town £628 was subscribed.

Dr. McCulloch was chosen its first president, and before the building was erected, teaching began. The first classes were opened, as near as we can ascertain, in the fall of 1817. A room was fitted up in one end of the house, in which the late Peter Crerar, Esq., resided, the other being occupied by the Rev. John McKinlay. Here plain pine desks were erected, so shaky, that on one occasion a Highland student, intent on taking notes, found it so difficult under the movements of his fellow-students, that, his patience being exhausted, he exclaimed, " Please master, they're shaking the dask on me." In this fashion was begun the first attempt at free liberal education in these Provinces. We give below a list of the first students.*

Soon after the building was completed, and the classes were transferred to it. After the practice of some of the Scottish Universities, students were now required to wear red gowns. These were made of light merino, and for the next twenty years these bright scarlet insignia of learning were one of the features of our town, reminding the Scotchman of the ancient seats of learning of his native land.

From this time, Dr. McCulloch's life was devoted to the interests of the institution. The largest part of the teaching devolved on him, and that under the most unfavourable circumstances. The late Jotham Blanchard thus wrote of his efforts, during the infancy of the institution :

" Of his daily labours and nightly vigils, after taking charge of the Institution, I am surely a competent witness. I was one of his first students, and have often seen him, at 8 o'clock of a winter morning, enter his desk in a state of exhaustion, which too plainly showed the labours of

---

* List of first students at Pictou Academy :—R. S. Patterson, John McLean, John L. Murdoch, Angus McGillivray, Hugh Ross, Hugh Dunbar, James McGregor, Michael McCulloch, Charles Fraser, Benjamin Dickson, William Dickson, David Fraser, Edward Harris, Jotham Blanchard, Thomas Forman, ——— Forman, Charles Archibald, David Sawers, John J. Sawyer, Duncan McDonald, John McDonald, Hugh Fraser, Archibald Patterson.

Perhaps these were not all present the first term, but they were in attendance with the first class, who passed through the Institution.

the night. To this those who are acquainted with the subject will give credence, when I state that his share of the course was, besides Greek and Hebrew, Logic, Moral Philosophy and Natural Philosophy. In each of these sciences, he drew out a system for himself, which was of course the results of much reading and much thought. When I add to this account of his daily labours, the repairs and additions which were necessary to a half-worn apparatus, and which none but himself could make, I am almost afraid my testimony will be doubted. And for the first five or six years of the institution, let it be remembered, he had charge of a congregation, and regularly preached twice a day, save when over-exertion ended in sickness."

His first co-laborer was the Rev. John McKinlay. He was a native of Stirlingshire, Scotland, who came to this country in the summer of 1817. Dr. McCulloch having given up the Grammar School, to take charge of the Academy, Mr. McKinlay succeeded him in the former. Teaching in it part of the day, and aided there by an assistant, he also became teacher of classics and mathematics in the Academy, a position for which he was well qualified, by the accuracy as well as the extent of his scholarship. He continued to hold this position till the year 1824, when Dr. McCulloch having resigned the charge of the congregation at Pictou, he was ordained as his successor on the 11th of August, 1824, and was succeeded in the Academy by Mr. Michael McCulloch, who had previously been the second teacher in the Grammar School.

But during the whole existence of the institution, Dr. McCulloch was its life and soul. As long as he continued in connection with it, he taught logic, moral and natural philosophy. Divers as were the branches devolving upon him, he taught them all efficiently. I have since had an opportunity of knowing something of the professors in

Edinburgh University, but never till I saw them did I know the greatness of Dr. McCulloch. We doubt not every professor there would have excelled him in his own particular field, but I believe there was no man in that institution, who could have made the same appearance in all the branches taught that he did. The same view was expressed to me by Dr. Dawson. He had a multifarious learning, so that he might be regarded as a whole *senatus academicus*. He could have taken any branch included in the faculties of Arts and Theology, and taught it in a respectable and efficient manner. I may add, that his intellect was of that peculiar clearness, that whatever he knew, he knew accurately and distinctly.

The teaching of the branches named however, was only a small part of the work which devolved upon him. Besides the charge of a congregation till the year 1824, he took an active part in the business of the Synod of the Presbyterian Church of Nova Scotia, and most of the public documents of the body came from his pen. As soon as the first class of students was sufficiently advanced, he was requested by the Synod to take charge of their studies in theology. To his other labors was added the instructing of these young men in Hebrew and theology. We may add here that he was a superior Hebrew scholar, and as such almost entirely self taught.

But his labours were chiefly increased by the opposition which the institution met with. This, as forming an important chapter in the history of Nova Scotia, as well as of the county, we must now notice. The trustees finding the amount insufficient to complete the building and provide other necessaries, in 1818 petitioned Lord Dalhousie, then Governor, " to recommend a grant of money from the public funds of the Province to assist them in erecting a suitable building, or for such other purposes as might be necessary in establishing said

Academy." His Lordship recommended the object to the Assembly by the following message :—

"The institution of an academy at Pictou, appears to me to promise advantages of education, highly favorable to the whole eastern part of this Province, and I therefore recommend the accompanying petition of the trustees of that academy to your favorable consideration."

Upon this message, the House, with only four dissentients, passed a resolution for £500 to the trustees, to be drawn for, as soon as they had expended £1,000 from private subscription; to this resolution the Council refused concurrence.

In 1819 a similar vote passed the House, and was sanctioned in Council. The trustees continued to make an annual application for money, and during the next four years—1820, 1821, 1822 and 1823—they received in all from the public funds £1,300.

For several reasons, the trustees about this time began to fear that the death of friends in the Council, and the increase of an influence in that body, which had always been opposed to the institution, might at some period deprive them of public support, and the possibility of this event they found injuriously to affect their arrangements. They therefore petitioned for a permanent endowment, and the Representative Branch, without a division, passed one to the extent of £400 a year. This bill the Council rejected.

In 1824 the Assembly passed another and similar bill, which was also lost in the Council. A vote of £400 for that year was then passed in the Lower and agreed to in the Upper House.

The Academy had now proved itself by its work. Several young men had completed their studies, and were coming forward to take their places in various professions, with good promise of usefulness. In that year, seven young men, having completed a course of study for the ministry in connection with the Presbyterian Church, were licensed to preach the gospel. As these were the first native preachers ever sent forth by that body, in any of these Provinces, and as they were all brought up in Pictou, we may give their names. They were John L. Murdoch, John McLean, R. S. Patterson, Angus McGillivray, Hugh Ross, Hugh Dunbar and Duncan McDonald. That autumn, three of these, viz. : Messrs. McLean, Murdoch and Patterson proceeded to Scotland, and preached with acceptance in the pulpits of various Dissenting ministers ; and having undergone an examination by Professors Walker, Sandford, Jardine, Miller, Milne and Meikleham, professors in the University of Glasgow, as to their scholastic attainments, they received the degree of A. M. from that institution.

The trustees of the Pictou Academy now felt, that they were entitled to appeal to the Legislature, on behalf of the institution, as no longer an experiment, but as having its character established. They accordingly in 1825, petitioned the Legislature for the removal of tests, for an enlargement of their powers and for a permanent grant. We have not a copy of the petition, but presume the enlargement of powers, meant the right of conferring degrees. The petition was referred in the Assembly to a committee, of which Charles R. Fairbanks, Esq., then Solicitor General, was chairman, which reported as follows :

" The Committee are of opinion that the Pictou Academy is a highly useful Institution, conducted on an excellent system, that of the Scotch Universities, and peculiarly adapted to meet the wants, and accords with the sentiments of the majority of the Province in regard to the higher branches of education. That its establishment and support has been and will continue to be a favorite object with the greater part of the Dissenters in the Province, on account of its total exemption from any disqualifications to students, originating in religious distinctions, and for the careful attention, which its conductors have manifested for the morals of those who attend it. That the attainment of a sound classical education, and of a competent knowledge of the other branches of science, commonly taught in the higher schools, is brought down to the means and ability of those, who, if the Academy did not exist, would be wholly unable to provide these advantages for their children. And lastly, that the Institution possesses decided advantages, in many respects to those students who are destined to the ministry in the Presbyterian and other Dissenting Churches, and is for this object, indispensably necessary, if these are to be supplied by the youth of the Province.

" Referring to the exclusive Scotch character of the population of the Eastern part of the Province, and to their known, and perhaps laudable, partiality and attachment to the Institutions of the country, whence they have originated, and regarding also the great and rapidly increasing population of that quarter, the Committee consider, there exists a fair claim on the part of Pictou, for support to this Academy, for which so decided an interest is there manifest, out of that General Revenue, to which they so largely contribute; and as from the evidence before them and other considerations, the Committee are obliged to believe, that this Institution will be attended by a class of persons, who, on various accounts, are, and will be, incapable of prosecuting their studies at Kings College, Windsor, or in the institution of doubtful and uncertain stability now forming in Halifax, they have deemed it their duty under the clearest convictions of the invaluable benefits, which Education confers on a country, to recommend the Pictou Academy to the continued support and fostering care of the General Assembly.

" And believing the honorary Collegiate Distinctions to be highly useful, as incitements to the emulation and diligence of students, and to be the means of extending the respectability, and character and influence of the institution ; while the incapacity to grant them possesses a tendency injurious, and perhaps discreditable to it, the Committee cannot perceive any substantial reason, for refusing to allow these privileges to the Academy.

" The Committee therefore report that in their opinion, it is expedient to provide, by an act of the General Assembly, for a permanent allowance to the Trustees of the Pictou Academy, of the sum of £400 from the Treasury, and for bestowing upon it, with full exemption from all tests now required of its Trustees, the name, distinctions and privileges of a college as known in Scotland. These the Committee believe will remove all impediments to the advancement and prosperity of this Seminary, give it stability and consideration, and justify its supporters in bestowing that assistance, which the doubt of its permanence now renders it prudent to withhold."

Upon this report the House first passed a vote of £400, which received the assent of the Council. It then proceeded to pass a permanent bill for a like sum, but, after two readings, it was delayed until the next session, on account of the absence of the Governor, and the supposed want of power in the President to give his assent. This supposed want of power was simply pretence. Wallace, the Administrator of the Government, never had any scruples about want of power, when it was a question of rewarding one of his creatures. Then he could exercise the powers of his position, in a way that the Governor himself would scarcely have done ; but a measure of those Pictou Dissenters, why it was simply flaunting the red flag in his face.

In 1826, the Assembly passed another permanent bill, to which the Council refused their assent. The Assembly then appointed a committee to search the Journals of the Council, who reported, that in favor of the bill there were four, Mr. Morris, Judges Stewart and Halliburton, and the Master of the Rolls, and against it, five, the Lord Bishop, and Messrs. Wallace, Jeffrey, Binney, and Prescott. The committee also reported, that the minority had entered a protest against the dismissal of the bill. This document we must present entire. But we may mention here, that during the Session, the Assembly passed the usual vote of £400, which received the assent of the Council.

REASONS OF PROTEST BY MINORITY OF COUNCIL,
DATED 22ND MARCH, 1826.

' 1. Because we think that the Dissenters in this Province, who compose more than four-fifths of its population, have entitled themselves to the favorable consideration of the Legislature, by their orderly, steady and loyal conduct, and the cheerful support which they have so long given to His Majesty's Government in Nova Scotia.

" 2. Because we think that when £400 sterling have been annually paid, for thirty-six years past, out of the revenue of this country, for the support of a college, which confines its academical honors to members of the Established Church, who pay but one-fifth of the revenue of this country, the Dissenters, who pay the other four-fifths, are entitled to at least an equal sum to support an institution in which their children can derive the benefit of a liberal education.

" 3. Because we do not think that the objection, which has been urged to the permanent establishment of such an institution in a remote part of the Province, as Pictou has been termed, ought to have any weight when the general wishes of the Dissenters have been expressed, by their Representatives in three successive Sessions in the House of Assembly, in favor of that situation, where the great body of the Dissenters reside, and where, out of a population of 12,000 persons, not 100 members of the Established Church could be found.

"4. Because we think the Bill, which His Majesty's Council have now determined to reject, is free from the objections to which the other Bills for endowing the Pictou Academy were liable, as the Institution is by this Bill placed sufficiently under the control of the Government, by empowering the Governor to nominate so large a portion of the trustees, and thereby securing the Province against the future introduction of teachers into that seminary, whose principles might be inimical to our political institutions.

"5. Because we are convinced that the public feeling, which has been so strongly expressed in favor of the Pictou Academy, will still continue to manifest itself, and defeat all the efforts of its opponents to destroy the institution; it will therefore continue to exist, notwithstanding the rejection of the present Bill,—but Government will not have that salutary influence over it which it would acquire if this Bill were passed into a law.

"6. Because, as members of the Established Church, we feel that the best interests of that Church will be consulted by manifesting a spirit of liberality to our fellow christians who dissent from us,—that even policy, independent of higher motives, dictates to us as a minority, the advantages of conciliating the Dissenters, and showing to them that we feel that the Church of England has nothing to fear from the diffusion of knowledge.

"7. Because we value highly that harmony and good understanding, which, without the compromise of principle, has so long prevailed among Christians of all denominations in this Province; and we fear, that the rejection of this Bill, while the annual allowance to the College at Windsor is continued, will excite a spirit of hostility to the Established Church among the Dissenters, which will seriously disturb the peace of the country, as upwards of 30 years experience has convinced all of us, who enter this Protest, that every attempt to give or retain exclusive privileges to the Church of England, has invariably operated to its disadvantage. If the clergymen of that church will exert themselves with tempered zeal, the purity of its precepts, the beauty of its liturgy, and the liberality of its sentiments, will insure its extension among the people of this Province, unless their feelings are so roused against it by any injudicious measures, on the part of the Government or the Legislature, to give to it advantages, to which so large a portion of the population think that it is not entitled.

CHARLES MORRIS,
JAMES STEWART,
BRENTON HALLIBURTON,
S. B. ROBIE.

We must now, however, refer to the nature and source of the opposition, which the institution encountered during these and subsequent years. This will appear, in part, from the preceding document; but we must explain, farther, that the Church of England Bishop—the second Bishop Inglis—not only had a seat in the Council, but was one of its most active members, not merely in matters which might interest him as a churchman, but in public affairs generally. Having often the ear of governors, he was a power behind the throne. He was a man of ability and an astute politician, but specially an able and persevering worker for the Church of England. None would have had any right to object to this, had his efforts been regulated by a due regard to the rights of others. But he was trained in the most intolerant school; and instead of relying for the progress of his church on the influence of her principles and practice, and the zeal and piety of her ministers and members, he devoted a large share of his energies to maintaining her exclusive privileges and political supremacy, as an established church.

The maintenance of Windsor College was essential to his object. Had he sought this without seeking any exclusive rights, it would never have met with a word of objection; but, notwithstanding that four-fifths of the population were excluded from its benefits, there was yet no word of removing the tests. As for giving them any share in the management of an institution established by Government for the benefit of all, why such a thought could scarcely ever be supposed to enter his mind ! / And now, when they dared to establish such an institution for themselves, his jealousy was excited against it, not only as likely to foster the evils of dissent, but as likely to form a rival to Kings College. His views are at least partially revealed, in the following extract of a letter published in the report of the Society for the Propagation of the Gospel, for the year 1823:—

" At Pictou an Academy, or college as it is called, has been built, at which there are now about twenty students. Much pains have been taken to make it attractive by its Philosophical Apparatus and lectures in the sciences, and the residence is agreeable to the students, as they lodge in private houses, at moderate expense and free from restraints. The Institution owes its rise partly to the difficulties and embarrassments, which have oppressed Kings College, Windsor, and partly to the zeal of the Presbyterian ministers, who have the sole charge of it. It is supported chiefly by an annual grant from the Provincial Legislature, and is *likely to rise or decay as the College at Windsor is depressed or advanced.*"

In these days of religious equality, and of good feeling among religious denominations, it is scarcely possible to realize the state of matters which existed then, the inferior position of Dissenters, the prejudices with which they were regarded even by sensible men, and the difficulties therefore which were thrown in the Doctor's way, at every stage of his efforts on behalf of the Pictou Academy. Many believed that a Dissenter must necessarily be disloyal. In the year 1809, on the visit of Sir George Prevost, the Governor, to Pictou, some parties made representations to him regarding the Doctor's loyalty, in consequence of which he felt it worth while to send to Government a lengthy defence, with a certificate signed by Hugh Dunoon and the other magistrates, that he regularly prayed for the King. In one of the debates of the Assembly, R. J. Uniacke, Jr., on some report he had heard of the principles of the Antiburghers, said that they ought to be looked after. When Wallace was administering the Government, representations were forwarded to him, accusing the trustees of disloyal principles. When they applied to him for a copy of the charges, he acknowledged having received such a paper, but refused to give a copy of it. This state of things as affecting the Academy, is thus described by T. C. Haliburton in one of his speeches :—

" There is much to regret, sir, in the state of public affairs in this province, and there are few colonies which present such a singular spectacle. There are a few individuals in Halifax, who direct public opinion, and who not only influence but control all publc measures. Seated in the capital, they govern the movements of all the different parts ; as they touch the springs the wires move, and simultaneously arise the puppets in the different counties and towns, play the part assigned to them, and re-echo the sounds which have been breathed into them. The smiles of episcopacy, the frowns of the treasury, and the patronage of official interest, have a powerful effect, when brought to bear upon any one object. There is also a wide difference between the success of any measure, when called for by the people, and when advocated by this party. Any project however absurd or extravagant, when required by the latter, to be carried into effect, has friends without number, but if the people solicit, it is viewed with caution ; you hear it whispered on all sides, it will offend such a person, it will not be acceptable in a certain quarter, and you are advised to be silent, as it may affect your personal interests, or draw down upon you a displeasure, which may retard your own advancement. The war cry of church and state has been raised against this persecuted institution, and it is said on all sides, it will militate against the interests of the Established Church and of Kings College at Windsor.

" I am a member of the Church of England, and admire and revere it ; I shall continue so, and though I disapprove of the intemperate zeal of some of its friends, I shall live and die a member of that church. I have also the honour of being a graduate of Kings College, and am a warm friend of that invaluable establishment. As such, sir, if there were any prejudices among the members of either, against the Pictou Academy, because it is the resort of children of dissenters, or if it was viewed by those with distrust, as a sectarian institution, I ought to know something of those prejudices. It is the misfortune of the church, and we all deeply lament it, that one or two unworthy members of it, have sought promotion through the paths of slander, and political intrigue, and have constantly represented Dissenters as disloyal and disaffected people. The value of these gentlemen has unfortunately been estimated on the other side of the water by their zeal ; and as they have uniformly reported sectarianism, as they are pleased to call it, synonymous with revolt and rebellion, the dependence of the colony has been absurdly thought, to be alone supported by these staunch friends ; and honor and promotion await their laudable exertions.

" I will never consent that this seminary of education for Dissenters, shall be crushed to gratify the bigotry of a few individuals in this town, who have originated, fostered, and supported, all the opposition to Pictou Academy. I do not mean to say, that they directly influence those gentlemen in this house, who oppose the bill, but their influence reaches to people who are not conscious of it themselves. They are in a situation to give a tone to public opinion ; few men take the trouble of forming just conclusions on any subject but adopt the sentiments of those, whose judgments they respect. In this manner they hint, ' ambitious Scotchmen at Pictou,' ' sour sectarians,' ' disloyal people,' ' opposed to church and state,' their hints circulate from one to another, men hear it, they know not where, adopt it, they know not how ; and finally give it as their own opinion ; until you find honest and honourable men, as you have heard to-day, pronouncing a judgment, evidently tinctured by the breath of poison, which they themselves are wholly unconscious of having inhaled.

"In one of the reports made to the Society for the Propagation of the Gospel in foreign parts, we find the following eulogium on Dissenters : 'It can be clearly substantiated, that in exact proportion to the influence of the established religion, will be the immovable loyalty of the inhabitants of the province.'—It would be difficult to find in public annals, such another abominable libel on Dissenters ; it is said the person who made it, was once your chaplain. Had I been a member of the Assembly at the time, I would have moved to have him publicly censured at the bar of the house ; he deserved to have been placed in the custody of the Sergeant-at-Arms, to have been deprived of his gown, and should have been admonished to ' go and sin no more.' "

At the starting of the Pictou Academy, the Bishop and his friends had succeeded in placing such restrictions upon it, as were likely to render it a small affair. But now that in spite of this, it was proving successful, his whole influence in the Legislature was employed against it. He was sure of Wallace's help to do any thing against the wishes of those Pictou Dissenters, and with two or three placemen he was able to command a majority in the Upper House to defeat its claims. It will be seen that in that year, (1826) it was only by his casting vote that the bill was rejected.*

In the course which the Bishop and his party adopted, he was acting against the opinions of the best, the ablest, and certainly the most liberal minded members of his own church. This will appear from the foregoing protest, all

---

* It is curious to note the position of these men to the public treasury. The Bishops salary was £2000, sterling we believe, largely through the liberality of the British Government, with £150 for travelling expenses, besides other perquisites, even it was said to a share of the royalty on coal ; Jeffrey had £2000 sterling ; Binney was Collector of Excise, and Wallace Treasurer, on what were large salaries for the times. Yet these four men, receiving among them annually fifteen or twenty times the whole sum asked for advanced education for four fifths of the population, could, with the aid of one other councillor, defeat the almost unanimous wishes of the country and their representatives in this regard. An objection was even made that Dr. McCulloch was receiving two hundred pounds a year!

the signers to which belonged to that body. The same spirit was shown in the Assembly, where Episcopalians, as T. C. Halliburton, were among the most earnest advocates of the Pictou Institution. And probably a majority of that church throughout the country, would have shown the same spirit. But the Bishop and his clique were unrelenting.

In the meantime the friends of the Institution had put forth vigorous efforts for its maintenance. The Synod of the Presbyterian Church had taken up its support in earnest, and subscriptions, liberal for the times and circumstances, were made in its congregations, though always the largest amount came from the members of the body in Pictou. Though anticipating, we may say here that up to the year 1830, about £5,000 was thus raised, of which about £3,000 was expended on building, library and philosophical apparatus. Much of this was from Ladies' Penny-a-week Societies, and sometimes from men, who, not having money, brought their produce to the stores, to pay their subscriptions. The very opposition which the Institution encountered, only intensified the earnestness of its friends.

That year Dr. McCulloch visited Scotland. He addressed the Synod of the United Secession Church, which unanimously recorded it as their opinion "that the Presbyterian Church of Nova Scotia and the Pictou Institution have strong claims on the sympathy and liberality of the Presbyterian and other churches in Britain, and of associations for religious purposes, and especially of the United Secession Church." They also issued a recommendation to the congregations under their inspection to make a collection, without delay, in aid of the funds of the Pictou Academical Institution, and they appointed a committee to prepare a short statement of the claims of the Institution, to be read from the pulpit of each congregation, when the collection was intimated, and also to consider what further measures might be adopted, for promoting the interests of our sister church in Nova Scotia.

A society was formed in Glasgow, entitled "The Glasgow Society for promoting the interests of Religion and liberal Education in the North American Colonies," and including in the committee of management, several influential laymen and ministers of different denominations. The students attending the Theological Hall of the United Secession Church, pledged themselves to raise the sum of £200, and as the result of these efforts, considerable sums were remitted in subsequent years in aid of the Institution.

By these means a library, deemed respectable at the time, was collected, and a philosophical apparatus, and later a chemical apparatus, partly the Doctor's own property, the first in these Lower Provinces, were added, and in spite of adverse influences, the institution was gaining strength.

Up to this period, the opposition to the Institution had come from the leaders of a dominant church, and was so clearly the expression of narrow-minded exclusiveness— so opposed even to the sense of justice of the best members of that body, that if nothing else had interfered, it must in the progress of events have been swept away by the rising tide of public sentiment. But now its friends were to be taken in flank, and the institution to encounter an opposition more intense, and ultimately more fatal, from men of the same religious name.

We are far from desirous of reviving old quarrels, but it is necessary to our history, to give the facts of the controversy in its new phase. The founders of the institution while desirous of establishing it on a liberal basis, and to give in it such a training as would fit our youth for usefulness in any sphere, attached special importance to it, as a means of educating young men for the Gospel ministry. Indeed the difficulty of obtaining ministers to supply the destitute parts of the Province, was one and perhaps the leading cause, which led them to found such an institution. The Presbyterian Synod had taken up its support on the same ground, and with this view Dr. McGregor and the other ministers had appealed to their people, and enlisted their sympathies on its behalf.

When the ministers of the Church of Scotland commenced the movement, forming an organization in connection with that body, they found the Institution ready to send out its first company of native preachers. They came with that contempt, which it was customary then and long after, for old country people to entertain for everything colonial. The idea of training some of the natives of the backwoods for ministers, seemed to them supremely ridiculous, and when they commenced preaching they decried them in the strongest terms. They also looked upon Seceders with that disdain, with which the members of the Established Church at that time generally regarded that body. They could not see that in this country they hold no more favoured position. But they were not long in seeing that the Presbyterian Church of Nova Scotia, composed principally of Seceders, who were in the Province before them, had a firm footing in the country, and that such an institution, by providing ministers in the Province, was giving them a great advantage, an advantage, it is true, equally open to all, but of which they had no intention of availing themselves. Hence they took a position against it, as favouring the Seceders in opposition to the Kirk of Scotland.

In the year 1826, Messrs. Fraser and McKenzie held an interview with the Trustees. At this meeting they objected to the teaching of the higher branches in the Institution, and proposed substantially, that it should be converted into Grammar School. To this of course the Trustees could not accede, and Messrs. F. and McK. left, intimating that as the Trustees would not adopt their views, they would henceforth meet with their most determined opposition.

Accordingly by their exertions, for the next few years petitions were forwarded to the Legislature, signed by their adherents asking for a change in the Institution, a change which amounted to an entire destruction of it, as far as the objects of its founders were concerned. They alleged that the teaching of the higher branches was a violation of its charter, and complained that English grammar, elocution, bookkeeping, navigation, geography and the elements of the classics, were not taught in the Institution, and they asked for their introduction.

To this it was replied, that the Institution had never been intended to teach these branches. It was shown that the original petition, on which the charter was granted, was distinctly for the establishment of an institution for teaching the branches not taught in the Grammar Schools of the Province—that the charter was given with this view —that the subscription list was in the same terms, and the money raised expressly for this purpose—that its bylaws, which defined the course of instruction, had been according to law submitted to the Governor, and approved by him—that the Legislature perfectly understood this, as for seven years after the establishment of the Institution, they had made a liberal grant to a Grammar School along side of it, and that a Committee of the Assembly, after a thorough examination, had expressed the highest approval of the system of education adopted. Therefore to employ the funds of the Institution in teaching the branches of an ordinary English education, was destroying the Institution, as far as its original purpose was concerned, and was entirely unnecessary as there was a Grammar School, in which these branches were efficiently taught within a few rods of the building.

It was objected farther that the Institution, in terms of

the charter, ought to be under the management of persons belonging either to the Church of England, or the established Church of Scotland, but that instead of this it had with a few exceptions, fallen into the hands of Seceders from the latter, who had adopted the present course of instruction, not suited to the circumstances of the Province, to forward their favourite design of raising ministers for their own connection, and that a divinity class had been introduced avowedly for this object, which gave the Academy a sectarian appearance.

To this it was replied, that according to the charter, the Trustees must be members of the Church of England or Presbyterians—that the Trustees had taken the test once in every three years though it had been forced upon them,—that the Trustees had always been anxious to have as full a representation of different religious bodies at their Board, as in their power—that they had elected Mr. Fraser a trustee soon after his arrival in this country—that the Rev. Archibald Gray, minister of the Church of Scotland, as well as Mr. Foreman, a member of his congregation, had been a trustee from the commencement of the Institution, had signed the first petition for its establishment, and every one presented by the trustees during his lifetime, and that they had also invited members of the Council to become Trustees—that as to education, the Institution was conducted on the principles of the Glasgow University, that the education given was restricted to such branches as were required for all the learned professions—and as to the Divinity class, that it was quite disconnected with the Institution. The Trustees had granted the use of a class room for the purpose, but agreed to extend the same favor to any other denomination. Dr. McCulloch further offered to remove this class to his own house. But his opponents demanded that he should relinquish it altogether. This he positively refused.

The opposition from these sources involved Dr. McCulloch in a vast amount of labour. His pen was constantly employed, in various ways, in representations to the Legislature, appeals to the Presbyterian congregations, and carrying on a scarcely interrupted controversy in the press. Besides these labours, he visited Halifax and the leading towns in the Lower Provinces, delivering popular lectures on science, especially chemistry, with a design of awakening an interest in education. These lectures were among the first of the kind in British North America. With the assistance of his family, he collected a Museum of Natural History, which was the finest in the Province at the time. Audubon pronounced the collection of native birds to be the finest, or among the finest he had ever seen. To the discredit of Nova Scotia, be it said, it was allowed to be sold abroad. By these controversies the feelings of parties were excited to a degree, which it is now scarcely possible to realize. The members of the Presbyterian Church looked upon the Institution, as that upon which the progress and prosperity of their church depended, and all their most sacred feelings were roused in its support ; while the adherents of the other body, taught to regard it as an institution against the Church of Scotland, through the strength of their best feelings toward their mother church, were roused to the most violent opposition. And the strife became deeply intensified, by becoming mixed with the political struggles of the day.

We must now return to the history of the question in the Legislature. In 1827, the trustees again petitioned for the abolition of tests, and for a permanent endowment; but owing to the absence of Mr. Smith, the Pictou member for the County of Halifax, and the person who took the lead in the business of the Academy in the House, no permanent bill was introduced. But now the majority of Council were emboldened to go farther than ever in their opposition to the Institution. There can be no doubt that the opposition in Pictou was fomented, perhaps instigated, by the official party in Halifax, but now they took advantage of this division, as an excuse for refusing all aid to the Academy. They could now turn round and say, agree among yourselves. Hitherto the opponents of the Institution, had generally after a fight allowed the vote to pass each year, though refusing to make it permanent. This was anything but satisfactory. For trustees to engage teachers, and teachers to accept engagements, on the faith of a grant, for which annually they were at the mercy of the deadly foes of the Institution, was a position which none would willingly occupy, and it was unjust, when Windsor had its grant secured permanently.

But now, when the House passed the usual vote of £400, the Council negatived it, and in a paper sent to the Assembly, gravely assigned as their reason, that "measures have been adopted by a majority of the trustees to excite a spirit of hostility to the Established Church," and to render Windsor College unpopular on account of its restrictions ; and they declared their determination not to consent to any grant, while the Institution remained under the management of the present trustees. The House then voted £300 for the current year's expenses, and £100 for the partial discharge of the pecuniary engagements of the trustees ; this also the Council rejected. A resolution was then passed in the Assembly placing £400 at the discretionary disposal of the Governor for the benefit of the Institution ; to this the Council assented. The Governor issued a commission, Judge Chipman chairman, to examine into the state and proceedings of the Institution, and the result after deliberate enquiry, was a most favourable report, which caused the Governor instantly to give his warrant for the money.

The history of the legislation of the following years we give in Mr. Blanchard's words: " In 1828, another permanent bill passed the Lower House, and was lost in Council. Next day the House passed another permanent bill, with some alteration of the provisions. To this, the Council sent down several amendments, or, more properly speaking, a very voluminous bill of quite a different nature. It is sufficient to mention, that Dr. McCulloch, the principal, was personally excluded from the trust. All the trustees were to be removed, and others appointed in their place by the Governor, and the Institution was to be reduced to the level of a Grammar School, or something lower. The House of course refused to concur in the amendments, and the bill was lost.

" The House then voted £500 to be placed at the discretionary disposal of the Governor, towards discharging the debts of the institution. This was sent to the Council and lost. There were only four nays. Next day the Assembly resolved, that if His Excellency the Governor should judge it proper to aid the Trustees, to the extent of £500 towards the payments of their debts, the House would provide for it at its next session. Next morning, however, the friends of the Institution, thought it was going too far thus to overlook the Council altogether, and upon reading their journals, moved the insertion of the words, " with the advice of His Majesty's Council." After the rising of the House, the Governor called upon the Council for their advice, and they advised to withhold the money, and it was accordingly withheld.

" To ascertain the proceedings of the Council with regard to the permanent bill, a Committee of the House searched their journals, and reported that sixteen petitions from various portions of the Province, had been presented to the Council, praying their assent to a permanent endowment of the Academy, and it was also reported that there were four nays in Council to the rejection of the bill.

"In 1829, a permanent endowment passed the House, but was lost in Council. The usual vote of £400 was also passed and also lost in Council. In 1830 a similar bill and a similar vote passed the Assembly, but were both lost in Council.

"In 1831, a Committee of ten was appointed in the Assembly, to report a bill respecting the Institution, and they introduced one which passed the House, and was sent to the Council. They returned it with several amendments, but these being connected with money, and so an infringement of the privileges of the House were not considered. The House then passed a resolution of £400; while the resolution was under discussion, several members who were opposed to the bill, expressed their consent to the vote for that year, and one of them proposed that the word "unanimous" should be prefixed to the resolution, in order that the Council might know the unanimity of the House, and be perhaps thereby induced to acquiesce in its wishes. The word "unanimous" was prefixed by the consent of a full House, but failed to produce the expected effect. The resolution came back disagreed to.

"In concluding this naked history of the Parliamentary proceedings relative to the Academy, the following facts are recurred to as particularly worthy of remark. The House of Assembly passed eight resolutions granting money to the Institution, which were negatived, or destroyed by amendments.

"During the fifteen years since the Academy was founded, there have been four General Assemblies, and in each of these, there was always a very large majority in favor of the Institution. The bills and votes for annual allowances often passed without a division, sometimes against minorities of four or five, and on the last occasion, unanimously."

These proceedings led to debates in the House, conducted with great ability. It was in these discussions, and those which arose out of them on the constitutional powers of the House of Assembly, that S. G. W. Archibald, then speaker, established his character as one of the finest orators of the day, while perhaps not less able, and little less eloquent, were the addresses of T. C. Haliburton and C. R. Fairbanks on the same side, while Alexander Stewart on the opposite did what could be done to cover a bad cause.

But the issue was what those in power little expected. The Council it is true had manifested their power in defeating the wishes of the House on a money question, strongly expressed for a dozen years, and their intolerant exclusiveness, in refusing an act of the commonest justice to the whole dissenting population, and the Bishop might feel as if he could smile defiance upon all foes. Indeed the Council seemed determined to exercise their power in very wantonness, for they now negatived a grant passed in the House for Horton Academy, and even refused the small sum of £30 to the Pictou Grammar School.

But these things were but the beginning of the end of the whole system of the irresponsible power of cliques and compacts. The discussions on the Pictou Academy raised the whole question of the Council's constitutional rights, and there were men now to claim for the Assembly that control of money matters, which, according to the British Constitution, belongs to the people's representatives. The temper too of both the House and country was being roused, by the manner in which the Council had exercised their powers, and men were now found boldly to cry out to have the whole concern swept away, or its Constitution radically changed. We will give one example of this. On the 19th March, 1829, Mr. Hartshorne, one of the minority, moved that the House appoint a committee to confer with the Council, and instruct them to concur with the Council's bill of the preceding year.

Then Mr. Haliburton, after indignantly repudiating the idea, that after seven bills had been passed by the Assembly and rejected by the Council, the views of the latter should now be thrust upon them, proposed that the resolution should lie on the table, and that instead, the members should agree to resolve themselves into a committee on the general state of the Province, and to prepare an address to His Majesty, humbly soliciting him to remove from his Council those who filled public offices, or to give them a Legislative Council, and afterward made the following remarks :—

" Will any man say that this is not necessary? or that it would not be a desirable amendment of our local government? Will any man say that we, the forty members here assembled from all parts of Nova Scotia, do not bring together a greater body of local and topographical knowledge than any similar number of men residing in Halifax? Or will it be denied that twelve or fourteen gentlemen appointed by the King, from different counties in Nova Scotia, to a Legislative Council, could not better subserve the interests of Nova Scotia than the same number of people in Halifax? It has been said that this country is a peaceable, quiet country, and is well governed. I admit that it has been a quiet and exemplary Province, but, sir, it is owing to the temperance, prudence, good sense and forbearance of this House, and the morality of the people for many years past. But as to our local government, the structure and frame of it is essentially defective. Is it possible that any man can assert that where the Legislative Council consists of the same persons as the Privy Council, and the latter is composed of all our public officers, whereby the servants of the public become its masters, that such a form of government is perfect, or that men so situated, unless equal to angels, could in the nature of things give satisfaction? Is it possible to affirm that a council separate and apart from the Privy Council, but appointed by the King from the country, would not be infinitely preferable?"

And then after a scathing rebuke of the Bishop for his treatment of Dr. McCulloch, as well as some other of his friends, he concluded :—

" I turn from them and him to this House and say, consider of this matter, and petition the King, either to remove the public officers fed and paid by the Province, from the Privy Council, or to grant us a Legislative Council. That there does exist a necessity for this measure, no man can doubt who understands the state of our affairs."

Though the Assembly was thus coming to learn and assert its rights, there was still wanting the firmness to take a firm stand in the maintenance of them. Had they done so on this question, and simply had this grant or given no supply at all, the contest would perhaps have come sooner, but the victory would have been more easily won, and with less loss and trouble to the country than in the contest which was soon after forced upon them. But the members had always objects, for which they wanted money votes, and the House resorted to a system of manœuvering, to obtain the consent of the Council even for grants necessary for public improvement, and the latter could always secure their measures, and in every collision with the House had hitherto been successful. And even the author of the above bold words, under the influence of a judgeship, quietly subsided into the most compliant of placemen. But freedom's battle was now begun, and when the next fight came, the Council found itself engaged in a vain struggle for life, and passed away, "unwept, unhonoured and unsung;" and in the new order of things, the people of Nova Scotia took good care that no Bishop managed our civil affairs.

It was perhaps unfortunate for the Academy, that it was mixed up with these political questions, but this position was forced upon its friends by the course of its opponents.

During the time that the annual grant was withheld, the friends of the Institution rallied nobly to its support. In one year friends in Halifax subscribed £500 towards it. At one meeting in Pictou a subscription was opened, which the next day amounted to £181.5s. ($725). Had it been a time of prosperity in the country, the zeal of its friends was such, that it might have been maintained without Government aid, and sometimes Dr. McCulloch proposed, that the Presbyterian Church should take the matter into their own hands, and leave the Legislature alone. But it was a time of great depression and scarcity of money, and subscriptions were insufficient, so that debt was accumulating upon the Trustees. The Council remaining obstinate, they resolved to lay their grievances before the Home Government. This had been threatened before, but the Bishop and the Council had hitherto had the ear of the authorities at Downing Street, and they smiled contempt on such a proposal. But they had to deal with men, who knew their rights and had the boldness to assert them, and who had confidence in the justice of the British Government. Accordingly in the year 1831, Jotham Blanchard, Esq., was sent to Britain as their agent. He addressed the United Secession Synod, which, after hearing his statements, resolved to strengthen his hands, by presenting an address to the King on behalf of his mission, and otherwise to aid in the promotion of his views. He then presented to Lord Goderich, then Colonial Secretary, a long memorial giving a full history of the Institution and its claims, which was confirmed by a variety of documents. Some members of Council were in London at the time, from whom His Lordship sought explanations, but they had to make the excuse, that they had not with them the documents necessary to reply to it.

The result was a despatch from the Colonial Secretary to the Lieut. Governor, of which the following are extracts:

"The arrival of Mr. Blanchard in this country, with a memorial upon the subject of the Pictou Academy, has called my particular attention to that question, which seems to be almost the only topic calculated to interrupt the harmony and good understanding, which in general prevails between the different branches of the Legislature, and throughout the Province at large.

"Unless, however, some means be found of adjusting the differences, which have arisen upon this subject, I fear it may swell into an affair of some magnitude, and threaten injurious consequences, which it might not be easy to avert. His Majesty's Government, therefore, feel most anxious that this cause of internal dissension should be removed, and that a bill should be passed, which might give to the Pictou Academy that permanent pecuniary assistance from the public revenue, to the grant of which the Assembly attaches so much importance; and I have no hesitation in submitting to you my opinion, that it would be most unfortunate, if the passing of such a bill should be frustrated, by attempting to annex to it conditions as to the constitution of the Body of Trustees, to which there is but little reason to expect that the Assembly would be prepared to agree.

"It may, I think, well be doubted whether, considering the nature of the institution and the great variety of religious opinions, which may be entertained by those who attend it, any benefit would result from placing the management of it, in the hands of a Board of Trustees composed of persons holding official situations under Government, who might thereby become, in the discharge of their duties, most inconveniently mixed up with questions in which they could not interfere with advantage. The veto which the Governor now possesses upon the appointment of new trustees when vacancies occur, seems to afford a sufficient guarantee against the introduction into the Board of improper persons; and although it can hardly be expected, that any board of management could at all times give unqualified satisfaction to every one, yet so long as the Assembly, representing all classes of persons in the community, should not deem any fresh Legislative interference necessary, it might fairly be inferred that the Institution was not improperly conducted; at all events, it is obvious at the present moment that the public at large are not desirous of any material change in its management.

"Whilst, however, I cannot say that I see reason to participate in the grounds upon which the Council have rejected a bill for a permanent grant, I should, of course, deem it to be more satisfactory if the measure were adopted in such a manner as to meet and conciliate the feelings and wishes of both parties. Your object will therefore be to endeavor to bring about, by the exercise of all proper means of persuasion on your part, such a state of feeling upon the subject as may lead to that result. All will then see that His Majesty's Government at home, and the individual who represents His Majesty in the colony, have no other object in view than the good of the Province, and the harmony and contentment of all classes of His Majesty's subjects."

This was a pretty severe rap on the knuckles of the old ladies, and most unexpected from such a quarter. With such instructions to the Governor, and such a plain expression of sentiment regarding the conduct of the Council, had the friends of the Institution now stood firm, we believe they might have gained all they had sought. But the House was disposed to compromise, and thus threw away the fruits of victory. Mr. Archibald now introduced a bill, the principal provisions of which were, that the Institution should be under the management of a board of twelve trustees, seven of the old trustees, to be elected by themselves, four to be appointed by the Governor, (these it was intended should be appointed from the party in Pictou opposed to the Institution), and the Roman Catholic Bishop. The Institution was to consist of a higher and lower department, the higher to teach the branches already taught, and the lower to teach those usually taught in the Grammar schools. The trustees however might obtain any suitable house in any part of Pictou, separate from the Academy for the lower division, which was to be commenced, as soon as funds had been raised by private subscription, tuition fees or otherwise to provide a salary of £100 a year. There was to be no Theological class taught in the Academy, but this was not to prevent any professor teaching any such class in any other part of the town of Pictou. The sum of £400 a year was granted for ten years, of which £250 was to be paid to Dr. McCulloch, as principal, while he continued in office, the rest to be for the benefit of the Academy, as the trustees might direct.

This bill was an attempt fairly to meet the views of all parties. The proportion of the trustees allowed to the old funds of the Institution, was what Mr. McKenzie professed himself willing to agree to. It preserved the teaching of the higher branches, which its friends had been so long endeavoring to build up, and it at the same time sought to meet the views of its opponents as to the lower branches. But Mr. McKenzie was no more reconciled than ever. It still made provision for maintaining the higher branches, and thus the Seceders might still educate ministers in Nova Scotia, and he accordingly appealed to the Council against the measure.

"The manifest aim," he complained to them, "and effect of the Bill, is the appropriation of £400 a year out of the public revenue, to gratify the ambitious views of a particular sect, to whom a combination of circumstances, and the injurious union of the three districts of this county, has given a temporary ascendancy and a degree of political influence, to which their relative numbers and strength in the districts of Halifax and Pictou, by no means entitles them. It is an ascendancy, which tramples under foot the just rights, and sets at naught the moderate demands of the other classes of His Majesty's loyal subjects and which in seeking to perpetuate itself, has led to consequences of which none are more fully aware than the members of your Hon. Board. The Academy has been declared by the Rev. Principal, to be subservient to the propagation of the gospel, that is, as is apparent from his printed memorial of 1826, containing this passage, ' to the education of preachers for the body of Presbyterian Seceders.' The political and religious influence of this sect is thus to be extended and confirmed at the public cost, and to the injury and depression of the Kirk of Scotland, and of all other denominations of Christians." *

Mr. McKenzie therefore demanded, that £100 or £150 of the grant, should be appropriated expressly to the teaching of the lower branches, of which proposal Dr. McCulloch said :

" He knows what education is. He knows what is taught in the Pictou Academy. If he thinks that the several branches are not taught, let him show it to this Honourable Board. If they are taught, I ask him to specify any seminary upon earth, where so much instruction is given for £400 Nova Scotia currency. Yet from this sum he would abstract £100 or £150, leaving at farthest £300 to keep the building in repair, to pay the interest of debt, amounting to, perhaps, £1200, and to maintain the teachers. The governors of Kings' expend, I believe, not less than £1,000 sterling on education. In the Pictou Academy the system of instruction is, at least, as extensive, and I

---

* To this, Dr. McCulloch's reply was easy, that as the benefits of the Institution were equally accessible to all, injustice was done to none, and that its course was restricted to those branches of education, which every civilized community has accounted necessary alike for all the learned profession. If he supported the Institution as subservient to the Propagation of the Gospel, because he believed the interests of religion required a native ministry, he sought to promote its efficiency equally in subservience to more general purposes. If Mr. McKenzie preferred teaching his people to depend on Scotland, he had no right to complain of others who pursued a different policy. At all events, instruction in Theology formed no part of the course of instruction at the Academy.

trust, not inferior in quality; but when it is proposed to grant to the latter £360, he eagerly solicits to take £90 for the lower branches. The result is certain destruction."

Mr. McKenzie also persisted in requiring the lower branches to be in the same building. To this its friends were strongly opposed. " There are," said Mr. Archibald in his speech, "but four rooms in the present building, one devoted to the library, one to the philosophical apparatus, a third is the class room, and the fourth contained the museum. To introduce a grammar school into the College is to destroy it. The higher branches could not be taught amidst the noise and tumult of a common school; and if young boys were allowed to range through the library, museum, and depositary of philosophical apparatus, these would shortly be destroyed."

All Mr. McKenzie's views were conceded. The clause allowing the lower branches to be taught in a separate building was struck out of the bill, and £100 of the annual grant was expressly appropriated to their support, and both departments were to be compressed into one building with four rooms. Mr. McKenzie pledged himself that he and his party would raise "penny for peny" with the others, to maintain the Institution; and he with Revds. D. A. Fraser, and John McRae, and David Crichton, Esq., were added to the trust.

From this time, the finger of decay may be said to have been upon it. It did some good work, and some of its best students were educated after this. The old friends still came forth from time to time liberally for its support. But there was a blight upon it. External war had been exchanged for internal strife. Hitherto the trustees were united, but now what the one party wished to build, the other laboured to destroy. The lower branches were taught for a time efficiently by George A. Blanchard, but afterward were committed to a friend and countryman of Mr. McKenzie, under whose management the teaching of them was discontinued, simply from want of pupils. In the meantime, the trustees were becoming embarrassed in bearing the expense of the upper department. In consequence, the second teacher's classes were closed. Friends became disheartened, and in the uncertainty as to the future of the Institution, young men were discouraged from preparing to enter it.

Under these circumstances, in the year 1838, an Act passed the Legislature, transferring Dr. McCulloch and £200 of the grant to Dalhousie College. The ministers of the Church of Scotland now argued that the founder of the Institution, in proposing as his model the University of Edinburgh, meant the institution to be in connection with the Church of Scotland, and though they could not exclude Dr. McCulloch, they succeeded, through their influence with Sir Colin Campbell, the Governor, in getting members of their own body appointed to the other chairs, to the exclusion of better men. Mr. Crawley, who was one of the rejected, immediately commenced the agitation, which resulted in the establishment of Acadia College.

The Pictou Academy was subsequently remodelled as an Academy or High School, and as such has been doing good work. And the higher education in Nova Scotia has ever since been inextricably muddled.

We have thus fully noticed the Pictou Academy and the discussions connected with it, as these involved important results to the Province:

In the first place, it was the means of training a goodly number of men for stations of usefulness, both in Church and State, which they have filled in a highly creditable manner, many of whom could not otherwise have had more than a common school education. Among those who gave themselves to the gospel ministry, we need only mention such men as John McLean, J. L. Murdoch, R. S. Patterson, John Campbell, Drs. Ross, McCulloch,

McGregor and Geddie. To law and politics, it gave among others, Sir T. D. Archibald, baron of the English Court of Exchequer ; Judge Ritchie, now of the Supreme Court of Canada, lately Chief Justice of New Brunswick ; Sir Hugh Hoyles, Chief Justice of Newfoundland ; A. G. Archibald, Governor of Nova Scotia ; Judge Young, of Charlottetown, P. E. I. ; Judge Blanchard, George R. Young, &c., &c. Among its students who followed the healing art, we may mention Dr. W. R. Grant, Professor of Anatomy in Pennsylvania Medical College, and among scientific men, Dr. J. W. Dawson, Principal of McGill College, Montreal.

Secondly. It largely advanced the cause of general education and diffused a taste for literature and science. The number of men it educated, with their general influence, the schools that they taught, the numbers of others partially taught, the popular scientific lectures of Dr. McCulloch, the general air which such an institution diffuses around it, and even the discussions to which it gave rise, made it the means of diffusing intelligence and a desire for knowledge, among all classes of the community, bebond any institution of its time. and we might almost say since. The illiberality of those who imposed tests upon it, in some measure limited its influence to Presbyterians. But persons of all denominations attended it, and by the discussions of which it was the subject, and in other ways, these bodies were excited to an interest in the same cause.

Thirdly. It was in the contests, of which it was the subject, that the equal rights of all classes to public education were secured. What sane man in our day would advocate the maintenance of only one institution, from which only one fifth of the population should derive any benefit ?

Lastly, As we have seen, it was in the same contests, that the movement began, in which the government of the country by irresponsible cliques was broken, and the Province secured the true force of representative institutions.

If the Institution cóst money, it will be difficult to find one that gave as much in return for as little. But its benefits cannot be reckoned by any money value. One John McLean was worth more to a Province a hundred fold, than all that it ever cost. If there was strife about it, this was only because it was attacked. All was harmony when it was founded, and for years after, but it was assailed and of course defended.

We may add here, that as Dr. McCulloch's exertions in connexion with the Institution, directed young men to the ministry, the county has ever since given a larger proportion of the best of her sons to that sacred employment, than any population of the same size in the Dominion, a circumstance probably in part owing to old Scotch training. We give a list in the Appendix, which contains the names of about a hundred, of whom seven have received the degree of Doctor of Divinity, and seven have been missionaries to the Heathen (Appendix J.) Probably there are others, whose names we have not ascertained.

We have already mentioned that Dr. McCulloch from an early date after his arrival, contributed to the *Acadian Recorder*. In that paper first appeared some of his writings, which have since been published separately. Among these was a tale or sketch of colonial life, called " William," which, with another of the same kind named " Melville," was published in Edinburgh in the year 1826, under the title " Colonial Gleanings." In that journal, several of his writings on local controversies, ecclesiastical and educational, first appeared. In it also in the years 1822–23, as we have already mentioned, he published a series of light and amusing sketches of the social habits of the people of Nova Scotia at the time, particularly in the rural districts. These were thrown off by him hastily as a sort of relaxation from engrossing labours, but they are so graphic, that in every part of the Province, persons were found that were regarded as the originals of the characters which he had delineated.

Some others of his writings were published in pamphlet form, among which may be mentioned the following:—

"The prosperity of the Church in troublous times." A sermon preached in Pictou. Halifax: 1814. pp. 24.

"Words of peace; being an address delivered to the congregation of Halifax, in connection with the Presbyterian Church of Nova Scotia, in consequence of some congregational disputes." Halifax: 1817. pp. 16. 18mo.

"The nature and uses of a liberal education." Illustrated. A lecture. 1818. pp. 34. 8vo.

After his death his theological lectures were published in the year 1849, at Glasgow, in a volume of 270 pages, 12mo., under the title, "Calvinism the doctrine of the Scriptures, or a Scriptural account of the ruin and recovery of fallen man, and a review of the principal objections, which have been advanced against the Calvinistic system."

Dr. McCulloch continued to labour in Dalhousie College, with some measure of success, but was ill supported in his work. The toils and anxieties of past years had broken down his constitution, and rendered him prematurely old, though his energetic spirit still bore him through bodily infirmities; yet all his energy was not sufficient to establish a new institution, against the weight of incapable colleagues, insufficient appliances, and half-hearted support. He died at Halifax in 1843, in the 67th year of his age. His remains rest in the old Pictou cemetery, where some years later his students erected a monument to his memory.

The Rev. John McKinlay, his associate in his first efforts to establish the Pictou Academy, continued from the year 1824 till his death, on the 20th October, 1850, to minister to the congregation in Pictou, now known as Prince Street Church. He was a man well read, a diligent student, a faithful pastor, and a man of peaceful disposition, and he passed away amid expressions of universal respect.

The Rev. Donald A. Fraser, in the year 1837, removed to Lunenburg, from which place, in 1842, he removed to St. Johns, Newfoundland, where he founded St. Andrew's Church and congregation there in connection with the Church of Scotland. There he died much respected, on the 7th February, 1845. It is but due to him to say, that removed from the scene of strife in Pictou, he saw matters in a different light, and acknowledged his mistake, in the part which he had taken in dividing the Presbyterian interest. He had not been long in Lunenburg till he found himself a Dissenter, haughtily treated as such, and obliged to employ the energies, which he had previously employed against his fellow Presbyterians and their Institution, against the arrogant pretensions of Episcopacy. On one of his first visits to Pictou, after leaving it, he said to a leading member of the Presbyterian Church, " You used to tell us that in the course we adopted in Pictou, we were making ourselves the tools of the Bishop and his party. I never saw it till I left Pictou, but I see it now." He spoke and wrote in favour of union, and strongly condemned the virulence, with which his late colleagues carried on their controversy against the Pictou Academy. But the waters of strife had been let out, and he was powerless to arrest them.

The Rev. Ken. J. McKenzie continued to minister to the Congregation of St. Andrews Church, Pictou, till the year 1837. His conduct grieved the hearts of the pious in his own congregation, and he then relinquished the ministry altogether. In the winter following, a vacancy occurred in the representation of the county, by the elevation of Mr. Smith to the Legislative Council, and he became a candidate for the position. He was opposed by Thomas Dickson, Esq., was defeated, and died a few months later.

We have omitted all reference to the personalities, which marked the deplorable controversies, in which these men were engaged, but as to the questions of issue, it would be foolish and wrong, not to employ the light of experience, to judge of the wisdom of the policy of the respective parties. We venture to say, that never did time, which tests all things, more thoroughly determine any question, than it has vindicated the wisdom of the course adopted by the Presbyterian ministers, in endeavouring to establish a collegiate Institution, with a special view to the training of a native ministry, and the unwisdom of the opposite. We need not enter into particulars. The history of the two churches since tells the tale.

CHAPTER XVII.

FROM THE FINANCIAL CRISIS OF 1825-26 TO THE DIVISION OF THE COUNTY, 1826-1836.

The financial crisis of 1825-6, described in a formre chapter, left the county in a very depressed condition. Still the various branches of business, which had previously occupied our public men, continued, though on a diminished scale,—and to those formerly mentioned as engaged in them, we may add the names of John Taylor and David Crichton, first in partnership, and afterward separately. Soon however these began to fail. Owing to the destructive manner in which the fisheries had been conducted, the fish visited our shores in greatly diminished numbers, and sometimes did not return to their old haunts at all, so that the trade fell off. About the year 1833, the fishing trade from Pictou ceased, and though there have been attempts made since to revive it, these have not proved successful. The West India trade was carried on under some disadvantages from Pictou, from the harbour being closed in winter, and from the fact of its not being a free port. As the fisheries failed and the supply of good lumber diminished, this trade came to an end about the same time.

Public attention was now directed to the obtaining for Pictou the privileges of a Free Port. Some explanation may be necessary on this subject for readers of the present day. In the colonization of America, all the European powers acted upon the idea, of making the colonies yield the utmost possible advantage to the mother country, and that often with little regard to the rights of the colonists. Great Britain, though distinguished for justice and magnaminity, as compared with other nations, long maintained a commercial system, narrow and selfish, alike unjust to the colonies and injurious to both. Up till the year 1825, her policy aimed at preventing her colonial dependencies having any trade, except with the mother country or with one another. But in that year, Mr. Huskisson, then President of the Board of Trade, passed his memorable act, by which the colonies were allowed to trade with foreign countries, which reciprocated the favour—an act which may be regarded as the emancipation of the colonies, and from which a new era in their history may be dated. Still the privilege was limited to some ports, known as Free Ports, of which Halifax was the only one in this Province. Under this system, if Pictou wished to export a cargo of any article, except fish, to a foreign country, it had to be sent to Halifax, unloaded there, reshipped and then cleared from that port. So all return cargoes of any description from foreign countries, had to be landed at Halifax, reloaded there, and thence cleared for Pictou, involving not only expense, but sometimes such a loss of time, as might prevent arrival for a whole winter. Thus the merchants here were virtually excluded from the trade with the Foreign West India Islands, South America, or the Mediterranean, which were the best markets for fish ; and as to the United States, a cargo of flour could only be imported by landing it at Halifax, and a few hundred chaldrons of coal that were sent thither, had to be trans-shipped in the same manner. Foreign vessels were also prevented from coming to Pictou.

We may suppose that the Halifax merchants had enough of human nature in them, to wish to retain for their port the monopoly which this afforded, and hence their opposition was long given to the extension of the Free Port system to the outports. The first movement to obtain this privilege for Pictou was now made, by the calling of a public meeting, which was held in the Court House, on the 8th January, 1828, when it was resolved unanimously, " That it is the opinion of this meeting, that it will be a great advantage to the trade, commerce, fisheries and agriculture of the Port and District of Pictou and the neighbouring harbours and places situate in the Gulf of St. Lawrence, that the Port of Pictou be placed on the footing of a Free Port, under the provisions of the Act of the Imperial Parliament, 6 George IV., chap. 114." It was also resolved to petition the King to that effect. The petition was accordingly forwarded, with another to the Lieutenant Governor, asking him to recommend its prayer. Before the petition reached Britain, however, Pictou and Sydney were declared Free Ports, through the influence of the General Mining Association. But the petition of the inhabitants, which prayed that Pictou should be made a free warehousing port, was refused, on the ground that it had already been made a Free Port. Even this, however, excited in the minds of people here the most glowing visions, as to the future prosperity of the place. " Stranger things have happened," said one, "than that the horses of the Governor General of India should yet travel on Pictou iron, paid for by direct importation of East India goods into Pictou Harbour," and the first arrival under the new system was thus hailed in the *Colonial Patriot*, of 21st May, 1828:—

" With much pleasure we record the arrival of the schooner Lovely Hope, from Boston, with a cargo of flour, corn, &c. This is the first arrival under the free port order, and it is creditable to the enterprise of Messrs. G. L. De Blois & Co., the merchants who so quickly availed themselves of the new system. * * We sincerely hope the cargo of this vessel will yield a liberal return to the consignees. We think the present an important era in the history of Pictou, and doubt not the Lovely Hope is the harbinger of much good to come."

The timber trade had now sunk to a low position. The finer qualities of wood were exhausted. All the pine fit for shipment was gone, and what little was left, was to be found only in the more remote parts of the county, was generally small in size and needed for home consumption. Pitch pine of value was not to be found. Instead of the splendid oak, which yielded abundant supplies of hogshead staves or timber for shipment, were to be found only a few small trees of second growth, scarcely sufficient to supply our own population with those articles, for which it was specially desired. There only remained the spruce and birch. For some years considerable was done in manufacturing the former into deals and battens, and in shipping them to Britain, but since about the year 1840 even that has come to an end, owing to the exhaustion of the supply. Birch timber continued to be drawn from the interior, and still forms an article of export.

The financial crisis of 1825-6 for a time nearly destroyed the ship-building business, but it soon began to revive and in subsequent years was carried on with much energy. But at that time, most if not all the vessels built were built to sell. After the close of the West India trade, there was scarcely a vessel of any size owned in the county and kept in regular employment. Even the building of them was carried on under disadvantages. The leading merchants had been left in debt by the events of those years, and others who began had but little capital. Hence the work was carried on by means of advances from parties in the old country. This involved expenses for commission, interest, &c. Then the vessels were sent to Britain for sale. Even if sold immediately, and at good prices, such were the charges, that very often the builder was as deeply in debt as when he began. But frequently they remained for a time unsold, with expenses eating up their value, and then they might be thrown upon the market at a time when prices were low, in which case they might not realize first cost. This business proved fatal to nearly all who were concerned in it, as we shall notice more particularly in our next chapter. A number of those however, who at this time began business with little or no capital and on the smallest scale, have since become the wealthy men of the county.

Owing to these circumstances, the closing years of the period we are now reviewing, were about the poorest the country has experienced since its first settlement. Farmers, owing to the credit system, and their giving so much of their attention to timber, were in the merchants books. Now that resource was gone. Their farms had been neglected, and from constant cropping did not yield as formerly. Crops too failed from other causes. The Hessian fly injured the wheat, and a disease affected the potatoes, so that they did not grow as formerly, sometimes the seed not coming up at all. There was little demand for farm produce, and no cash market. Merchants received it in exchange for goods at low prices. The pork, butter, etc., thus received was shipped principally to Halifax, Miramichi, or Newfoundland, and cattle were sometimes driven across Mount Thom to Halifax, frequently realizing but a small return. The ship timber, or other produce of the forest, which they traded with the merchant, were paid for in goods dear in price and often trashy in quality, so that the farmers were so destitute of cash, that it used to be said they could only look for as much as would pay their taxes and stipends, though too commonly they sought to pay the latter, either in some other way or not at all.

We now turn from business matters to notice some other matters connected with this period. In the year 1827, the first newspaper published in the Province out of Halifax, was started in Pictou. It was called the *Colonial Patriot*, and was published by William Milne, in partnership with J. S. Cunnabell of Halifax, but its establishment was mainly owing to Jotham Blanchard, Esq., who for several years edited it anonymously. The important part which the paper played in our Provincial history, as well as the character and services of its editor, entitle them to special notice in this work. Mr. Blanchard was born at Peterboro, N. H., on the 13th March, 1800. His grandfather, Jotham, usually known as Col. Blanchard, had left the United States at the close of the Revolutionary war, from loyalty to the British crown, and settled at Truro. His son, Jonathan, remained behind and married there, and Jotham was his eldest child. When he was fifteen months old, his parents removed with him to Truro, where the old people with their family were still residing. Here from accident or sickness he incurred permanent lameness, and probably from the same cause his constitution was feeble and ill fitted for the labour, to which he was impelled by his active mind. After he was able to go about, his father provided a pony for him to attend school, or go where called. Some years later, the family removed to the West River, where the father purchased George McConnell's farm and put up the large house, so long known as the Ten Mile House, where he for several years kept an inn. Afterward they removed to Pictou town, where Jotham completed his education at the Pictou Academy, being one of the first class of students at that institution. He studied law under Thomas Dickson, Esq., and was admitted to the bar on the 18th October, 1821. In his profession, he soon established his character as an able lawyer and an eloquent pleader, but at the same time was. noted as always discouraging litigation, at a time when there was so much disposition to it.*

The first number of the *Patriot* was issued on the 7th December, 1827, and had for its motto, "*Pro rege, pro patria.*" In exposition of this, the editor said :—

"In politics we shall side with the most liberal system. Our motto, if rightly understood, conveys our sentiments. We reverence the British Constitution, and honor the king as its head, but feel assured that the best way of showing true regard for the king is by advancing the interests of his subjects. All governments are designed for the general good of the people, and that government deserves most praise, which most effectually succeeds in this object; and we boldly assert, that he who pretends to support the dignity of the government and the honor of the crown, at the expense of the general happiness, alike commits treason against the king and his subjects;—he betrays the people and dishonors their sovereign.

" Respecting our Provincial politics, we can only say that we shall advocate what we consider sound and just principles ; and if we find the government or any branch of it deviating from these, we shall not fail to proclaim it. This determination, we are well aware, would be ridiculed by the members of Government, were it to travel so far as to meet their eyes, but neither their scorn nor our own weakness shall deter us from this course of conduct, being convinced that it forms no excuse for permitting obnoxious measures to pass

---

* An instance of this was given by a gentleman then holding a humble position in the printing office. He was one day waiting upon Mr. B. for "copy." While the latter was driving his pen with great vigor, a countryman came into the office. Scarcely lifting his head, Mr. B. asked his errand. The man replied that a certain person had sued him for debt. "And do you owe him?" said Mr. B., while the pen went with undiminished rapidity. The man mumbled an uncertain reply. "Do you owe him?" said Mr. B., more sternly. "Well, perhaps I do," the man drawled out. "Then go and pay him," was the reply, while the pen never stopped in its career. The man slowly retired, glancing back with a mixture of wonder and curiosity, at this new specimen of legal advice.

in silence, that observation upon them is attended with no immediate results. It is an important point, to keep the eyes of the people open to their own interests, and thus convince the Government that they know when their rights are overlooked. This is the safest and surest mode of preventing and rectifying mal-administration, though we must confess that in the latter case the process is tedious. It has, however, proved successful in Britain, where ancient prejudices and their abettors have been forced to yield to the increase of knowledge, and the consequent march of liberal principles.

"With our cotemporary editors, we shall carefully cultivate the most friendly feelings, but our public duty is paramount to private inclination, and if we find them betraying the people's rights, or inculcating excessive servility, we must not be backward in exposing their errors, and reminding them of their duty as sentinels of the public interests.

" The peculiarities of our religious tenets we do not think proper at present to divulge. . . . While men do not cherish religious views subversive of the order of society, or inimical to the great and leading principles of our glorious constitution, we think it the very acme of injustice that there should be civil distinctions on account of religious opinions. Influenced by these sentiments, we shall never hesitate to strike in our feeble lance against any man—be he friend or foe, for us or against us—whom we shall find prostrating' the landmarks of his neighbor's rights.*

"Having witnessed the beneficial effects resulting from an unshackled press in Britain, we shall always advocate the same system here.

" We will discuss the interests of Pictou. We shall at all times, however, when opportunity permits, be happy to raise our voice in behalf of the whole Province of Nova Scotia, without reference to east or west, north or south ; and even beyond the limits of our own Province, our humble efforts shall always be at the command of our sister colonies, when we think their just rights attacked or disregarded, or in danger of being compromised by the negligence or inertness of the great body of the people, or the adroitness or power of the few.

" Our infant establishment is the first of the kind in the country, and we do hope that the friends of general improvement in all parts of it will, by the kindness of their smiles, brighten us into a vigorous existence. The town has advocates in abundance, and papers in abundance,—we shall endeavor to advocate the peculiar interests of the country."

These sentiments seem innocent enough, and in the present day would alarm no person, but they covered principles, which at that time were considered by those in power as dangerous, if not altogether subversive of

---

* In explanation of this paragraph, it is only necessary to remind our readers, that Catholic emancipation was at that time still one of the great questions of the day.

society. Referring to this in the second number, the editor says:—

"Before setting out with so open an avowal of our principles, we perfectly knew that the voice of slander would follow our track, and that we should be charged with disloyalty and radicalism. This has been the refuge of all the supporters of existing abuses and new oppressions, since the world began. Pharaoh, no doubt, considered Moses a great radical. William Tell was a radical; the sturdy barons who forced Magna Charta from King John were villainous radicals; so were Luther and John Knox; and they were a radical crew, to be sure, who drove the last of the Stuarts from England's throne. In later times Chatham, and Burke, and Fox, and Brougham have all been charged with disloyalty and radicalism by the advocates of gray-haired abuses. If we, then, of the *Colonial Patriot,* suffer from the same species of slander, we shall suffer in good company, and we prefer suffering in a good cause to prosperity in a bad one."

These were the days when the Council of XII., combining executive and legislative functions, sitting in secret, all, with scarcely an exception, Churchmen and residents of Halifax, and nearly all placemen,* most, if not all of them, decent men in their way, but trained in the narrowest school of political sentiment, full of the highest notions of arbitrary power, ruled the country with undisputed authority. Successive Governors had been but tools in their hands, and the House of Assembly, in any attempt hitherto made to assert its independence, had been obliged, whenever it came into collision with their high mightinesses at the other end of the building, to succumb by a threat of the latter of refusing to do business with that branch of the Legislature. The Council too seemed to feel under no obligation, to adopt any measures for the improvement of the Province, so that for anything of that kind the country was indebted either to the persistent efforts of the Assembly, or to the independent judgment and energy of such a Governor, as Sir J. C. Sherbrooke or Sir James Kempt, yet were jealous of anything that seemed in the remotest degree to affect

---

* In subsequent collisions with the Council, it was stated that ten out of the twelve were paid officials of Government.

their own dignity, and resented it as subversive of the British Constitution or treason to the Sovereign. Only the winter previous, they had rejected a measure of the House for increased aid to common schools, and when such an unpatriotic, if not unconstitutional, exercise of power, provoked the author of the measure, T. C. Haliburton, to describe them as twelve old women, one in lawn sleeves (alluding to the Bishop), the House was called to account, and from fear of consequences, and contrary to their own judgment, meekly bowed to reprimand the author of the speech. Moreover, the majority of the Council having scarcely been outside the town of Halifax,* the country was to them of so little account, that any attempt on the part of the inhabitants to discuss their proceedings, they would have regarded almost as we might suppose a farmer, to regard a criticism on his style of farming, from the sheep in his back pasture. Indeed, the Attorney-General described the members of the House, particularly referring to those from the country, as the Caribous.

Such were the circumstances in which the *Colonial Patriot* was issued, as the advocate of liberal politics. The newspapers of Halifax were devoted to the news of the day, containing only some common-place remarks on public events, and discreetly silent regarding official doings. But Mr. B. had entered keenly into the political discussions of the mother country, on the subject of popular rights, on which at that time feeling there was running high. Impressed with the much greater subserviance of the people in general to the few in power, which existed here, he threw his whole soul into the work of securing for the popular will, that control over public affairs, for which the Reformers in Britain in another shape were contending. Those measures of reform in

* It was asserted afterward in one of the newspapers of the day, that some of them had never crossed Sackville Bridge, ten miles out of Halifax.

colonial administration, which the popular party in Canada and Nova Scotia afterward succeeded in carrying, the "*Colonial Patriot*" was the first paper in the Lower Provinces to advocate.

Mr. B. wielded the pen of a ready writer. He wrote rapidly, but his writings were marked by great vigour and independence. He had the assistance, however, of other pens, lay and clerical, and the paper soon began to excite public attention. The political questions of the day were then mixed up with the Pictou Academy dispute, which was in fact the battle ground of party, and the *Patriot* was ever the fearless advocate of the institution.

The principles of the paper and the free spirit in which it assailed Government abuses, soon brought it into notice. Its radical, or as they were then deemed, revolutionary views, were received in some places with horror. We recollect of hearing of an old Scotch minister, a Seceder too, who, hearing Mr. B. advocate in his earnest way his political views, lifted up his hands in holy amazement, and exclaimed, "daring innovator." In Halifax particularly, the *Patriot* created no small stir, especially in official circles. The style of writing in it would not appear very violent, as compared with the political writing to which we are now accustomed, nor its sentiments very extreme, but at the time they were so unusual, and such was the general sycophancy to men in power, that they produced quite a sensation. It was read nevertheless.

But the paper was to receive attention from higher quarters. A few weeks after its commencement, an article, which appeared in it from the pen of a correspondent, was regarded as rank treason by the powers that be. It is said that the matter was seriously debated in the Council of XII., and that the feeling was general, if not unanimous, in favour of bringing the author to condign punishment. Milne received notice of an intended prosecution for libel, and the Attorney-General's son, R. J. Uniacke, Jun., entering the House with the paper in his hand, and, as Blanchard described him, " with all the greatness of a full-blown bladder," declared those connected with it to be dangerous persons—that they had violated parliamentary privileges, and that he would never move another resolution in the House, unless it would avenge the insult by calling them to account. It was proposed not only to prosecute the proprietor for libel, but to bring him in custody to the bar of the House. The writer of the article had submitted it to Mr. B. as a lawyer, instructing him not to publish it, if it were libellous, and the latter was satisfied that there was no danger on that point, but for some time Milne expected that the House might take the last step proposed. But the majority of the House stood firm against Uniacke's denunciations, which were no doubt inspired by Government.

There was much anxiety on the part of the authorities, to find out the authorship of the obnoxious article, which they were disposed to attribute to Dr. McCulloch. But as the real author *afterward* freely acknowledged his work, and when liberal principles had triumphed, rather took credit for it, we violate no confidence in saying, that it was written by the late Rev. Thomas Trotter, of Antigonish. We may add, that instead of containing anything violent, it would now be considered calm and logical. Its offence was, that it questioned the constitutional right of the Council, to act as they were doing regarding money questions.

But a circumstance, which gave Mr. Blanchard and the paper special notoriety, was the publication of what was called " the Canadian letter." There being at that time much political agitation in the Upper Provinces, their condition and affairs occupied a prominent place in the *Patriot's* discussions. The very first number strongly condemned as unconstitutional, the course taken by Lord Dalhousie, in rejecting Mr. Papineau as speaker, when elected by the Assembly, and Mr. B. continued warmly to support the course taken by that body in adhering to their choice ; and maintained that if the people were true to themselves, they must triumph in the end. He was for a time a warm admirer of Mr. Papineau, though, like the rest of the Nova Scotia Reformers, he would had he lived, have condemned the course taken by him and his compatriots in the outbreak of 1837.

Not long after the report of the proceedings of the Canadian Parliament, in which Mr. P. was a second time rejected, reached this Province, an extract from a private letter from a gentleman in Nova Scotia, was published in the " *Canadian Spectator*," in which the spirit of the popular party was applauded, assurance was given that whatever the enslaved press of the Province might say upon the subject, the great majority of the people, who knew the merits of the conflict, thought well of the objects they had in view, and in general of the means they took to accomplish them. It was stated, that while in the Legislature of this Province, there was a growing spirit of independence, there was still far too much servility to those in power, and though the existing state of things in Canada was much to be deprecated, it was desirable that some of the same spirit should come our way. "A moderate quantity of it now might supercede the necessity of more hereafter. As prevention is preferable to remedy, I am in hopes a little of it will creep our way, before a greater share of it will be required." And what was no doubt considered more dreadful, in reply to the accusation of the popular party being the disturbers of the peace, he maintained that "Lord Dalhousie, by stretching doubtful prerogatives to their utmost limits, and unnecessarily irritating the people, has made himself the public disturber."

The extract was copied into the Halifax papers and the writer of it was denounced as a political libeller, not fit to crawl on free soil, and his opinions characterized as disloyal and dangerous. As "the writer of the Canadian letter," which had been addressed to Mr. Leslie, member for Montreal, Mr. Blanchard defended the extract, but denied the legitimacy of the inferences drawn from it. Such was the feeling excited in high circles, that Mr. B. did not trust the office with the knowledge of the authorship of what he wrote, but employed a friend as scribe, in whose handwriting the manuscript went to the printer.

Mr. Joseph Howe, at that time editor and publisher of the *Nova Scotian*, was prominent among the assailants of the principles, which "the writer of the Canadian letter" advocated, and a somewhat fierce controversy was maintained for a time, which did more for the elucidation of the principles of liberal government, and their introduction into this Province, than any thing that had hitherto transpired. Mr. Howe was then a young man, just beginning his career as a journalist. His early writings gave indications of the talents he possessed, although he had not reflected deeply on political questions. He was naturally connected with the official party, his father having been both Queens Printer and Postmaster General, and his elder brother succeeding to both offices; and indeed was regarded as the chosen champion of the party. But the result of his controversy with Mr. B. and the other writers, who came to the aid of the latter, was that he become a convert to the views, which at that time he denounced, but in the advocacy of which he afterward became so prominent and so celebrated. He has been known to say, that he received his first impressions of liberal politics from Jotham Blanchard. He did not approve of them at first, but the more he thought upon them, the better he liked them, till he embraced them fully, and devoted his life to their establishment.*

The maintaining of a country paper at that period was no easy matter. Even for many years after, it was with difficulty that a publisher could make ends meet, but of course at that time the difficulties were much greater. The population likely to support it was but small, the country was not in a very prosperous condition, the habits of payment were very irregular, the publisher in Pictou was not a practical printer, and patriotism was not then the paying business it has since become. At all events a long time had not elapsed, till the publisher found himself in jail for debt. In an editorial the situation was thus humourously described :

"We do not know what our readers may think of it, but for our own part, we can honestly declare, that it has affected us more than if we had heard of the incarceration of every other expounder of news in the Province. For subscribers to be in our debt is bad enough, but for patriots like us, who have been grumbling for them immeasurably, to be shut up in the prison house, because they have not paid us, while they are going at large, is almost beyond the endurement of flesh and blood. Anybody but ourselves would have long ago delivered them to the judge; and sure are we, that had they fallen into the hands of such a prompt and righteous dispenser of justice as our old Treasurer, he would have made them down with their dollars on the spot, and given them a good pounding to boot.

"If our subscribers cannot pay us, we give them this notice, that they must find us an equivalent. Let them only recommend us to His Majesty's Council, and get us into some moderate office, which will help us out of our scrapes. We are not ambitious men, we assure them. With such a salary as Mr. Jeffrey's we will be perfectly contented to be publicans and crave nobody. Though we may now and then take a race after the smugglers for fun, not one of our subscribers need lengthen his steps.

"Some of our subscribers seem to think that if, like our old Treasurer, they say that they have no money in their chest, we are very well off. But we do assure them that we are not very well off. Such a thing was never known of patriots since the world began. Had we twelve thousand pounds lying past us to the good, our subscribers would have something like reason upon their

---

* Entering the *Patriot* office when on a visit to Pictou at the time of the election of 1830, he laughingly remarked, " The Pictou scribblers (so he used to call the writers in the *Patriot*) have converted me from the error of my ways."

side; but upon the word of honest patriots, we positively declare that we have not half that sum in our possession, and to the best of our knowledge and belief, are not likely to have it before next meeting of Assembly.*

"When we were dragged to jail, we had no doubt of a speedy deliverance. We were perfectly confident that as soon as our confinement was known, there would be a rushing to see us, which would far outdo anything of the kind that had ever occurred in the Province. We said to ourselves, that if the large bullock and Mr. Barry had each a thousand visitors, patriots, such as we, must have ten thousand at least. But except a few of our creditors, who called to enquire when we would pay them, not a creature came near us.

"It is very little to the credit of our subscribers, that we are in the hands of the Sheriff. We will not therefore allow them any longer to affront themselves. And we give them this notice, that there are only two ways: either they must send us their money, or come and live with us. In the last case we shall have them under our own eye, and if, between the treadmill and breaking stones for the streets by way of relaxation, we do not work it out of them, we shall have ourselves to blame. By these means we will collect as much as will pay all our considerate creditors; and when we find that there is nothing more to be got, we will send for the individual who for the pure purpose of annoyance, has been persuaded to put us in jail, and like honest gentlemen that we are, surrendering to him all that we have, that is to say, all our debts which have become bad through his placing us in confinement, we will walk out, and prosecute the patriot trade with redoubled vigor. If he expects the favor of councillors and their creatures by crushing us, he is likely to find that he has caught a Tartar. We were not born so far north for nothing, and we assure our friends that after coming all the way from Aberdeen † for their benefit, they will not find us so easily put down."

But the darkest cloud we are told has a silver lining. Above all, no circumstances can be so desperate as to be beyond woman's sympathy, and the darkest scenes of life will be bright with the light which shines not on sea or shore, if cheered by her smile. So did the poor printer find it. The Sheriff had a fair daughter, whose pity was moved by his hapless condition. Our prosaic history cannot adequately tell how pity passed into deeper feelings, but at all events so well was the enforced leisure of the prison employed, that when he again went forth to

---

* The allusion in this paragraph is to the accounts of the Treasurer, which showed a balance of twelve thousand pounds in favor of the Province, while applicants sometimes were roughly turned away with the declaration that there was no money.

† Mr. Milne's native place.

liberty, it was under bonds which, we suppose now after nearly fifty years of wedded bliss he has no desire to see dissolved.

To relieve him from his embarrassment, and to continue the publication of the paper, which had now become popular among the friends of liberal politics, a number of gentlemen combined, and subscribing the requisite funds, took the concern under their own management. For four or five years longer, the paper continued the same political course, Mr. Blanchard acting as editor the greater part of the time.

The course which Mr. Blanchard pursued excited against him much personal hostility. In the press he was accused of assailing all that is respectable, and subverting the very foundations of society, and his private character was attacked in ways that would outdo even our present political newspapers. To this however was added burning in effigy, a proceeding which Mr. Blanchard noticed in the following manner, duly honouring the more prominent actors in the scene, by giving their names :

" A number of the merchants and other respectable inhabitants did us the honour on Thursday night, to burn us in effigy in the middle of the town. Mr. ———, merchant and all his clerks, and (naming some others,) and about 100 other most respectable gentlemen, all assembled and performed in a most gentlemanly style the noble feat of burning our effigy. The blaze was so good, that many persons thought a house was on fire, particularly as these gentlemen were so careful of the property of the town, that they bawled fire most vociferously, to warn all, that gentlemen were employing that element for gentlemanly purposes.

" We cannot adequately express our gratitude to these numerous merchants and other gentlemen for this signal honour. To be ranked with Popes, and Kings, and Dukes and Governors is an honour, which does not come the way every day to editors.

" We must regret one or two mishaps that occurred. Mr. ———, whose clerk had the honour of carrying the effigy (and it was an honour even to be the jackass or packhorse of us, the Editors of the *Patriot*) mistook ——— for an intruder, and beloaboured him very severely, and occasioned the loss of his hat. However, in so laudable a work as honouring us, a few wounds and the loss of a hat were trivial misfortunes. If ——— will call upon us, we shall sympathize with his sufferings, and contribute to the purchase of a hat. We do not wish to be honoured free of expense."

This was followed by a gentleman publicly spitting in his face. Mr. Blanchard being feeble, and his assailant being attended by others, supposed ready to proceed to personal violence, was unable to resent the result, though a friend did so the next day.

Considering the hostility, of which Mr. Blanchard was the object, we are naturally led to enquire, whether there was any cause for this in his personal character. After careful enquiry we must say, that, as far as we can learn, the opposition arose entirely from his political course. His life was pure, he was a genial companion and a firm friend. But the fact is, that at that time the very idea of criticising the proceedings of those in power, was not only so new, but was so contrary to the arbitrary principles then prevalent, that it was held as sedition and rebellion. " It has long been a crime," he says in one article, " to stand up in the Assembly and advocate the rights of the people, or to say that they have rights. It is a crime to establish a paper under the hated name of Patriot. It is a crime to subscribe for such a paper. It is a crime to treat of public men and measures according to their deserts. It is a crime to call public functionaries to account, and to hint that tax gatherers and smuggler seizers may not be immaculate in official duties, and infallible in legislative conduct." And this was all the more intolerable, when it came from one " from such a remote part of the Province as Pictou." In fact, his real offence was the political position which he assumed. When Mr. Howe advocated the same views afterward, he was assailed with equal bitterness, and with similar accusations.

In the year 1830, came the great conflict between the Council and the Assembly. The history of this does not belong to our present work. But the state of the question may be given. In the year 1826, the Assembly had passed a revenue bill, by which the duty on foreign brandy was raised from 1s. to 1s. and 4d. This duty was at first collected, but the officials having discovered a flaw in the wording of the act, had not collected the extra duty for several years. So little control had the Assembly over the financial affairs of the Province, that this was not discovered till now. They immediately determined to have the error rectified, and sent a bill to the Council imposing the extra 4d. This was sent back rejected, at four o'clock on the day on which the revenue bills expired; and next morning Honourable Councillors, in the midst of a blinding snow storm, were busy taking out of warehouse large quantities of spirits, which they had there in bond. The House of Assembly immediately passed another bill, which was also rejected by the Council, who now assumed an attitude on the subject, to which the Assembly felt they could not submit, without sacrificing the time honoured rights, which belonged to them as the Representatives of the people. The result was that no revenue was collected that year. The Assembly was dissolved, and a new election took place that fall amid much excitement. In Halifax city, the Council carried things their own way, but the county was the subject of a keen contest. Mr. Blanchard became a candidate along with Messrs. William Lawson, S. G. W. Archibald, and George Smith, as the friends of the Assembly, while the Government candidates were, Messrs. Hugh Hartshorne, J. A. Barry, J. L. Starr and Henry Blackadar.

There was not the same party discipline as now, when men, even Christian men, vote for the candidate of their party, whatever his capacity or even whatever his character. Every candidate had to depend largely on his personal influence and popularity. The Government too were nearly all powerful in the city of Halifax. For a young man from the country, like Mr. B., without wealth or influence in the capital, personally almost unknown there, to seek election as representative of the Metropolitan county, was a bold undertaking, and almost enough to give the old ladies fits. But as a candidate, he made a good impression upon independent men. His note books still in existence shew him to have cultivated his mind, by diligent study of the writings of the best poets and orators of Britain and Ireland. He had also been interested in the modern political discussions of the mother country, and now his speeches attracted attention. But such was the hostility of the Government party to him, that in Halifax he was insulted, which however only rendered his friends in the country more determined in their efforts on his behalf. Even in Pictou they would not allow him to be heard on the hustings, while the gentleman who proposed him, Adams Archibald, Esq., of Musquodoboit, one of the greatest natural geniuses the Province ever produced, was soon after dismissed from the commission of the peace.

The election for the County of Halifax caused much excitement throughout the Province. In Pictou the political question was mixed with the religious division that had been growing up, and with the feelings that had been excited regarding the Pictou Academy, so that party feeling reached an unprecedented height; and this election, ever since known as the big election, witnessed deplorable scenes of violence, pitched battles being fought, sticks freely used and one man killed.

Mr. Blanchard was returned with the other popular candidates, and for five years proved an energetic member of the House. It is generally said, that he disappointed expectation. This may be true in part, but it is easily accounted for. Perhaps the expectations of his friends were too high. We may add that the House proved rather a subservient one. True there were only eight returned as Government supporters ; but when the Council swallowed the revenue bill, which they had rejected the year before, members of the House seemed inclined to rest and be thankful. The loss to the Province by the late collision with the Council, seemed to make them tremble at the thought of another. And then and more especially his health failed.

Still while his strength remained, his voice was ever raised on behalf of any measure, which promised to advance the public interests. The subject which engaged his most energetic efforts, was the Pictou Academy. The Government still continuing hostile, he was as we have seen, in the year 1831 sent by its friends to Britain to lay its claims before the Home Government.

He also succeeded in carrying some important measures, among which was an act for the relief of honest insolvent debtors. Up to this date, any one creditor could retain a debtor in gaol after the surrender of all his property, by supplying eight pounds of bread a week for his maintenance ; and persons were found ready to use this power, in the hope of leading the friends of the unfortunate to pay the claim, in order to obtain his release. This power was now taken away, and two magistrates had power to order the discharge of an insolvent, where without fraud he gave up all his property. He also advocated the abolition of imprisonment for debt altogether, but the country was not prepared for such a measure. He had studied the works of Brougham and others of the school of English law reformers, and advocated some of their measures here, among others the conferring of equity jurisdiction upon the Supreme Court. But it required twenty years to prepare for the introduction of this grand improvement in legal procedure. Out of the House he still supported measures for public improvement. Among these may be mentioned the establishment of circulating libraries. After his return from Britain, where he had seen the system in operation, he spent a good deal of effort in endeavoring to have it introduced into our rural districts, but not with much permanent result.

His labours were too great for his bodily strength. In the session of 1836, the last of that House, he travelled to Halifax in a covered sleigh, in which a small stove was fitted up for his accommodation, and was able to attend to local county business at his rooms, but was unable to occupy his place in the House. In the year 1838, his mind also gave way, and he sunk into a state of mental imbecility, from which he never recovered. He died 13th July, 1840.

We may mention here that Alexander Lawson was an apprentice in the *Patriot* office, and afterward established and still conducts the Yarmouth *Herald*, the first successful venture in newspaper printing, in the Western part of Nova Scotia, and long the only supporter of the popular party in that section of the country.

We may here give the subsequent history of the newspaper press in Pictou. On the 31st August, 1832, Mr. Milne commenced publishing from the *Patriot* office, a small weekly paper for the young, called *The Juvenile Entertainer*, at the rate of 5s. per annum. It continued for a year or two to give selections of interesting reading for the young, and was a creditable effort for the time, being the first of the kind in the Province.

The Government party in the year 1831 established a paper in opposition to the *Patriot*, called the *Pictou Observer*, of which the Rev. Kenneth John McKenzie was the editor, or in which he was at least the ablest writer. The *Patriot* expired about the year 1838, and the *Observer* followed it to the same bourne.

In the year 1836, Mr. James Dawson purchased the press and types of the old *Patriot*, and commenced a paper called the *Bee*, and soon after the *Observer* was resuscitated by Mr. Roderick McDonald, a native of Stornoway, who had taught the lower branches in the Pictou Academy. He removed to Ontario, and the paper became defunct. In the year 1840, the *Bee* was bought out by Mr. John Stiles who established in its place the *Mechanic and Farmer*. In 1842 the *Presbyterian Banner* was established under the editorship of the Rev. James, now Dr. Ross. But in 1843 both these papers were merged in the *Eastern Chronicle*, which has continued to the present day. In this office was trained E. M. McDonald, who became its editor and proprietor, and afterward Queen's Printer, and with Hon. William Garvie, established the Halifax *Citizen* and became member of the Dominion Legislature, and died Collector of Customs for Halifax. The *Observer*, after a short suspension, was revived by Mr. A. McCoubray, of St. Johns, Newfoundland, Martin I. Wilkins, Esq., being its editor or chief contributor, but again became defunct. In its place was established the *Colonial Standard*, which still continues.

We have already described the rum drinking of former times, but have now to notice the commencement of a movement for the suppression of its evils. The necessity of some measure of the kind may be inferred from the following facts: In the year 1825, there were imported into the Province 753,786 gallons of rum, besides 30,000 gallons of wine, and several thousand gallons of gin and brandy, to which the quantities smuggled, and what was made in the Province required to be added. When we consider that the population of the Province was estimated at 120,000, it will be seen, that even allowing for what was exported, the consumption might well be regarded as truly alarming. Again, in the year 1830, there were entered at the Pictou custom house 73,994 gallons of ardent spirits, and it was calculated that what of this was exported, would be equalled by the product of domestic distillation. The population of the county, by the census of 1827, was scarcely 14,000. Allowing for increase, the consumption would still be about five gallons for every man, woman, and child, the cost of which could not be less than $60,000, or $20 for every family, and $4 for every individual.

The evils of this had been long felt. But hitherto good men believed, that the use of ardent spirits in moderation was beneficial and even necessary, but now was started the idea of total abstinence from them as a beverage, and to the West River belongs the honor of having formed the first society on this basis in Nova Scotia, the second in British America, one in Ontario having been organized a few months earlier. The *Boston Recorder* had been circulated for some time in the settlement, and had rendered the people there familiar with the subject. The first movement however for the formation of a society, took place at one of the quarterly meetings of the Agricultural Society. These meetings had lost their interest, and the attendance at them was small. When therefore the members met in October, 1827, there being little doing, Mr. George McDonald moved that they form a Temperance Society. The Rev. Duncan Ross immediately seconded the proposal. The only other supporter at the meeting was Mr. Donald McLeod, and from these three the movement originated. The next to join them was Mr. David McLeod. These four held several private meetings, and at length arrangements were made for the public organization of the Society, which took place at the next meeting of the Agricultural Society, in January, 1828, when 12 persons signed a temperance pledge.* The following is a copy of it:—

---

* We have given the above dates, as we received them from the late George McDonald. The claim of the West River Society to be the first in the Province was for a time disputed, on behalf of the Beaver River Society, in the County of Yarmouth. The matter was discussed forty years ago, when the parties were alive and the records in existence, and it was then clearly *proved* that the former was the first. We regret, that the Society's book has disappeared within a short period, so that we are indebted to tradition for the above dates, more particularly as Rev. Mr. Campbell, in his history of Yarmouth County, in ignorance of these facts, has revived the claim of the Beaver River

" We whose names are hereunto annexed, believing that the use of ardent spirits is not only useless but hurtful to the social, civil and religious interests of men, agree that we will not use them, unless in case of bodily hurt or sickness, that we will not, as an article of luxury or living, traffic in them, nor will we provide them for the entertainment of our friends, or for persons in our employment, and in all suitable ways we will discountenance the use of them throughout the country."

The movement excited great opposition, and the members had to encounter no small amount of ridicule, if not worse. It was then considered impossible to do any work, particularly any job requiring a number of men, without rum, and an opportunity came, to test the principles and power in this respect of the friends of the new movement. One of them had the frame of a barn to raise. At that time it was customary, to make the timber of frames very heavy, and in raising them, first to lift the whole of one side, and then of the other, by main strength. For this of course, in the case of a building of any size, there would be required a large number of men. On this occasion, all the neighbours as usual assembled, and all the Temperance men for some distance round. The others however refused all assistance, if there was to be no liquor; and the friends of the new movement, having said, that in that case they would raise it without their aid, were left to try their strength. But on attempting to raise the side, they found themselves unable to move it, and after they were fairly beaten, and endured no end of jibes from the other

---

Society. There is still however sufficient evidence to show its groundlessness. Mr. Ebenezer McLeod, who was for some time Secretary of the West River Society, not only remembers the old discussion, but from his recollection is able to affirm, that that Society was formed as early as the date given by Mr. McDonald. The Rev. Dr. Blaikie of Boston testifies, that he was teaching at West River in 1828, and in that year joined the society, which was already organized. Again in the *Colonial Patriot* of date 17th September, 1828, the editor, urging the formation of such societies, says, " We are happy to state that one has been organized at the West River in this district, and would recommend to the office bearers the propriety of publishing its constitution." But by Mr. Campbell's own statement, the Beaver River Society was not formed before April 25th, 1829.

party, the latter laid hold, put the whole up at the double quick, and then had their dram from a supply which they had privately brought.

On another occasion, at the raising of the frame of a mill at Six Mile Brook, the two parties quarrelled, and as neither would yield, and neither was strong enough to do the work alone, they separated for that day, without its being accomplished.

This state of things did not continue, for in the *Colonial Patriot* of 17th September 1828, we find the following:

"On Friday last, the frame of a large dwelling house, the property of George McDonald, was erected without the use of *rum*. In lieu of it, ale and beer were used, so that the work was completed in a superior manner, while neither abusive language nor profane swearing was heard, no black eyes nor drunken men seen, but peace and friendship pervading the concourse. That this change of custom will be followed in future, (at least to a great degree) may be reasonably expected, since it tends not only to promote the harmony, health and respectability, of those who assemble on such occasions, but the interests of the builder. Ten or twelve years ago, he must have used almost as many gallons of the mighty rum, in erecting a frame of similar dimensions, and for this not unfrequently have his name stationed on the wrong side of some ledger, whence it may not be so readily erased, as some purchasers of spirits allow themselves to believe."

Not content with the promotion of the cause in his own congregation, Mr. Ross advocated it in the public press, pled in private with his brethren in the ministry on its behalf, and preached and lectured on the subject in their congregations, as he had opportunity. It was not, however, till the year 1830, that the first movement was made for the formation of a society in the town. It began with a sermon on the subject preached by him in the old Presbyterian Church. The discourse gave considerable offence, and even as the audience retired, some gave audible expression to their dissatisfaction, in such sayings as, "he might have given us something else than the like of that," &c. This was followed by a private meeting at the house of Mr. James Dawson, when he, the Rev. James Robson, James Hepburn, Francis Beattie and three or four others, associated themselves under a temperance pledge. They held several private meetings, and after some time agreed to call a public meeting, for the purpose of more formally organizing their Society. This took place on the 15th March following, in the old court house. It was well attended, but largely by people opposed to the movement, among whom were a number of rowdyish characters, who occupied the back part of the room, and who had been put forward by the rum interest to make disturbance. They had got a well known negro, named John Peters, well primed with liquor, as their chosen instrument to spoil the meeting. Accordingly, when Mr. Dawson had spoken in advocacy of the proposed Society, they set John forward to have his say. "Fine man, Missa Dawson, go into West Ingy trade—bring hun'eds puncheons of rum, make plenty money," &c. This rejoiced the rabble, who supported the speech by a volley of eggs at Mr. Dawson. A merchant of the place then arose and spoke at considerable length against the proposal, when the laugh was rather turned on him and his friends, by their chosen champion, Peters, exclaiming, "I secken Missa ——'s motion, dem's my sentiments." The friends of temperance, however, succeeded in adopting a constitution for their Society and opening a book for subscribers, and a few days after it was announced that it had received forty-four signatures. We have no list of names, nor of ths first office-bearers, but we know that from this time it received the support of some of the most respectable members of the community, among whom Jotham Blanchard deserves special mention.

The following from the *Patriot* of June 26, 1830, however, shows that the cause had been making progress :—

"We barely noticed, several weeks ago, a launch of a vessel from the yard of John Gordon, Jun., another native Nova Scotian. We were well pleased with the name (Patriot), of course, but we were better pleased to learn, that she was built and launched without the use of ardent spirits. We have since learned that she was sailed to Newfoundland and sold, and still no spirits used."

About a year elapsed before another public meeting was attempted, when a lecture was announced to take place in the old Grammar School house, by Jotham Blanchard. But he had not above two dozen of hearers, as he expressed it, not as many as the pages he had written. No further attempt was made at any public demonstration, till October of the following year (1833), when the Rev. John McLean delivered a lecture in the old court house. The attendance was large, the audience respectable and orderly, and from the eloquence of the speaker and the strength of his facts and arguments, the lecture made a profound impression. A vote was passed requesting its publication, which took place a few months after. From that time temperance has had a firm hold in the town, though we can recollect a time after this, when there was still scarcely a shop in town which had not over its door the words "spirituous liquors by license."

We may add that for the purpose of combined effort on behalf of the object, a Central Society, composed of representatives from the various Temperance Societies in the County, was formed on the 7th March, 1832, and called the Pictou Temperance Union.

We must now refer to the mail arrangements and improvements in travel made at this period. Ezra Witter, who had removed from the western part of the Province, and settled at Bible Hill, Truro, where he engaged in carriage building, commenced about the year 1815 carrying the mail from Halifax to Truro; and in conjunction with him, Jacob Lynds, carried it from Truro to Pictou. For some years they used a chaise drawn by a single horse, but afterward drove a double seated waggon, carrying three or four passengers, drawn by two horses, making one trip each way every week, the journey being performed in two days or two and a half. They continued in this way till 1828. The following is their advertisement in that year:

EASTERN STAGE.
To run once a week between Halifax and Pictou,
By E. WITTER and J. LYNDS.

THE public are respectfully informed that until the middle of November, the subscribers intends to run a weekly Stage, which will accommodate four passengers between Halifax and Pictou. It will start from Mr. Boyle's in Halifax, every Tuesday morning, at seven o'clock, reach Truro on Wednesday at 7 A. M. and arrive in Pictou at 8 in the evening. It will leave Pictou one hour after the arrival of the packet from Prince Edward Island, and arrive in town on Saturday afternoon. The fare to or from Pictou will be £2, and every exertion will be used to insure comfort and security to passengers, and their baggage, of which each will be entitled to carry 20 lbs.—Apply in Pictou to Mr. Robert Dawson—in Halifax to Mr. A. Boyle, where any other information will be given.

June 18.

In that year a company was formed of persons in Pictou and Truro, with one or two in Halifax, called the Eastern Stage Coach Company, to run a line of coaches between Halifax and Pictou. In the following year (1829) they began with a heavy double seated waggon, drawn by two horses, which made the journey in two days and then, with the same horses and driver, returned on the two following. They then erected a frame over this, which was covered with canvas on top and had curtains at the sides. The next year the company was enlarged, and their carriages were drawn by three or sometimes four horses, though often in changing, it was only transferring them from the pole to the lead. They also put on double sets of horses, and they now left Halifax and Pictou, on the same day, making three trips a week each way. Proper coaches were put on, though not we believe till a year or two later.

This arrangement continued under various proprietors, and sometimes with opposition lines, till the year 1842, when Hiram Hyde purchased the establishment, then owned by Arnison and Trenaman, and commenced daily coaches, leaving Pictou at 4 o'clock each week day morning, and Halifax at 6 o'clock, and making the journey in fourteen to eighteen hours, according to the roads and other circumstances. This arrangement continued till the building of the railroad.

When the stage began to run, the road went over the highest hills, but about this time commenced the system of making level lines of road. The credit of inaugurating the new era of road making, is due to Sir James Kempt, then Lieutenant-Governor. He was not long in Nova Scotia, till he began to use his influence to change the whole system of road expenditure. He employed Mr. George Whitman, as surveyor on the eastern road, and when he reported that he had found a line with a rise of not more than one foot in thirty, people laughed at him. The commencement of level roads was made along the Grand Lake in 1828. Previously the road had gone round the basin, but now the road was taken to Dartmouth, shortening the distance some miles. About the same time the work of levelling began in Pictou. Sir James passing over Mount Thom, his eye at once saw the valley below, where a level line might be easily obtained. Soon after the work of alteration began, and about the year 1832, the present line, by which it was said the Governor "circumvented Mount Thom," was completed. It still, however, crossed the Six Mile Brook at Gass's place on what was called the Kempt Bridge, in honour of him, and came out at the West River at Mrs. Brown's place. The work of alteration continued, till in the year 1840 the whole line was completed.

We may add here that it was still some time before the process was completed, in regard to the road toward the east. About the year 1847, the road from New Glasgow round Frasers Mountain was made; in the year 1850 the road at the foot of Green Hill was completed, and in 1851 the road to Antigonish by the Marshy Hope Valley was opened. Previously the travellers in that direction had only the choice of a long round by the Gulf Shore, or a road ten miles shorter, but scarcely fit for carriages, over the Antigonish Mountains.

On the 3rd March, 1830, died Dr. J. McGregor, and we may say, that no man was ever more warmly loved while he lived, nor more deeply mourned when he died. Hundreds of homes were filled with weeping at the intelligence of his departure, and far beyond the bounds of the county, multitudes mourned him as a father and friend. On the Saturday following, devout men carried him to his burial and made great lamentation over him. His funeral was the largest that had ever been in the county, and with all the increase of population, probably larger than any since, it being estimated that there were 2,000 present. A monument was erected to his memory, with the following inscription, a copy of which was kept framed in many houses throughout the county. But it has been replaced by another :

"AS A TRIBUTE

OF AFFECTIONATE REGARD FOR THE MEMORY OF THE LATE

JAMES MACGREGOR, D. D.,

The first Presbyterian minister of this district, who departed this life, March 3, 1830, in the 71st year of his age, and the 46th of his ministry, this tombstone was erected by a number of those, who cherish a grateful remembrance of his apostolic zeal and labours of love.

When the early settlers of Pictou could afford to a minister of the gospel little else than a participation of their hardships, he cast in his lot with the destitute, became to them a pattern of patient endurance, and cheered them with the tidings of salvation. Like Him whom he served, he went about doing good. Neither toil nor privation deterred him from his Master's work, and the pleasure of the Lord prospered in his hand. He lived to witness the success of his labours in the erection of numerous churches, and in the establishment of a Seminary, from which these churches could be provided with religious instructors. Though so highly honoured of the Lord, few have exceeded him in Christian humility ; save in the cross of our Lord Jesus Christ, he gloried in nothing ; and as a public teacher, combining instruction with example, he approved himself to be a follower of them who through faith and patience now inherit the promises."

The year 1831 was marked by the commencement of steam navigation from the port of Pictou, and, indeed, on the coast of British America. The pioneer boat in this trade was built at Three Rivers, on the Lower St. Lawrence, for a company formed in Quebec the previous year and was called the Royal William. She was of 1,000 tons burden, and had engines of 180 horse power. This was considered enormous in those days, and in all the ports she visited she was regarded as a wonder. She was intended to ply between Quebec and various ports in the Lower Provinces, in fact to do the work that the Gulf Ports and other lines of steamers are now doing, and was aided by the Canadian Government. She made her first trip in August, arriving in Halifax on the 31st, in seven days from Quebec, having been detained in Miramichi two days. Crowds assembled on the wharves, with almost the feelings that the appearance of the Great Eastern would now excite. She arrived in Pictou on the 3rd September, and we still remember the excitement which her presence created. She made several trips that season, ending her voyage in Halifax, as required by the act giving her a subsidy. Her first summer's work showed the folly of her builders. Not only was she far larger than was needed, but she was fitted up in a style of elegance, that would compare with the floating palaces of the Hudson or the Sound. On her first arrival, the editor of the *Patriot* pointed out the mistake that had been committed, and while advocating the enterprise, urged that the company should get a boat one quarter of the size, and fitted up in a substantial but plain style.

The next year it was arranged, that she should run regularly to Pictou, the Legislature having agreed to give the subsidy on voyages terminating here. But on her first trip she left while cholera was raging in Quebec, and when she arrived in Miramichi, she had the disease on board, and was sent to quarantine, where the engineer died. Afterward she only made one or two trips that seanson.

The next spring she was sold. Her original cost was £17,000, but now she did not bring one-third of the amount. Her new owners sent her one or two trips on the old route, but finally determined to send her to Britain. She arrived here on the 13th August, on her way thither and cleared again on the 17th for London,* where she safely arrived, being the first steamship to make the entire passage across the Atlantic under steam. Previously several vessels had crossed partially by the aid of steam, but these made their way principally by sails, steam being used only when a wind was wanting, and even then only at a low rate of speed. But now a Canadian built ship, sailing from Pictou, first proved the practicability of ocean steam navigation, and introduced a new era in the trade of the world.

In the year 1832, the General Mining Association purchased the steamer "Pocahontas," which commenced to ply between Pictou and Charlotte Town, sometimes going as far as Miramichi. She was commanded by David Davidson and made her first trip on the 11th May. In the year following, they sent a large steamer called the "Cape Breton," which commenced to ply between Pictou and Miramichi, on which route she was employed for some years.

Another institution formed near the close of this period deserves notice here. We refer to the Pictou Literary

---

* "Cleared, 17th. Ship Royal William, McDougall, London, Coal, Natural curiosities and spars, by W. Mortimer."—*Patriot*, August 20th, 1833.

and Scientific Society. It originated with the following paper :

"We, the undersigned, agree to meet at Mr. Blanchard's class room, in the Pictou Academy, on Monday evening, December 8th, 1834, at seven o'clock, to make arrangements respecting the formation of a literary society, such as may be considered most beneficial to the interests of all concerned.

"W. J. Anderson, G. A. Blanchard, W. B. Chandler, Daniel Dickson, David Matheson, Joseph Chipman, James Fogo, Jas. W. McCulloch, Wm. Burton, Edward Roach, Jas. Purves, George S. Harris, Wm. Gordon, John B. Davison, James Primrose, David Crichton, C. Martin, James Johnston, Robert Corbet, Michael McCulloch, G. M. Johnston, A. P. Ross, Charles Elliott."

Accordingly a meeting was held at the time appointed, James Primrose, Esq., in the chair, and George S. Harris, Secretary, when it was resolved that "the meeting form themselves into a society, to be called the Pictou Literary and Scientific Society." The object was stated to be "the mutual improvement of its members in the sciences and general literature," and it was agreed that this object may be best attained by the delivery of lectures or essays on literary and scientific subjects, which afterward may form topics of discussion.

The first lecture was delivered on the 16th of the same month, by Dr. W. J. Anderson, on phrenology. There were some present who had read the discussions in the *Edinburgh Review* on the subject, and the lecture was followed by an animated debate, which was continued at the next meeting, when Dr. Martin gave an address on the brain. That winter lectures were delivered fortnightly, ten in all.

The Society continued in existence for twenty-one years. During this time, it had every winter a course of lectures, sometimes fortnightly and sometimes weekly. From the Pictou Academy there had been diffused a taste for literature and science, and many of the lectures were of a high character. Several clergymen, such as Dr. McCulloch, Mr. Trotter, Mr. McKinlay, and Mr. Elliott lectured with more or less frequency. Conspicuous among the lay lecturers were J. D. B. Fraser and J. W. Dawson. The former generally lectured on chemistry or some kindred subjects, and he showed a skill in experiments, which rendered his lectures highly interesting and popular. Mr. (now Dr.) Dawson delivered his first lecture in April, 1836, the subject being geology. Though then a young man, he already gave evidence of that attention to natural science, in which he has since attained so much distinction. Afterward he frequently lectured on that and other branches of natural science. The medical men, such as Drs. Anderson, Chipman, and Martin, lectured on scientific subjects kindred to their profession; members of the legal fraternity, such as Daniel Dickson, James Fogo, George A. Blanchard, John McKinlay, and Hiram Blanchard, discoursed on a variety of general subjects, while mercantile men, such as T. G. Taylor and Charles Robson, and others, contributed their share to the usefulness of the Institution. Altogether, the lectures were in a style superior to anything in the Province. By those who had an opportunity of judging, they were pronounced in general of a higher character, even than those delivered in the Halifax Mechanics' Institute. They were frequently followed by discussions, often animated, sometimes even exciting, giving rise to displays of wit or oratory, or eliciting valuable information. The society afforded many an evening's instructive entertainment. But from various causes, interest in it declined, and it finally expired, its last meeting having been held on the 12th April, 1855.

To this account of a creditable effort for the diffusion of the light of knowledge, we may add as a close to this chapter, that the lighthouse at the Beaches was finished in the year 1833, that the lantern was raised to its place in August of that year, and it was first lighted on the 1st of March, 1834.

CHAPTER XVIII.

MINES AND MINING INDUSTRIES OF THE COUNTY.

Since the failure of the timber trade, perhaps nothing has been so important to the progress of the county, as the Coal Mining, carried on first by the General Mining Association, and later by other companies, which have made Pictou up till this time the greatest coal producer in British America, it having been only during a few years surpassed by Cape Breton. We have already mentioned the discovery of coal in 1798, Dr. McGregor's exhibiting a fire of it to the candidates at the election of 1799, and the first efforts at coal mining. We shall now proceed to give the history and present condition of this industry in the county.

In the year 1807, John McKay, son of the Squire, usually known as Collier, obtained a license to dig for the inhabitants, and at a later date, to export. He and his father commenced working a small three feet seam on the farm of the latter, but it soon became exhausted. They then searched further and found what has since been known as the "Big Seam," though they did not know its value. John continued to work at this for some time, selling it at the pit's mouth and sending it down the river in lighters. A demand sprang up for it during the war, to supply the garrison, navy and inhabitants of Halifax. In the year 1815, we find 650 chaldrons exported. After the peace, the price fell to half its former rate. Owing to this and perhaps other causes, McKay failed, and was imprisoned, and his property seized by Hartshorne of Halifax, who had been supplying him. The workmen being unpaid, the latter tried to compromise with them, but they persisted in claiming full payment of what was due. Mr. Adam

Carr, who was one of them, joined with Mortimer, and by his influence, the Government were induced to let the mines to the highest bidder, and in that way they obtained the lease in the year 1818. They worked together till Mortimer's death in the following year, when on the 3rd November, the lease was transferred to George Smith and William Liddell, on the following terms, the Mine on the west side the river for a rent of £260 and 3s. per chaldron for all raised over 400, and that on the East side the river, for £110. We may mention, that this last has never been found productive of good coal. It is the same that a few years ago was opened by the German Company. Smith and Carr worked in partnership, but after a time separated, when the latter got the whole into his possession, and continued to mine, raising the coal by horse power, selling it at the pit mouth, and carting it to the river, where it was sent away in lighters.

Of these years, we may give a statement of the amount of coal raised, as reported to Government.

```
1818.............................. 2820 chaldrons.
1820.............................. 2609     "
1821.............................. 1370     "
1822.............................. 2004     "
1823.............................. 1725     "
1824 ............................. 2261     "
1825.............................. 2801     "
1827.............................. 2523     "
```

In the year 1825, the home Government leased all the reserved mines of Nova Scotia for sixty years to the Duke of York, excepting, of course, those which had been already leased to other parties. Sir James Kempt, in laying before the Council correspondence on the subject, intimated that he was authorized to state, that the reserved profits of the mines would be applied to the benefit of the country. This was a transaction which no person in the present day will defend, and which subsequent British ministers have acknowledged themselves unable to justify. It had this compensating effect, however, that it introduced into the country a wealthy company, at a time when the same capital could not have been easily obtained. The Duke's lease was transferred to Messrs. Rundell, Bridge and Rundell, the celebrated London Jewellers, in payment of his debts, and from them to the General Mining Association, in which, I believe, they were large shareholders. The company had been formed, as the name imports, for mining purposes generally, and, I have been informed, did attempt the working of mines in South America. But for a length of time, their attention has been confined to the coal mines of Sydney and Pictou.

On obtaining their lease, they sent an agent to the Province to explore for mines, and, on his report, resolved to commence operations at the East River. They purchased Mr. Carr's lease, and having about the same time become possessors of the rights of the lessees of the Sydney Mines, they thus came into possession of all the mines and minerals in the Province, with the exception of what might be found on a few old grants, on which there had been no reserve. Early in the summer of 1827, they sent out Mr. Richard Smith, intending to commence operations both in coal and iron mining. In June a vessel arrived in Pictou, bringing machinery and implements, with colliers, engineers and mechanics.

On the 11th June, the Lieutenant-Governor issued a proclamation, calling on the officers of Government, magistrates and proprietors of land, to give every facility to Mr. Smith in carrying on his operations. He accordingly made all necessary arrangements for working on a large scale. He purchased the farms of Dr. McGregor, William McKay, and Colin McKay, commenced sinking new shafts, 212 feet, and erecting the proper machinery for working on a large scale and in a more scientific manner than hitherto. On the 6th September, their first coal was raised, and in the month of December, he had a steam engine in operation, the first ever erected in the Province. The event was thus noticed in the local paper of the day :—

" The same day on which our first number appeared (December 7th, 1827) another event happened which we may with great propriety, hail as the harbinger of illimitable prosperity to Pictou, of great utility to the whole Province, and we might fairly add to the British North American colonies.' On Friday last, for the first time in Nova Scotia, the immense power of steam was brought into successful requisition at the Albion Mines on the East River. Let us rejoice that this district is the favoured scene of its first operation. The engine is of 20 horse power, and the perfection of its first operations evidence the skill of the engineer. The Company's works will now proceed with redoubled celerity and vigour. Their progress, though retarded by the selfishness and overreaching disposition of individuals, has surpassed the imagination of individuals."

Before describing the Company's operations farther, we must give a brief account of the position and structure of the coal fields of this county. As formerly mentioned, the southern portion of the county is occupied by rocks of Silurian or other formations of the older geologic eras. At their northern base are found Lower Carboniferous rocks, with limestone and gypsum. Then come the Newer Carboniferous, or coal measures, occupying the whole front of the county. These, however, are divided by a remarkable formation, known as the Great Conglomerate, which extends in an east and west direction, crossing the East River below New Glasgow, the Middle River at the bridge near Alma, and the West River near Durham, and forming the eminences of Frasers Mountain and Green Hill. To the north of this range some small seams of coal have been found, at Carriboo, Merigomish and the south side of Pictou Harbour, but none large enough for profitable working, and it is yet a question whether any may be expected. Dr. Dawson regards the rocks as of the upper coal measures, or Permo-carboniferous, and therefore not productive, though he expresses a hope that good workable beds may yet be found at greater depth.

At present, however, all the valuable coal seams known, are found on the south side of the Conglomerate, and near its base in the great coal basin of the East River, and its extensions eastward and westward. Of these, the most important is that commonly known as the big seam or the main seam, the enormous size of which is one of the most remarkable phenomena of the field, in which respect we do not know that it is paralleled in the world. It is from this that most of the coal yet mined has been taken. The whole thickness of the seam vertically is 40 feet, or a little over, but in a line perpendicular to the dip 38 or 39. This, however, is not all good coal, there being several bands of ironstone through it, and some portions of the coal are inferior. Dr. Dawson, who is moderate in his calculations, says that at least 24 feet of good coal may be taken out of it.

A cubic foot of this coal, according to the same authority, weighs about 82 lbs., rather less than 28 feet being equal to a ton of coal. Hence, a square mile of this seam would yield in round numbers 23,000,000 tons.—This coal is a highly bituminous baking coal, and is shown by Professor Johnston's trials to possess high qualities as a steam producer, one pound being capable of converting 7.45 to 7.48 pounds of water into steam.—The greatest objection to it is, that it contains a considerable quantity of light, bulky ashes. Hence, it is not so much esteemed for domestic use, as the better qualities of Sydney coal. But otherwise it possesses very high qualities. It burns long, gives a large amount of heat, is free from sulphur, and remains alight much longer than most other coal.

Next in importance to the main seam, is what is commonly called the deep seam, about 150 feet below the first. It is altogether about 25 feet in vertical thickness, but it also is divided by ironstone and impure coal into three layers of good coal, making together, according to Dr. Dawson, about twelve feet in thickness. The quality

of some portions of this seam, is superior even to that of
the main seam, but these layers prevent its being mined
so economically, but only its nearness to so large a seam
prevents its being worked extensively.

Next in value, though not next in order, is what is
known as the McGregor seam, which lies at a depth of
about 280 feet below the deep seam. It is about twelve
feet thick. The two upper veins, amounting together to
nearly six feet, have been worked, and found to be of
good quality, though requiring care in removing the
shaly band which separates them.

About five feet above this, there is a small seam, between
three and four feet thick, of good quality. Between this
and the deep seam are two other small seams, each about
four feet in thickness. These would be valuable elsewhere, but in the presence of such large seams, we need
not expect them to be worked for a length of time.

About 240 feet below the McGregor seam, is a peculiar
bed known as the "Stellar"* or oil coal, so called from
its peculiar scintillations in burning, which some time
ago attracted attention for its yield of oil. The following
is its arrangement and composition:—

	ft.	in.
Inferior bituminous coal	1	2
Oil coal	1	8
Bituminous shale	2	0

The central portions of this have been found to yield
120 gallons of oil to the ton; it and the shale together, 75
gallons.

These seams all lie conformably. At the southern
outcrop on which working commenced, they dip to the
northeast at an angle of about 20 degrees, and the strike
is about north-west.

Lately another seam has been discovered overlying the

---

* Hence the name Stellarton has been given to the village adjacent.

main seam. It is found on the northern part of the coal field, probably to the base of the Conglomerate. It is said to be five feet nine inches thick, of pure coal of good quality, but so far as we are aware, the seam has not been explored, nor the coal analysed.

Having thus described the position and characteristics of the different seams, we return to the operations of the General Mining Association.

To get their coal to market, they constructed a railroad from their works to a point a little below New Glasgow, on which they hauled the coal by horses. Here shoots were erected, and vessels drawing not more than six feet of water were loaded. To load larger vessels, they constructed lighters, in which the coal was conveyed to the Loading Ground, as it is called, at the mouth of the River.

In the *Patriot* of January 28, 1829, the progress of these works is thus noted :—

" The progress of the Mining Company appears to be daily becoming more important. The foundry is in successful operation, and railways (rails) are now casting, and will be ready for laying down in the spring, for the purpose of facilitating the conveyance of the coals to the navigable part of the river. A considerable addition to the number of lighters to be employed on the river, will be made in the course of the winter, and a wharf or place of deposit at this town, which will contain several thousand chaldrons of coal for exportation, is contracted for. It is also gratifying to learn, that orders have been received by the late packet to build two steamboats (the machinery for which will arrive in spring), one of thirty horse-power, for the river navigation, and one of 100 horse-power for the purpose of coasting and carrying the coals to market."

The foundry was under the charge of a man named Onions, but did not do much till the arrival of W. H. Davies, Esq., in 1830, to take charge of it, and who may be regarded as the father of the iron foundry business in Nova Scotia. The boilers for the foundry and pit engines were put together, at John McKay's blacksmith shop, near where Russell's now stands, the plates and rivets having been brought out from England. They were then pushed over the wharf, to the amazement of most people, who expected such immense articles to go to the bottom. But they floated lightly, and were towed up the East River, as far as the water would bear them, when they were landed on the intervale and dragged on rollers to their place.

A wharf was built at Pictou, long known as the miners wharf, and for a time coal kept there, but a depot there was found unnecessary and unprofitable. As mentioned above, the hull of a small steamer, intended to be used for towing lighters and vessels, was launched from the shipyard of Mr. George Foster, Fishers Grant, on the 19th August following. She was called the "Richard Smith." She was towed up the river, where she remained all winter receiving her machinery, and did not commence her work till the following summer. The *Patriot* of July 17th, 1830, contains the following announcement: "We stop the press to announce that the Steamboat 'Richard Smith' has just appeared in the harbour for the first time." It being the first time a steamer was seen on these waters, the whole town turned out to see the marvellous spectacle. She was at first commanded by Capt. McKenzie and continued for some time to ply on the harbour.

This plan of loading being slow and tedious, they next resolved on deepening the river. For this purpose they obtained an act of the Legislature, giving them full authority over the river, so that no vessel drawing over six feet of water was to enter without their permission, and only by paying toll to them. But, in passing the act, the Assembly, which had resented the act of the British Government, in transferring our mines and minerals, added a clause to the effect, that the bill was not to be construed, as admitting the right of the home authorities, to dispose of our mines in the way they had done. In consequence of this, the act was disallowed at Downing Street, and, at the same time, a feeling of opposition rising in the country against such a monopoly, the company did not renew their application for similar power. They continued, therefore, to ship their coal in the manner described for several years, but the demand was greater than they could supply, and the long delay of vessels in receiving their cargoes, was a great discouragement to the trade. In the meantime, the use of locomotives on railroads had been tried successfully in England. Accordingly it was resolved to build a railroad from the East River to the Loading Ground, for the conveyance of their coal in that way. The road was laid out and operations commenced in the year 1836.

The surveys and plans were made by Peter Crerar, Esq. When they were sent to Britain, it was proposed to send out an engineer to superintend the construction of it. But on his plans being submitted to a competent engineer, the latter said that they needed no better superintendent than the man who prepared them. In consequence, the supervision of the work was entrusted to him, although he had never seen a railroad, and he accomplished it satisfactorily. It was opened in the year 1839, when the first locomotives in British America ran upon it. There were three of them, built by Timothy Hackworth, who competed with Stevenson at the first trial of locomotive engines in England. They were of great power, but slow. They continued doing their duty regularly till lately, when one of them was taken down, but the other two are still at work.

The opening of the railroad was made the occasion of general rejoicing. The two steamers, Pocahontas and Albion, with lighters attached, each carried from Pictou about 1,000 persons to New Glasgow, whence they were taken by train to the mines. Crowds of people on horseback and on foot were here assembled from all parts of the county. Here a procession was formed of the various trades, the Masonic lodges, the Pictou Volunteer Artillery Company, and visitors mounted, with bands of music and pipers at intervals, and various banners, marched to New Glasgow and back again, when the Artillery Company fired a salute. A train of waggons, fitted up to receive passengers, had been attached to each engine, and, being filled with the crowd, now made the first trip to New Glasgow and back again, giving a new sensation to multitudes.

On their return, a feast was given to the employees of the Company, for which 1,100 lbs. of beef and mutton, with corresponding quantities of other articles, were provided; a dinner was given to invited guests, and the night was spent in general festivity.

This railroad, we may mention, was six miles long, and so nearly straight that the least radius of any of its curves was 1300 feet. Its width was 18 feet. The estimated quantity of excavation was 400,000 cubic yards. At the terminus was a wharf 1500 feet long by 24 feet broad, commanding a fall of 17 feet above high water level at the shoots. The rails were of malleable iron, and the estimated cost $160,000.

The first operations of the General Mining Association were on the low ground, close by the East River, where an engine pit was sunk to the depth of 400 feet, and about 250 yards from the outcrop. Here they mined for some time 12 feet of the upper part of the main seam, the lower being regarded as inferior, over a tract of about 800 yards to the west and 250 yards to the east, and covering an area of about 40 acres. In working to the eastward, however, the coal was found to deteriorate in quality. On the 29th December, 1832, at an early hour in the morning, the works were discovered to be on fire. On the day preceding, nearly 100 miners and 14 valuable horses had been at work in their several places in the pits. The men retired from their work between 5 and 6 p. m., leaving the horses as usual in the places provided for their accommodation under ground. On the following morning, when the men assembled for work, to their consternation they found several of the shafts emitting dense volumes of smoke. Immediately the Mines' bailiff, with two other persons, descended one of the ventilating shafts, when the works were discovered to be on fire in several places, and all the horses dead from suffocation.

The intensity of the fire obliged them to make a hasty retreat to the surface, and immediately on their reporting the state of things, the manager instantly set all hands to work to cover the mouths of the pits, hoping by preventing the circulation of air, to stifle the flames. The pits were thus left covered for several weeks. When they were again opened, it was found that the fire had done extensive injury, and was still slowly burning. Having originated in the lower rooms, it was fortunately confined to that part of the works, and to the prompt closing of the pits must be attributed the saving of the upper. But as this had proved ineffectual for the total suppression of the fire, the managers were reluctantly compelled, toward the end of April following, to introduce the waters of the East River. This proved successful, but it required the power of their steam machinery for pumping, working night and day till the 14th December, to clear the pits of water.

An examination of the works led to the belief that the fire was the work of malice and design. An investigation took place under the direction of the Solicitor-General, William Hill, Esq., when a mass of testimony was taken, which left no doubt of the fact. Large rewards were offered for the discovery of the guilty, but they were never detected.

Several other fires occurred, but one in October, 1839, exceeded all the rest in severity. The heat was so intense that it melted the iron chains which were used for hoisting the coal out of the pits. These workings were in consequence abandoned, and have since been known as the Burnt Mines.

Farther to the dip, other shafts were sunk, now known as the Old Bye Pits, and others 960 yards to the west, known as the Dalhouse Pits ; and also one nearer the outcrop, known as the " Cage Pit," was sunk to the Deep Seam, which it reached at a depth of about 300 feet.

From the first of these, the workings were considerably extended east and west, the upper part of the main seam only being mined. In some workings to the dip of these, an accident occurred in May, 1861, which rendered it necessary to let in the water to extinguish the fire. An attempt was made to get into these workings in 1862. But such was their condition, and another fire having occurred in 1863, they were abandoned, and this district has received the name of the " Crushed Mines."

From the Dalhousie pits the main seam continued to be worked in its entire thickness, the lower portion being much improved in quality, and from the Cage Pit the deep seam still continues to be mined in its entire height.

During the year 1866, a new shaft was sunk to the face of the west workings. A steam engine for hoisting was erected, and a railway between the pit and the main line constructed. But from some unexplained cause, this pit, known as the Foster Pit, was found to be on fire in May, 1869. The place in which it was first seen was not being worked, but was near those in operation. Immediate steps were taken to extinguish the fire, but the rapid accumulation of smoke so overpowered the workmen, that they were obliged to resort to the plan of excluding the air, by closing the top of the shafts, and all other places by which it could enter the mine. The coal at this part of the seam had been found to deteriorate, and from the state of the mine in consequence of the accident, it has since been abandoned. This so affected the Dalhousie Pit, that it too was abandoned shortly after.

The last pit sunk by the Association is known as the Foord pit, which, in the costliness and efficiency of its equipments, is said to be unequalled in America. We may therefore give a particular description of it. The hoisting shaft strikes the main seam at a depth of 960 feet, but to the bottom of the seam it is 1,000 feet. Its size is 12 feet by 9 feet 6 inches, inside the lining; and it is divided into two compartments, with cross stays and slides, passing perpendicular to the sides of the shaft, and bolted to the cross stays, where the cages work in. The cages are double decked, each cage holding four boxes, and each box containing 12 cwt. coal. The winding engine is a double horizontal one, with cylinders of 36 inches in diameter and 5 feet stroke, and nominally of 160 horse-power, though capable of working considerably above this. The winding drum is 18 feet diameter, with two six inch iron wire ropes, which pass from the drum over two large pulley wheels 14 feet in diameter, which is elevated on a strong wooden frame, 30 feet above the top of the shaft. The ropes pass over the pulleys and connect to the cages. When the engine is put in motion, one of the cages goes down with empty boxes, and the other comes up with full ones. On reaching the surface, the boxes are passed to the screens, where the slack is separated, and the coal passes into the cars ready for shipment.

The pumping pit is 40 feet deeper. It is provided with an engine, known as the Cornish pumping engine, with a cylinder 52 inches in diameter, and a stroke of 9 feet, and nominally 260 horse-power. There are three sets of pumps, 18 inches in diameter, two what are called bucket pumps, the third known as a ram or forcing pump. At each stroke 100 gallons of water are brought to the surface, and the engine works 7 strokes a minute and 10 hours each day.

The winding and pumping engines are supplied with

steam from 10 large boilers, each 35 feet long and 5 feet 6 inches in diameter. When the mine is in full working order, it will produce 1,000 tons of coal per day.

The ventilation is produced by a Guibal fan, 30 feet in diameter and 2 feet wide, an instrument closely resembling a steamer's paddle wheel. It is placed at the mouth of what is known as the fan pit, which is 600 feet deep, and is driven by an engine of 70-horse power. The engine is supplied with steam from two boilers, each 25 feet long and 5 feet 6 inches in diameter. The air goes down by the winding and pumping shafts, circulates through all the works, making a course, it has been calculated, of 7 miles in length, and is expelled by this fan, which produces a current of air equal to 60,000 cubic feet in a minute. This is tested in the pits every day.

The Foord pit coal is noted for its excellent quality, for generating steam, for making gas and for making coke. There are at present 42 coke ovens, each 11 feet in diameter, making coke night and day from the slack coal, and a large addition to their number is contemplated. The coke is of superior quality for smelting iron ore, and is now used for that purpose at the Londonderry Mines.

From the Foord pit a drift level, 600 yards long, has been run to the deep seam, and by it and the Cage pit that seam is now mined.

The system of working pursued from the commencement of the colliery has been continued, with some modifications in the size of the pillars, which from the thickness of the seam, and its declination, often proved inadequate, and led to crushing of the workings. The bords are driven eighteen wide, and parallel to the levels. They are turned out of balance ways or headings, which are put up to the full rise at intervals of 150 yards, the width of the pillars between being eight or ten yards. These balance ways are used to bring the coal down to the horse road, on the principle of a self-acting incline;

the only difference being that the loaded bogie raises the
empty tub to the respective bord ends, and it is in its
turn taken back by a tub of coal, which exceeds it in
weight. The bords are driven in opposite directions
from these inclines, to shorten the putting.

The Company's works gathered around them a large
population. They own about four hundred houses,
which are occupied by their employees. These, with
the residences and places of business of others who have
been attracted hither, form a large village, which now
contains five churches: two Presbyterian, one Episcopal,
one Wesleyan Methodist and one Roman Catholic. The
population around these works necessarily made a demand
for farm produce, and afforded a ready cash market for
it, and this has been a great convenience to the rural
districts around.

In the year 1872, the General Mining Association sold
all their rights in the mines at Pictou to a new company,
known as the Halifax Company, of which Sir George
Elliott is chairman.

In the year 1856, the monopoly of the General Mining
Association was abolished, they retaining in Pictou four
square miles where they might select. The area, as
chosen by themselves, extends from the Albion Mines to
the upper part of New Glasgow, a distance of about two
miles, embracing the ground on both sides of the river,
but extending a greater distance to the west than to the
east of it. Exploration for coal immediately became
active, and in this work the late James D. B. Fraser, Esq.,
of Pictou, deserves special notice. He took out rights
of search in the neighborhood of the General Mining
Association's area, to the west and south. From the
strike of the large seams toward the west, it was to be
expected that they would appear to the westward toward
the Middle River, but for a time even scientific men were
baffled in tracing their course, and some came to the

conclusion that they became exhausted in that direction. Mr. Fraser spent a good deal of time and money in his explorations, but failed to find the Big Seam.

Finding, however, the Stellar coal, he ran two slopes into it, one 215 feet, the other 204, and commenced shipping it, along with its accompanying oil shale, to Boston, where there was manufactured from it oil of good illuminating quality. But the discovery of the oil wells of Canada and the United States, so lowered the price, that it was found impossible to compete with them. The work was therefore abandoned, and until either the supply from the oil wells diminishes, or other uses are discovered for oil, this vein is not likely to be again worked. He also formed a company, of which the principal shareholders are in New York, called the Acadia Company, which commenced working the McGregor seam, at the place originally worked by the Doctor. They spent a considerable sum in erecting buildings, and providing the plant necessary for carrying on extensive operations, when a fine seam of coal, since known as the Acadia Seam, was discovered about two miles to the south west of the Albion Seam, where the Nova Scotia company's works now are, which geologists regard as the equivalent of the main seam. We believe the credit is due to Mr. James Fraser, Mount William, of being the real discoverer, though a Connecticut yankee, named Truman French, reaped the fruits. Mr. John Campbell, by careful exploration, conducted in a scientific manner, traced it farther to the south. It was now found that about a mile and a half to the westward of the East River, the seams suddenly turned to the southward, and the line of outcrop continued for more than a mile in that direction, forming a sort of bay, which now forms the area of the Intercolonial Company. It was then found again to strike to the north-west towards the Middle River; then turning again toward New Glasgow, it has been again found with high dips to the southward nearly opposite the town. "The East River coal area," says Dr. Dawson, "between that river and the Middle River, would thus appear to constitute an irregular trough, with a deep bay to the southward." The seams in this direction, though regarded by geologists as the continuation of the seams on the East River, are considerably changed. Thus the Acadia seam has a thickness of about twenty feet, of which from sixteen to eighteen is good coal.

It will thus be seen, that as compared with the main seam at the Albion, it is diminished in thickness, but improved in quality. On the Intercolonial area, the "deep seam" has also been discovered with a thickness at right angles to the bed of eleven feet, and other beds, supposed to be the equivalents of the other beds on the East River.

The discovery of the continuation of the coal seams towards the Middle River formed a new era in coal mining in this county. The Acadia Company, abandoning nearly, if not quite entirely, all operations on the McGregor seam, on the area south of the General Mining Association's, commenced operations on the main seam on their area to the west, and for some years exceeded even the old company in the amount of coal raised. They built a railroad connecting their works with the Government line, and have sent their coal for shipment over it to Fishers Grant, a distance of about 13 miles.

Their slopes are now 1,575 feet deep, on an incline of 22 degrees. Levels have been driven to the boundary lines on either side of their area, and the seam is found to be remarkably regular, not a single dislocation or disturbance having been encountered in any direction. At present, the Acadia colliery employs 180 men and 20 boys, and produces 400 tons of coal per day, which is greatly below its actual working capacity.

The quality of their coal is excellent, having been tested for a great variety of purposes, and with the most satisfactory results. The mine is admirably equipped with all the best and most approved appliances for securing the safety of life and property in mines, and the works have been remarkably free from accidents.

Mr. John Campbell, who had first traced the coal seams to the southward, and obtained a lease of an area in that direction, sold his rights to a company in Montreal, of which G. A. Drummond, Esq., was president, called the Intercolonial Company. They immediately commenced developing their property. In the year 1868, two slopes were sunk to the dip of the large seam, usually known as the Acadia seam, and a pair of winding engines erected at their mouth. About 14,000 tons were mined the same year, and a large amount of preparatory work done. A railway about six miles long was constructed to the Middle River, where they had built wharves, and provided all the conveniences necessary for shipping coal in quantities. The railway was opened on the 1st of October. The ballasting, however, was not then completed, and from the lateness of the season, only between two and three thousand tons could be shipped. They have since erected a short line from their works to the Government road, by which they are enabled to send coal to Halifax and places along the line of the Intercolonial Railroad.

In the year 1869, the colliery, under the management of the late James Dunn, Esq., was fully equipped with everything necessary for the production, transportation and shipment of coal, and under the improved markets of the following years, the company's business rapidly increased, so that in 1872 their sales amounted to 105,545 tons, their shipments ranking second in the Province, the Acadia Company alone exceeding them.

In 1873, the markets still further improved, and elaborate preparations were made in the mines for a heavy production. A large stock of coal was banked on the surface and about 7,000 tons stowed in the upper workings of the mine. In all a greater quantity was on hand than that possessed by any other company, when the spring trade opened, with every prospect of a successful year's business. But just as the shipping season opened, the terrible explosion took place, by which many lives were lost, the pit set on fire, much of their plant destroyed and their operations suspended. The following account of it is taken from the report of the Inspector of Mines:—

" Early in May the shipping had already become vigorous, when a strike of the colliers for certain privileges and higher rates of wages closed the workings. After a week's intermission, an agreement was made with the men and they resumed work on the 13th. About noon on that day, a shot fired in one of the low levels on the south side of the pit ignited the coal. Every exertion was made, to put out the fire, but the peculiarly broken condition of the face of the level prevented the men from attacking the flame, where the burning gas directly issued in great volume from the solid coal. The fire spread rapidly, and as it was soon evident that the chances of subduing it were small, an order was issued that all the hands, who were disinclined to assist at the fire, should leave the pit. Many had previously left, having been driven out of their bords, by the smoke. The boys, all except one, had gone up, and of the rest, all but about a dozen men who remained with Richardson, the overman, at the fire, left the lowest landing to walk up the slope. Richardson and his men, who so heroically remained to battle with the fire, so long as there was the slightest hope of success, must soon have followed to endeavor to check as speedily as possible the progress of the flames, and save the pit by closing all openings. No attempt to do this was, however, made, for before many of the men who were in the slope had time to escape, an explosion of gas, unexampled on this continent for violence, occurred, dealing on all sides death and destruction. The force of the explosion was so great, that the wooden rope rollers were torn from the track and hurled out of the slope, as from the mouth of a cannon, falling in the woods some two hundred yards back of the bankhead. Great baulks of timber 14 feet long, by 9 inches through, were cast up out of the Campbell pit to so great a height that on falling, they struck the ground with such force as to fracture them, and the rush of air swept away as would a hurricane the exposed roof of the bankhead. Many explosions took place during the afternoon, and the second occurring about two hours after the first, killed four volunteers, who were nobly endeavoring to rescue some men then known to be alive at the bottom of the pumping pit. By the second explosion, the ventilation was thoroughly destroyed, and as hopes could no longer be entertained that any life still existed in the mine, all the preparations to explore the workings were then abandoned, and attention alone directed to saving property. The violence and frequency of the explosions, struck terror into the hearts of all who rushed to the scene, and paralyzed the efforts of those who sought to close the openings. All the available water was turned in to cut off the lower workings, and effectually seal the bottom of the pumping pit. Still the fire raged, despite of every exertion, for 36 hours, and the flames shot up with a fierce roar to the height of from thirty to forty feet from the many openings along the crop. Two days passed before the men engaged in filling the openings had effectually sealed this fiery grave of fifty-five of their comrades.

" The workings remained closed until the end of October, when one of the slopes was opened, and air allowed to circulate between it and an opening made by a fall near the rise. At the end of a fortnight, and just when appearances seemed to warrant preparations being made to re-open the workings in a regular manner, the return air showed unquestionable signs, that the fresh air was finding its way into places, where the heat was still sufficiently intense to cause combustion of the coal or the bituminous shales of the roof. In consequence the pit was again closed."

The total number of lives lost was sixty, among whom was Mr. Dunn, the manager, of whom 31 were married men, 28 single men and 1 boy, leaving 29 widows, 80 orphan children, besides parents dependant on the lost. Contributions to the amount of about $23,000 were made in various parts of the Dominion and the United States for their relief.

To keep a small business going, a pit some 70 feet deep was sunk to the south of No. 2 slope. In the fall of 1873, a new manager, Mr. Robert Simpson, M. E., arrived from Glasgow and under his supervision a new slope was driven to the south of the old workings, and winding machinery there erected. Subsequently he conducted the re-opening of the two original slopes—an operation involving great skill and expense, but one successfully consummated, the most of the exploded workings being recovered in 1875 and safety in restoring the remainder assured. The most of the water has now (1876) been pumped out, the workings cleared of *debris*, and reconstructed thoroughly, so that the colliery now, with its three working inclines, is in a better position for a large out-put than ever it was before. A fan 20 feet diameter, 6 feet wide, on the Guibal principle, for the ventilation of the underground workings, was erected in 1875 and has proved a complete success.

Mr. French, who had obtained the lease of an area of three-and-half square miles, where the extension of the seams in this direction was first discovered, worked for a time spending money uselessly, but his rights were transferred to a company, composed of persons principally resident in New Haven, Connecticut. In 1869, they commenced building a railway from their mine to the Middle River, a distance of six miles ; and in July, 1871, they had it completed, with shipping wharf, and commenced shipping coal. On this railroad the most noticeable feature is the high bridge across McCulloch's Brook. It is a trestle work, built of Southern pine, imported for the purpose. It is 400 feet long, consisting of four spans of 100 feet each. The middle span is 78 feet above the bed of the brook.

The works of these three collieries being in immediate proximity, a village has sprung up named Westville, of which the growth has been more rapid, than that of any place known to me in these provinces. The land here, owing principally to heavy fires, was so barren that one man, who owned fifty acres, after clearing some of it, offered the whole for the cow; and a lot of 100 acres, on part of which is now situated the Acadia Company's works, was willed to the Pictou Academy, in payment of a subscription of five pounds. In 1866, I visited the spot. Part was covered with wood, but part seemed too barren even for that. Having been severely burnt over, it produced only small bushes. Some men were then engaged in erecting a hut, of round poles cut almost on its site. In 1875, a census was taken, when the village was found to contain a population of 2,500, with three churches—two Presbyterian, of large size, and a small Methodist; and a Roman Catholic chapel is now building. But a great mistake was made at the outset, in the ground not having been properly laid out. The consequence is that the buildings have been placed in most admired disorder.

These four companies are all that are in successful operation on the west side of the East River. Another company, known as the Montreal Company, established by Mr. Robert G. Haliburton, sank a shaft just opposite New Glasgow, on an area owned by them, which thus lies at the north side of the coal field, and near the base of the Conglomerate. Here they found the coal of good quality, but lying at a very steep angle, and abounding in inflammable gas. But nothing has since been done to develope the property.

Before the commencement of the General Mining Association's operations, a seam was opened on the east side of the river, and after the abolition of the Mining Companys operations, it was again opened by the Pictou Mining Company. The coal was found to be inferior in quality, and a continuance of the deterioration having been ascertained by a shaft sunk farther to the east, all operations were abandoned, though Mr. Rutherford expresses his opinion, that it may not continue far into the dip.

Considerable labor and means have been expended in endeavors to trace the course of the seams further east. The result has been the discovery of several beds of coal; but the field, on examination, has been found so intricate, the measures so disturbed and broken, that their extent and position, as well as their relation to the other seams, are as yet involved in some uncertainty. We shall, however, give a brief summary of the facts ascertained.

Immediately behind New Glasgow two seams have been opened, the lower known as the Stewart seam, upwards of three feet in thickness, and the upper as the Richardson, 2 feet 9 inches, both of which are regarded geologically as overlaying the main seam. The last of them has been partially mined by a company, known as the "Crown Brick Coal and Pottery Company," which was formed for the purpose of working an extensive deposit of fire clay found here. Though the seam was small, the coal was found to be of excellent quality. The company, however, has been for some time in a state of snspended animation.

About a mile further east, two seams have been discovered about 3 1-2 and 4 1-2 feet thick, and another larger. But here a large fault is found to cross the field, and the whole measures are so broken, that very little has been done in the way of mining upon them.

Farther east, at what was known as the Marsh, four young men named McBeans, two of them brothers, and cousins of the other two, also brothers, took out rights of search. They were at the time possessed of but limited means, but they spent time and labour and what means they had, in exploring their area, and in opening some veins found on it. Their enterprise in due time met with its reward. The examination of the field by Sir William Logan, proved that their lease covered valuable seams of considerable extent. It was accordingly purchased by a Company in Montreal, known as the " Vale Coal Iron and Manufacturing Company," of which Sir Hugh Allan is president. Since 1872, under the able management of J. B. Moore, Esq., the vice-president, and J. P. Lawson, engineer, they have provided and erected everything necessary for mining and shipping coal on an extensive scale.

This colliery is situate about six miles to the eastward of New Glasgow, on a seam formerly known as the McBean area. It contains three square miles, or 1,920 acres. There are five known workable seams of coal on it, which are found in descending order, as follows: The uppermost of the series is the " Captains seam," a good coal well liked for domestic purposes. It measures three feet six inches in vertical thickness. Seventy-five feet below is the " Mill Race seam," so named from being first discovered in the mill race below Jas. McDonald's saw mill. It is not quite so good coal. It is three feet thick, with impurities. Over sixty feet below is the " Geo. McKay seam." This is a good seam of coal, well liked both for steam and domestic purposes. It measured in three openings, 3 feet 9 inches, 4 feet, and 4 feet 10 inches in vertical thickness. A small seam of oil shale of uncertain size and value, about eight inches in the centre of the seam, is very rich, but it gets poor as you go from the centre. Next is " The Six Feet seam." It is not quite so good coal, but is purer to the dip.

About 1,450 feet across the measures behind the above seam, is the " McBean seam." It is a good coal, both for steam and domestic purposes. It measures seven feet of vertical thickness. Two small seams are found, about 200 feet across the measures from the McBean seam. They are too small to work. The vale colliery is placed on the McBean seam, into which two slopes, one sixteen feet wide (the main slope), the other eight feet wide (a travelling and pumping slope), are driven on the dip of the seam, from which the levels are driven in the coal. A pair of winding engines, 12 inch cylinder and 18 inch stroke, built at the Acadia foundry, New Glasgow, have been erected in front of the Main Slope, and a double acting steam plunger plump, also made at the Acadia foundry, is placed at the foot of the pumping slope, which throws the water to the surface and drains the mine.

A railroad six miles long, leading from the colliery and joining the Intercolonial Railroad at New Glasgow, with all the necessary sidings, has been constructed by the Company. From New Glasgow the coal is conveyed over the Intercolonial Railway to the Pictou Landing, where it is shipped.

The works of the Vale Colliery were started in the woods, in the fall of 1872. A few trains of coal were run over the railroad to Halifax in the fall and winter of 1873–4 ; and in 1874, the out-put of merchantable coal was about 39,000 tons, the dull state of the markets keeping the mine idle one-half of the months of September and October, two-thirds of November and all December.

Workshops for carpenters and blacksmiths, and an

office and store, have been erected at a convenient distance from the works. A number of miners' houses have also been erected by the Company, in all about ninety buildings. To the north of the Company's property, the land is regularly laid out in building lots. A good many of these have been sold, and quite a neat village has sprung up, with stores, halls and dwellings, and a number are in course of erection. As the situation is picturesque, the Vale will be one of the prettiest villages in the eastern part of the Province. During the year 1876, a Presbyterian church has been built and a pastor ordained.

Messrs. Mitchell & Barton have an acre or two of the McBean seam, at the north-east corner of the McBean lease. There is only coal enough there to supply the inhabitants with their winter's fuel for a short time.

Between the Vale Colliery works and New Glasgow, and extending southwardly for some distance is all good coal measures, and doubtless containing valuable seams; and it is more than probable that should the coal trade revive, other valuable works will be started upon them.

It will thus be seen that we have now five large companies in vigorous operation, in mining and shipping coal. It only remains to give a few statistics, showing the amount of their work.

The following is a statement in tons of the sales by counties for the last four years:

	1872.	1873.	1874.	1875.
Cumberland	14,153	26,345	49,599	60,744
Pictou	388,417	334,984	357,920	337,102
Cape Breton	380,273	520,189	337,000	304,702
Other counties	31,070	588	4,588	4,047
Total	785,914	881,106	749,127	706,795

STATEMENT OF THE QUANTITY IN TONS SOLD BY EACH OF THE COMPANIES SINCE 1865.

	1865	1866	1867	1868	1869	1870	1871	1872	1873	1874	1875	1876
Albion or Halifax	203334	194301	120385	96042	86639	79240	77133	98865	107453	110431	115488	120527
Acadia	6964	10630	10912	25804	52567	73983	104007	129063	109975	100658	61983	56401
Intercolonial	40	603	443	914	58271	72603	51467	105545	36524	56214	62252	46914
Nova Scotia	12	160	41	683	294	650	12818	60590	79955	51065	50378	17808
McBean or Vale			10	85	427	172	276	600	194	38131	46767	33968
Mitchell and Barton							61	354	843	1219	234	
McDonald and McKay	104											
Crown, Brick, & P. Co				172	71		18					
George McKay	46	176	321	27		58						
German or Picton Co		57	83		50							
Montreal and Pictou			443	90								
Merigomish Co		00										
Total	210500	205627	132639	123872	198210	226406	245800	388417	384984	357920	337102	275618

In the above the round and the slack coal are counted together. The latter will be about one-ninth of the whole. To these quantities must be added an average of ten per cent. for colliery consumption, to show the whole amount raised.

STATEMENT OF THE NUMBER AND CLASSES OF PERSONS EMPLOYED AT EACH COLLIERY DURING THE YEAR 1875.

	UNDERGROUND.				SURFACE.				CONSTRUCTION.		TOTAL.		HORSES.	
	Cutters.	Laborers.	Boys.	Day's labor.	Mechanics.	Laborers.	Boys.	Day's work.	Persons.	Day's labor.	Persons.	Day's labor.	Above.	Below.
Halifax Co	275	62	69	83674	47	118	38	61388			607	145062	22	32
Acadia	95	27	20	31216	27	66	5	24692			240	65308	15	5
Intercolonial	84	31	18	31656	29	49	15	27508	69	6255	298	64449	8	2
Nova Scotia	78	25	14	34441	13	39	7	17947			176	52353	5	6
Vale	124	28	11	33645	27	37	5	16205	15	1705	247	51555	3	5
Mitchell & Co	3	2	1	265		1		68			7	333		1
Total	657	175	133	214927	143	310	70	147808	84	6960	1575	369605	53	50

When the General Mining Association commenced operations, they designed to work the iron as well as the coal deposits known to exist on the East River. They quarried ore from a bed now known as the Blanchard bed, and collected a quantity of Limonite about the banks of the river, near Springville. They also erected a blast furnace at the Albion mines, for the purpose of smelting these ores. But those in charge, accustomed to English ores and English fuel, did not understand how to manage ores of a different character. They declared that the ore was too rich, and, the company not having discovered the bed of Limonite, the work was abandoned.

Some iron, however, was produced, which, in combined hardness and toughness, excelled anything known. When quartz crushing began at the gold mines, and iron possessing these qualities was specially required for stampers, parties gathered up the lumps that had been thrown away at the old blast furnace at the Albion mines, and they found it superior for the purpose to any iron that could be obtained from any other quarter.

During the last few years, attention has again been directed to the subject, and careful explorations have been carried on, under the direction of competent scientific men. The result has been, to show the existence in this county of a variety of iron deposits, of great extent, and superior quality. Geologically, these lie among the Upper Silurian and Lower Carboniferous rocks, which we have formerly mentioned, as traversing the interior and southern portions of the county. We shall briefly notice the principal of these. Among the most important is a great bed of Red Hematite, which is most extensively developed at Blanchard, near the East Branch of the East River of Pictou, and on the upper part of Sutherlands River. The ore bed is an enormous deposit, varying in width from fifteen to thirty feet, and where it has been opened up, affords from ten to twenty feet in thickness of good ore. This bed has been traced for several miles, and rises into some of the higher elevations of the country. At Sutherlands River, it is found at an elevation of 400 feet above its bed, and its position will allow the extraction of millions of tons above water level, by the simplest operation of the miner. Though not one of the richest ores in the district, its great quantity and accessibility render it of great value. The analyses made of it, show a percentage of metal varying from 43 to 54 per cent. The foreign matter is principally Silica and the proportions of Phosphorus and Sulphur are very small. The principal exposures of this bed, are distant only twelve miles from the great collieries of the East River of Pictou, and less than ten miles from the Halifax and Pictou Railway, while the extension of the latter eastward will pass close by its outcrop at Sutherlands River.

At Sutherlands River, about three miles from Merigomish Harbour, is a valuable deposit of Spathic Iron ore or Siderite, occurring in Lower Carboniferous sandstones, and varying in thickness from six feet six inches to ten feet six inches. It affords from 42 to 43 per cent of iron, and contains from 2 to 8 per cent of manganese. This bed is only four miles distant from the "Vale" colliery.

At the junction of the Lower Carboniferous and Upper Silurian rocks, in the valley of the East River, near Springville, is a vein of Limonite of exceeding richness and value. It varies in width from five to twenty-one feet, and the ore is of the finest quality, affording from 62 to 65 per cent of metallic iron. A similar vein has been opened near Glengarry station.

Besides these, a large vein of Specular iron ore, similar to that at Londonderry, occurring in similar conditions, and supposed to be a continuation of it, has been traced from New Lairg, near Glengarry, eastward to near the East Branch of the East River. About a mile to the West of this stream, it has been examined, and thence explored for two miles, following the course of a high hill, and its width was found to vary from five to twenty feet. The ore is a nearly pure peroxide of iron, containing from 64 to 69 per cent of metal, and great quantities could be easily taken out from the outcrop of the vein.

There are also other veins of less importance. Clay Ironstones also occur in many parts of the coal field, but no attention has hitherto been given to them as sources of iron ore. It may be anticipated that should the richer ores be worked, they may be rendered available in connection with them.

It will thus be seen, that these explorations have shown that from Glengarry to Merigomish, a course of over 20 miles, there extends a series of iron ore deposits, of good quality and more than usual dimensions. The ferriferous rocks extend westerly into Colchester, and though these have not been explored, yet small veins of Specular ore are found on the upper part of the Middle and West Rivers. It is probable also, that they will be found in the opposite direction toward Antigonish county.

As the presence of a cheap flux is important for the manufacture of iron, we may add that limestone is found in every part of this section of country. Abundance of fire clay also of superior quality, is found in various places. Moulding sand also is plentiful on the East River and its tributaries. The best known deposit is near the mouth of McLellans Brook, which has for years supplied our local foundries.

It will thus be seen, that in its rich iron ore, in the immediate vicinity of coal, Pictou possesses the elements of national prosperity. The course hitherto pursued of raising coal to export, is simply a waste of our natural resources. Let it be employed in developing the treasures of the rocks, and the county, we may say, the Province, will enter upon a boundless career of progress.

CHAPTER XIX.

FROM THE DIVISION OF THE COUNTY TILL THE PRESENT
TIME, 1836—1876.

In the year 1836, the Act erecting Pictou into a separate county, came into operation. By that Act, it received two representatives for the county and one fo rthe township of Pictou. The first election under the new arrangement took place that season, when by a compromise between parties, George Smith and John Holmes, Esquires, were returned for the former, and, after a contest, Henry Hatton, Esquire, for the latter. On the remodelling of the Council in 1838, Mr. Smith was elevated to a seat in that body, and Thomas Dickson was elected in his place. At this first county election, Mr. Holmes first came into public life. From that time till incapacitated by old age, a few months before his death, in 1876, he occupied a prominent place in our county and provincial politics, having been for several years member of the House of Assembly, then a member of the Legislative Council, and, at the adoption of the Confederation Act, one of the first senators from Nova Scotia. His public course was that of an honest, thorough-going true blue Tory. At a very late period of his life, and, we suppose, to the last, he declared his admiration of the Government of the old Council of XII., and his detestation of responsible government. His father having taken the lead in forming a body in connection with the Church of Scotland, he succeeded to his influence, and both in the civil and ecclesiastical movements of the members of that body, wielded the influence of a Highland Chief in the days of clanship.

We append a list of members of the different branches of our Legislature to the present time. (Appendix K.)

The present period presents few events calling for special notice. In the year 1843, the disruption of the Church of Scotland took place, and was followed the next year by a similar division in Nova Scotia. Of the Presbytery of Pictou, in connection with the Church of Scotland, only one minister, the Rev. John Stewart, then of St. Andrew's Church, New Glasgow, adhered to the Free Church. Of the rest, all returned to Scotland to occupy the vacant parish churches, with the exception of the Rev. Alex. McGillivray, of McLennan's Mount, who, it was said, by accident missed a presentation. Congregations were formed in various parts of the county in connection with the Free Church. That portion of St. Andrew's Church, New Glasgow, which adhered to their minister, formed the congregation of Knox Church, which has since amalgamated with the congregation of Primitive Church in that town. The people of Blue Mountain and the Garden of Eden generally, and a majority of the people of Barney's River, joined the Free Church, and obtained as their first minister the Rev. D. B. Blair, in the year 1848. Congregations were also formed at Pictou, Rogers Hill, West Branch River John, Earltown and Saltsprings. The Rev. Alex. Sutherland, who had been brought up at Rogers Hill, but had completed his studies in Edinburgh, returned from Scotland, and became minister of Earltown and West Branch River John in 1846. Shortly after, the Rev. Murdoch Sutherland became minister of Pictou and Rogers Hill. He was greatly esteemed, but his career was short.

The large majority of the adherents of the Church of Scotland remained in their old connexion, and for ten years received very little ministerial service. The folly of depending on Scotland for ministers, was now apparent, and, as the body was not in a position to educate young men in this country, they sent a number of promising natives to be educated in Scotland, and from their return, the revival of that body may be dated. The first of these were the Revds. Alex. McLean, Alex. McKay, George M. Grant, William McMillan, Simon McGregor and John Cameron, all natives of Pictou.

In the year 1867, just one hundred years after the arrival of the first settlers, the railroad from Halifax to Pictou was completed. It had been for some time open to Truro, and this had somewhat changed the trade, especially of the rural districts of the county, large quantities of agricultural produce being sent over land to Halifax, thus making improved markets for our farmers. The effect of the completion of it, by the increased facilities which it affords for communication with the rest of the continent, it is unnecessary to point out.

We shall now briefly review the various branches of business in the county during this period.

At the commencement of the period, ship building was carried on with considerable activity, and so continued for a time, so that in one year, forty vessels were registered as built in Pictou, and its outports, including Tatamagouche. But these were, with scarce an exception, built to sell, and under the ruinous system described in our last chapter. The result was that about the year 1841, the leading ship-builders, and a number who were engaged in it on a smaller scale, became bankrupt

Since that time a new system has been pursued. Persons build vessels to own and sail them, and the business has proved highly profitable, and, until the present depression, was rapidly progressive. Nearly every trader in Pictou, New Glasgow and the outports, as well as many tradesmen, have at least shares in ships, and a large fleet of vessels of superior character, is now owned in this county and is found in the carrying trade of the world.

The following is a statement of the vessels registered at Pictou:—

8 Ships measuring		8.428
24 Barques	"	14.337
4 Barquentines measuring		1.040
1 Brig	"	236
19 Brigantines	"	4.956
31 Schooners	"	1.573
6 Steamers	"	162

Total vessels... 93   Total tonnage.... 30.732

This however does not represent the whole tonnage of the Port, several vessels partially or entirely owned here being registered elsewhere. "The County of Pictou" for example, one of our largest traders and entirely owned in the county, being registered at Glasgow.

Another change in the system of shipbuilding must be noticed. Building vessels for sale was simply building vessels of inferior quality. So much was this the case, that our colonial built vessels got a bad reputation, from which they have scarcely yet recovered, and of all colonial vessels, those of Nova Scotia and Prince Edward Island were the lowest in the scale. There were other evils connected with the system, not the least of which was the frauds connected with insurance. To hear of one of these vessels, perhaps, on her first voyage being wrecked, often gave its owner pleasure, which he scarcely affected to conceal.

With the building of vessels for the use of the owners, commenced an era of improvement in the quality of the vessels built, so that now the character of our vessels will compare with those either of the other Provinces, or of any portion of the world.

In this good work, Captain George McKenzie, of New Glasgow, deserves special mention. If Captain Lowden was the father of the old ship building trade in Pictou, Captain McKenzie was the father of the modern system, and though we did not intend particularly to refer to the living, or to those who, though departed, belong to the present period, yet we think the position he held in reference to this important part of the business of Pictou, as well as the character of the man himself, entitles him to a full notice.

He was born in Halifax, in December, 1798. His father died in 1802, and his mother removed with her five children to Fishers Grant in the same year. When he grew up to youthful manhood, he turned his attention to shipping and ship building, for which he appeared to have a natural talent In 1821, he and John Reid, of Little Harbor, built a schooner of 45 tons at Boat Harbor, in which he shortly afterwards took a trip to the West Indies. She was called the "James William." The two men hewed the timber, took it from the woods, and did all the work of building themselves. An event happened when he was about nineteen years of age, which had the effect of bringing him into prominence among ship builders. A vessel built at some point near the Beaches was being launched, when she stuck. Those engaged about her spent a great deal of labour, used all the mechanical appliances at their command, and exhausted their ingenuity in efforts to get her off, and finally gave up the work in despair, and from her position, it was feared that she would break up before spring, when George McKenzie volunteered to get her off and succeeded in doing so at the next tide. From that time he was a "marked man." In 1824 he went with Robert McKay, of Pictou, to superintend work in his yard and continued with him about three years. In superintending the launching of a vessel for Mr. McKay at River John, he had the misfortune to get one of his thighs broken. He then sailed for some time as master, both to the West Indies and to Britain. In the year 1829, we find him in command of a brig called the "Two Sisters," owned by his brother-in-law, James Carmichael, Esq., and his brother John, then doing business in partnership. In her he went up to Glasgow, and she was then noticed as the largest vessel, that up to that date had gone so far up the Clyde. On her return, it was noticed in the Pictou paper, as something good, that she had made the round trip in twelve weeks. He then settled down in business in New Glasgow, where he first built a schooner of 100 tons, and then a trader for Almon, of Halifax. The "Sally," a barque of 350 tons, one-half being owned by Henry Hatton, of Pictou, was his next venture. From this he advanced to building vessels of 600 tons, and then to others of 800, which were thought wonders for a time, but not content with this, he soon was building still larger. In 1850 he launched the "Hamilton Campbell Kidston" of 1,400 tons. In launching her, the launch ways spread, and her stern took the mud, when half way down the launchways. The spectators beheld the accident with dismay, women wept, but the captain was as calm as a summer eve. He quietly walked round her as if nothing had happened, and then told his men to go home and return at such an hour, as they could do nothing till the tide changed. At the appointed time he set them to work, and in a short time had her safely afloat. Considering the position of the vessel, the shallowness of the water, and the appliances available, this was almost as great a feat as the launching of the "Great Eastern." When she arrived in Glasgow, her appearance there created quite a sensation, she being the largest vessel that up to that time had sailed up the Clyde. In 1854 he built the "Sebastopol" and the "Magna Charta," the latter attracting great attention as the largest vessel built in the Province. Others of his vessels were well known, as the "Sesostris," the "Catherine Glen," the "John McKenzie" and the "George." The "County of Pictou," built in 1865, was the last vessel built by him. She is still afloat, and has been remarkably successful.

So well was he known in old Glasgow, that a number of gentlemen, connected with the trade of that port, presented him with a service of plate as a token of their esteem.

Captain McKenzie was one of the first to see the evils of the old system of building vessels to sell. And, although from his character as a ship-builder, he obtained contracts for building at remunerative rates, he gave particular attention to running his vessels, some of which as the "Sesostris" and the "John McKenzie" were well known, and thus led the way in the formation of that large carrying trade, which forms such an important part of the industries of the county.

We believe too that it will be universally conceded, that Captain Mckenzie took the lead in building vessels of a superior class, and that largely to him is owing the vast improvement in the character of our ships, of which now we have no reason to be ashamed, comparing them with those either of other Provinces or of any portion of the world.

During a period of forty years, Capt. McKenzie went to sea more or less, and in this he soon developed his character. As a commander, he was daring, clear headed, calm even under the most difficult circumstances, prompt in deciding upon his plans, and energetic in having them executed. When he commenced going to sea it was the ordinary practice to carry only a moderate amount of canvas and to take in sail at night, if the wind was any wise fresh, so as not to have to rouse the men during the night to lessen sail, should it blow harder, and if there was any appearance of a blow, even the upper yards were lowered. Capt. McKenzie soon pursued a different course. He availed himself of all the sail his vessel could carry, and thought not of lowering his yards or taking in sail, because night was coming on. Doubtless by this time, others were adopting the same system, but at first it appeared so strange to safe going old fogies, that some times they thought him mad. Though commanding vessels of all sizes from the coasting schooner to the 1400 ton ship, and in all latitudes, and that for the period of about forty years, he was only wrecked once, and then lost but one man. What is however perhaps more worthy of notice was, that in all that time he never lost a man overboard. Once indeed when in Harbour, (at Savannah Harbor we believe) a man fell overboard. The Captain, though he could not swim himself, jumped overboard after him, which induced others to follow his example, and they were both saved. But at sea, in taking in sail, or any of the other ways in which this accident occurs, he never lost a man, and as he heard of such cases, he could not avoid giving expressions to feelings of mingled disgust and indignation, as he could only regard them as resulting from bad management.

Much of his energy he infused into those around him. Full of patriotism, he was not only anxious to advance the welfare of the Province, but it was his delight to bring forward Nova Scotians, and particularly young Pictonians to command his ships or to fill stations of usefulness. The sons of the farmer or the widow, in his hands were soon in command of his big ships, and proved successful commanders, ever speaking of him with affection.

His manner in command was slightly brusque and imperative, but no man was ever more distinguished by the spirit of kindness and readiness to help others. In fact he was too open handed, too free in buying, too ready in distributing, to have been a thorough going money making man. In consequence of this, as well as from heavy losses experienced at different times, while always having a competence, he did better for others than for himself, and many a one has reason to remember his seasonable help. He died in the winter of 1876.

At the commencement of this period, there was still an export of deals and battens, which continued for a few years, but now not only is pine imported, but there is not enough of spruce lumber produced, for the consumption of the county, and a considerable quantity, especially in the form of flooring, is annually imported. The sawmills, of which by the census there are 78, are employed principally in sawing hemlock into boards and shingles, with a little spruce and some hardwood plank. The only article of wood now exported is squared birch timber, and as a great part of the southern portion of the county is still covered with forest, on a soil of which a large amount is unfit for cultivation, this trade may continue for some time, and at all events there is here a perennial source of supply of hardwood, to meet the ordinary purposes of our own population.

The failure of the timber trade, as well as other circumstances, has led to greater attention to their farms, on the part of our rural population. The farming is not any where yet of a high order, but it is very different from what it was forty years ago. There is every where attention to improved modes of culture, and the introduction of better stock. Machinery is being generally used. There was not at the beginning of this period a threshing machine in the county. My father introduced one about the year 1840. It was literally a one horse power, on the tread-mill principle. There was not another at that time in the county.* Now these are universally in use, as well as mowing machines and other improved implements. At the same time the markets have been so favourable for farm produce, that, notwithstanding the failure of the potato crop, and the very general destruction of the wheat crop for some years by the wheat

---

* Mr. Donald McLeod, of West River, erected one some years before on a small stream on his farm. The power did very well, but the threshing part did not succeed.

midge, in nothing has the county been more distinguished during this period than by the progress, which the farming population have made in comfort and independence. At the beginning of this period, there were few farmers, who were not in the merchants' books, generally for considerable amounts. The credit system was still almost universal. The farmer was thought to do well, and was reckoned a good customer, who in the fall brought a sufficient amount of pork and other produce to reduce his account to a moderate figure. But now this mode of trading has largely passed away. Farmers now generally sell their produce for cash, or exchange it for necessaries at the store, and can generally do so at prices, at which they would once have thought themselves set up altogether.

Their improved circumstances appear in their dress, their dwellings, and the conveniences by which they are surrounded. At the beginning of this period milled cloth was but little used, and while the better class of farmers had at least a Sunday dress of English cloth, the large majority were clad in homespun, undressed and dyed at home. We need not describe the change produced by the general use of fulled cloth. The younger generation will scarcely remember the old blue-dye coats, so characteristic of many parts of the country, and occasionally so odoriferous under a summer shower; and we can scarcely tell the time when we have seen a countryman come to town with his feet encased in raw hide moccasins.

At the beginning of this period, there were a few old-fashioned chaises, but these only among the better class of farmers in the older settlements. In the new they were almost unknown, and the roads were scarcely fit for them to traverse. Hence we might see a countryman come to town with a conveyance, which we shall not attempt to describe, his horse garnished with a straw collar, a straw' saddle, all kept in place by hair ropes. Now it is a poor farmer who cannot drive his pair well harnessed at farm work, or who does not own a comfortable carriage for riding to church on Sabbath, or for travel on week days. By the census, it exceeded every other county in the Province in the number of light carriages and vehicles for transport, the number of the former being 4,596, of the latter 7,246.

This improvement has been most apparent in the newer settlements. In these the people then generally lived in log houses with few conveniences, but now, largely through their greater economy, and partly through their being on good soil, they are as independent in their circumstances as the people of any part of the country.

Farming is still the leading industry of the county, the number engaged in it according to the census of 1871, being more than equal to the number employed in all other lines of business. Thus we find that 5154 are classed as agricultural, while the rest are classed as follows :

```
Commercial ..................................... 699
Domestic........................................ 452
Industrial ...................................... 2,611
Professional .................................... 346
Not classified.................................. 692
                                                -----
                                                4,798
```

Of these 488 were miners, 369 carpenters, 298 mariners, 257 blacksmiths, 202 shipbuilders, 172 teachers, 105 tailors, 37 foundrymen, 34 clergymen and 5 booksellers, there being 20 in Halifax, and only 11 in all the rest of the Province, eight counties having none.

We may here give a statement of its agricultural products as compared with other counties. By the census of 1871 there were produced the previous year as follows:

Bush. wheat.	Oats.	Other grain.	Potatoes.	Tons hay.
76,426	409,868	64,937	415,524	32,334

In wheat and oats, it largely exceeded any other county, Cumberland coming next to it in the former, with 47,395, and Inverness in the latter, with 276,330. In other grain, it was exceeded by Lunenburg, which while only raising 2,661 bushels of wheat, and 22,447 of oats, produced 67,957 of barley and 13,109 of rye, by Cumberland, which produced 64,023 of buckwheat, and Colchester, which produced 43,995 of the same grain and 18,294 of barley. In potatoes and hay, it was exceeded by Kings, Annapolis, Cumberland and Colchester. In the production of butter, it exceeded every other county, the amount being 804,661 lbs., Colchester coming next with 625,026. In cheese it was exceeded by Kings, Annapolis, Antigonish and Inverness, though it is probable that the proportions will have since been altered, as four cheese factories have been erected in the county, one at West River, one at Gairloch, one at River John and one at Barney's River. In cloth, it largely excels every other county, the amount being 183,008 yards, Inverness coming next with 138,996.

The following is a statement of farm stock owned in the county :

Horses.	Milch cows.	Other horned cattle.	Sheep.	Swine.
6,787	14,958	12,560	43,416	4,343

In horses and milch cows it ranks first, Colchester coming next in the former, and Inverness in the latter. In other horned cattle and sheep, it was slightly exceeded by Inverness, and in swine by Kings, Inverness and Antigonish.

Altogether we may set down Pictou as the first agricultural county in the Province, the only one which can compete with it being Kings. In fruit growing, however, it is only seventh, being exceeded by Kings, Annapolis, Lunenburg, Digby, Hants and Cumberland.

In manufactures, the only one in which as a county we can take credit for great progress, is tanning leather.

We had tanners from an early period, but the business has been so developed during this period, that now Pictou manufactures almost as much leather, as the other seventeen counties of the Province put together. We give the figures :—

	Value of Raw Material.	Value of Articles Produced.
Pictou	$202,702	$346,974
Rest of the Province	210,793	423,019

Fair progress has been made in woollen manufactures. The first application of any but hand power to this, was by the erection of a carding machine at Middle River by a Mr. Humphrey, of New Brunswick, in a mill owned by the late Isaac Archibald, about the year 1822. The year after it was erected, Mr. Archibald bought it and carried on the business himself. The first fulling mill was established three or four years after, by James Farnham, of Truro, and Edward and Stillman Lippencott, under the name of Lippencott, Farnham & Co., at what is now Roddicks Mills, below Durham, then owned by the Rev. Duncan Ross and George McDonald. They carried on dyeing in connection with it. But they had only the privilege of the waste water, and, after a time, finding that insufficient for their purposes, they, in the year 1829, removed to the Six Mile Brook, and set up a mill, where F. Miller & Co.'s establishment is now, at which place the business of cloth dressing has ever since been conducted. The first spinning machine and power loom were set up by Mr. James Grant, near Springville. There are now, besides carding machines and fulling mills, four other establishments, where the whole business of manufacturing cloth from the wool is carried on, Messrs. George Kerr & Sons', Middle River; Messrs. Frasers', Rocklin, Middle River; Messrs. McDonalds', Hopewell, and Messrs. F. Miller & Co.'s, Six Mile Brook. These are mainly driven by water, but some of them have steam as auxiliary.

Manufactures of wooden ware are yet in their infancy. John Fraser, of Green Hill, has a rake factory, and Messrs. Nute Bros., and Messrs. Cumming Bros., both of New Glasgow, have established factories, driven by steam, for the production of furniture, and Messrs. Samuel Archibald & Co. have an establishment for the manufacture of rakes and other implements, and also of furniture, at Watervale, West River. The first successful application of steam to mills in the county was at the Clarence Mills, Pictou, established by James Primrose, Esquire. It was employed in driving a grist mill, a carding machine, and in sawing, planing, and other work in wood.

In the manufacture of iron, besides the Albion Mines Foundry, previously established, three others are now in successful operation, Messrs. Davies', in Pictou, Messrs. Frasers' and the Acadia Foundry, in New Glasgow ; and the Nova Scotia Forge Company by the application of steam power, is doing a large business in the production of wrought iron for various purposes, and Mr. Conolly at Middle River has a small axe factory, which has produced implements of the finest quality.

Little is done in developing the beautiful free stone so abundant in the county. There was at the time of the census only one establishment for the manufacture of grindstones, Mr. Robert McNeil's, at Quarry Island, Merigomish, the products of which was valued at $4,500. Some stone is quarried for buildings, and, particularly at the Eight Mile Brook, some is taken out for monuments and other work requiring a fine polish.

Though we deem the manufacturing interest in Pictou small, compared with what it might be, yet it is larger than that of any county in the Province, except Halifax, which includes the city. By the census, the value of the products of all its industrial establishments was $1,273,018, while that of Hants, which came next, was only $836,503.

The period we have been considering has been marked

by a large emigration from the county, the majority of the young of both sexes in the rural districts, going abroad as they reach maturity. In the year 1848 or '49, a number were attracted to Australia, who were generally unsuccessful. Since that time, California has been the great attraction for young men, and service in the New England towns for young women. But our Pictou boys are to be found in every part of the world. Some, under the law, by which not only prophets, but other good men, are not without honour save in their own country, reaching the upper rungs of the ladder, the majority by industry attaining a competence, which they might have done at home, and many, alas, making shipwreck of earthly prospects, and even of conscience and a good name.

We add a statement of the population at different periods:—

```
Census of 1817 .......................... 6,737
   "    1827 ........................... 13,949
   "    1838 ........................... 21,449
   "    1861 ........................... 28,785
   "    1871 ........................... 32,114
```

We think, however, that there was manifestly an imperfection in the first return ———, that with the immigration of previous years, the population must have been greater, and that with the state of business from 1817 to 1827, there could not have been such an increase in that period as indicated by the census. Such was the progress, that at the time of that census, it was only exceeded in population by Halifax and Annapolis, which, however, then included Digby, and ever since it has been the largest in population of the rural counties.

We have had occasion in the course of our history, to refer particularly to the Presbyterian Church and Presbyterian ministers. This has not been from any denominational prejudice, but simply because they have been so closely connected with the past progress of the county, that a history of Pictou without reference to them, would be like the play of Hamlet, with the part of Hamlet omitted. It is proper, however, to give some notice of the early history of the other Churches in the county.

Among the 82nd men were several Roman Catholics, who settled in Merigomish, and, as we have seen, in the years 1791 and 1802, a large number of Scotch Catholics arrived, who settled along the Gulf shore, part of them in this county, and part of them in Antigonish. They received their first church service from the Rev. Mr. Jones, of Halifax, who performed a mission among them, but the first resident priest was the Rev. James McDonald, who lived at the Gulf, but ministered on both sides of the county line. We do not know the date of his arrival, but he was here as early as 1793. He offended his people by advising them to attend Dr. McGregor's preaching, and otherwise showing disrespect for his own Church, so that he was obliged to take refuge from their wrath in Walter Murray's house. They then gave out that he was crazy, and he was sent up to Quebec to a monastery, from which he never returned.

He was succeeded by the Rev. Alexander McDonald, who settled at Arisaig about 1800, and officiated among the same people from that time till his death. Through the influence of Mortimer, he was made a magistrate. He died in Halifax on 15th April, 1816, in the 62nd year of his age, and his people carried his remains through the woods all the way to Arisaig. In 1810, the first regular church was erected at Arisaig, though they had a place of meeting previously.

He was succeeded by the Rev. Colin Grant. In the year 1834, the church at Baillies Brook was erected, and in the year 1869, that settlement was formed into a separate parish, with the Rev. Simon McGregor, D.D., as their first priest.

The chapel in town was commenced in the year 1823, when the frame was erected, without, however, any rafters. It stood on the west side of Chapel Street. The next year the rafters were put on, but in a gale soon after were blown down. The most zealous Catholic in town at that time was an Irishman, named Thomas Jones, who kept a grog shop. About forty pounds were now collected, principally among Protestants, to aid them in finishing the building. But Jones having got the amount into his hands, withdrew the light of his countenance from Pictou altogether, and departed to some more congenial region. From this, together with the smallness of their numbers, and their humble circumstances, their church long remained unfinished.

The first priest stationed in Pictou was the Rev. Mr. Boland, who arrived in town in the autumn of the year 1828.

The present church at Merigomish was built in the year 1865, but they had a small one for many years previous.

The few Episcopalians among the early settlers in Pictou generally fell in with the Presbyterian ministers. Their first organization was owing principally to the late Dr. Johnston and Robert Hatton, Senr. The latter was a lawyer, a native of Dublin, who came to Pictou with his family in the year 1813. Through his influence, Col. Cochrane had presented the Society for the Propagation of the Gospel, with the lot on which the church now stands, then valued at £150. He himself put up the frame in the year 1824. About three years after the church was finished, mainly through the exertions of his son, Henry, and consecrated on the 16th April, 1829, by Bishop Inglis. In the year 1847, it was lengthened by an addition of seventeen feet and a new spire erected. A transept was added in 1866. In the year 1832, Pictou District was erected into a parish under the name of St.

James by order of the Governor in Council on the petition of the clergyman, church wardens and vestry.

Among the first clergymen who visited them, was the Rev. Dr. Gray, then of Sackville, we believe. He was here in 1814, as we learn from a marriage celebrated by him in that year. Previous to 1830, they were visited by the Rev. W. B. King, then a teacher in Windsor College, during vacations, and by the Rev. Mr. Burnyeat, who visited them two or three time a year, holding service in the old court house.

The Rev. Charles Elliott, who came to this Province in the year 1829, under the appointment of the Society for the Propagation of the Gospel, was employed for some time as a travelling missionary, and Pictou was included in his district. In the year 1832, however, he finally settled here. He was a B. A. of St. Edmunds Hall, Oxford, was admitted a deacon, in the Chapel Royal of St. James Palace, on the 14th June, 1829, by the Bishop of London. He was ordained a priest in Nova Scotia in St. Johns Church, Cornwallis, by Bishop Inglis, on the 27th June, 1830. He was admitted Rector of St. James Church, Pictou, on the 23rd April, 1834, by the same Bishop, the first church wardens being Henry Hatton and Robert Hockin. The whole county was then his parish, and he preached regularly once a month at the Albion Mines, and also at River John, and visited Wallace, Pugwash and Tatamagouche, and occasionally even Cape Breton. He was a man of amiable disposition and gentle manner, and labouring diligently but quietly in his own calling, he gained the affection of his own church and the respect of the community.

On the 12th July, 1849, a meeting was held at the Albion Mines, at which it was decided to erect a church for public worship, according to the rites and ceremonies of the Church of England. For this purpose, £175 were subscribed by the inhabitants, and £125 given by the

General Mining Association. The building was begun soon after, and completed about July, 1851. That district was erected into a separate parish under the name of Christ Church, Albion Mines, by a decree of the Right Reverend Hibbert, Lord Bishop of Nova Scotia, bearing date 24th March, 1852, at or about which time, the Rev. Joseph Forsythe, missionary of the Society for the Propagation of the Gospel, took charge, W. H. Davies and Henry Poole, Esquires, being the first churchwardens. The Church and burial ground were consecrated in the latter part of September of that year.

In the year 1865, the Rev. Mr. Kaulback was appointed first curate of River John, and in October of that year, Mr. Elliott went to England, where he died at his residence, Falkland House, Painswick, Gloucestershire, on the 27th September, 1871.

The first Baptist Society in the county was organized on the principles of the Scotch Baptists, or Disciples, as they call themselves, who are distinguished from others of the name, by rejecting the office of the ministry, all the members using their gifts for edification, and by the observance of the ordinance of the Lords Supper every Lords day. The society was founded by James Murray, who came to Pictou in 1811, and afterward moved to River John. Here on the 18th June 1815, the day of the battle of Waterloo, he baptized two persons, and on the same day dispensed the communion. Since that day with the exception of a few very stormy days, they have not failed to meet on the first day of the week to break bread. They now number 40 members, and are the only society of this order on the north coast of the Province.

The first society of the regular Baptists was formed in Merigomish, in the year 1838. Two years previous, the Rev. George Richardson passing through the settlement, preached at Thomson Carmichaels, Barneys River. Mrs. Alex. Meldrum was present, and attributed her conversion to the sermon. On his return, a few months later, she was baptized by him. In the following spring he again came to Barneys River, when her husband, Mr. Peter McEwan and Mrs. Fogan were also baptized. These with Mr. and Mrs. Carmichael, Mr. and Mrs. Crandall, then resident in the settlement, formed the first Baptist Church, Mr. Peter McEwan being ordained deacon. From that time to the present, about 40 have been added to the church, but from deaths and removals, the membership is now only 14. They have a meeting-house at Barneys River, commenced in the summer of 1874, but they have never had a settled minister.

The first Wesleyan Methodist Society in the county originated with some dissentients in the congregation of the Rev. Mr. Mitchell, River John. They were organized into a society by the Rev. Mr. Snowball, then on the Wallace Circuit, in the year 1822. Previous to that they had been visited by local preachers, the first of whom was Mr. Andrew Hurley. They built their first church in 1824, and since that time, River John has been one of the regular Methodist circuits.

About the year 1845, a society and congregation was formed at the Albion Mines, by the Rev. Richard Weddall, consisting principally of miners, who had come from England. Wesleyan ministers had previously preached to them. A church and parsonage were subsequently built, and more recently a small church at Westville in connection with the same circuit.

A society has also been formed in Pictou, principally composed of parties who had previously been connected with the Evangelical Union, as they call themselves in Scotland, but usually known as Morrisonians. They had previously been under the pastoral care of the Rev. Alex. McArthur, who adopted the sentiments of Swedenborg. After several changes, they connected themselves with the Wesleyans.

For many years there was in town a small society of Friends, consisting, however, almost entirely of the family of Jas. Kitchin, an Englishman, who kept a watchmaker's shop on Water Street. In the case of each watch which he repaired, he placed a paper with the following :—

> " Behold, oh, mortal man!
> How quick the moments fly;
> Our life is ever on the wing,
> Prepare, prepare to die." —JAMES KITCHIN.

With this solemn warning of this old worthy, we might appropriately close our work, but we may add a statement of the relative numbers of the different religious bodies in the county at the date of the last census :

Presbyterian Church of the Lower Provinces	14,105	Wesleyan Methodist	797
Church of Scotland	12,250	Baptist	345
Roman Catholics	2,965	All others	193
Church of England	1,470		
			32,125

Our work is done. It has taken time and trouble. But it has been pleasant, and we trust not unprofitable. Especially do we feel satisfaction, in being the instrument of rendering an act of justice to the sturdy pioneers, who first invaded our forests and prepared homes for us in the wilderness. The present and future generations, in this county, and beyond it, have great reason to be profoundly grateful for the sturdy energy and moral worth of the mass of those, who first peopled our county, as well as for the intelligence and public spirit of those, who were the leaders of society at its formation, and particularly for the high talents and the devoted Christian zeal of James McGregor, Duncan Ross and Thomas McCulloch, who first planted the gospel among us, and who moulded the moral and religious, and we may add, largely the intellectual character of our population. It only remains that Pictonians at home and abroad, while thankful to God for what they were, and profiting even by their errors, preserve the noble heritage of steady habits, and sound religious principles, received from them, and transmit the same to the race that is to come.

APPENDIX.

A.

LIST OF GRANTEES, BY GRANT OF 26TH AUGUST, 1783, WITH THE NUMBER OF ACRES RECEIVED, AND NOTICES OF THE SITUATION OF THEIR LOTS.

ON WEST RIVER.

David Stewart 300 acres ; John McKenzie, 500 ; Hugh Fraser, 400 ; William McLellan, — ; James McDonald, 200 ; James McLellan, 100 ; Charles Blaikie, 300, and in an after division 250 acres, 550 in all ; Robert Patterson, 300, and in an after division, 180, 500 in all ; James McCabe, 300 ; Alex. Cameron, —. All these lots are still occupied by descendants of the grantees, with the exception of Charles Blaikie's, which was situated opposite Durham, and Robert Patterson's, which was farther down the river.

ON MIDDLE RIVER, EAST SIDE.

Alex. Fraser, 100 acres, where Samuel Fraser now resides ; Alex. Ross, jr., 100 acres, just below ; then above following up the stream, John Smith, 350 ; Robert Marshall, 350 ; James McCulloch, 240 ; Alex. Ross, 300, an after division to Alex. Fraser, sr., 400 (on the rear of which Westville is now situated) ; Alex. Fraser, jr., 100 ; John Crockett, 500 ; Simon Fraser, 500 ; Donald McDonald, 350 ; David Urquhart, 250 ; Kenneth Fraser, 450 ; James McLeod, 150.

ON EAST RIVER, EAST SIDE.

Walter Murray, 280 acres (adjoining Indian burying-ground); and 70 acres in an after division. Then following up stream : James McKay, 70 ; Donald McKay, jr., 80 ; John Sutherland, 180, and 70 in an after division ; Rod. McKay, sr., 300, and an after division, 50 ; James Hays, — ; Hugh McKay, 100 ; Alex. McKay, 100 ; Heirs of Donald McLellan, 260, (then a blank, where New Glasgow is now situated) ; Hugh Fraser, 400, and an after division, 100 ; Wm. McLeod, 80 ; John McLellan, 200 ; Thomas Turnbull, 220, and in an after division, 180 ; Wm. McLeod, 210, and in an after division, 60 ; Alex. McLean, — ; Colin McKenzie, 370.

ON EAST RIVER, WEST SIDE.

Donald Cameron, 100 acres, at Loading Ground ; James Grant, 400, at Basin ; Colin McKay, 400 ; Wm. McKay, 550 ; Donald Cameron, 100 ; Donald McKay, sr., 450 ; Donald Cameron, a gore lot ; Anthony Culton, 500. These extended from below the mines to some distance above them.

B.

LIST OF THE NUMBER OF FAMILIES IN THE DISTRICT OF PICTOU, VIZ.

" Jonas Earl, Robt. Watson, Robt. Watson, jr., Daniel Earl, Daniel Earl, jr., Jas. Watson, Isaiah Horton, Patrick Berry, Wm. Aikin, John Fulton, James Fulton, John Patterson, George McConnell, Mat. Harris, Robt. Harris, John Rogers, Wm. McKenzie, Wm. McCracken, Abram Slater, Moses Blaisdell, Wm. Kennedy, Jas. Davidson, John McCabe, Bar. McGee, John Wall, Colin McKenzie, Alex. Ross, Donald McDonald, Wm. McLeod, Walter Murray, Thos. Fraser, Alex. Fraser, Wm. McKay, Hugh Fraser, Alex. Faulkner, Colin McKay, Colin Douglass, James Campbell, Thomas Troop, James Hawthorn, Joseph Glen, John McLennan, Ken. McClutcheon, Hugh Fraser, John Ross, George Morrison, Robt. Jones, Don. Cameron, Rod. McKay, Robt. Sims, Peter Hawthorn, John McLellan.—November 8, 1775. (Signed) John Harris." A number of these, set down as families, however were unmarried men at this time. Upon this a petition was presented to the Governor to issue a writ for the election of a representative, but the request was not granted.

C.

LIST OF PASSENGERS IN SHIP "HECTOR," WITH NOTICES OF THEIR HISTORY AND SETTLEMENT, SO FAR AS KNOWN.*

SHIPPED AT GLASGOW.

MR. SCOTT AND FAMILY. Unknown.

GEORGE MORRISON AND FAMILY. From Banff, obtained grant on west side of Barneys River, where he settled. An island there still called Morrisons Island. Left one daughter, married to David Ballantyne, Cape George.

JOHN PATTERSON. Fully referred to in the history.

GEORGE McCONNELL. Settled on West River, at Ten Mile House. His descendants numerous in this and adjacent counties.

ANDREW MAIN AND FAMILY. A native of Dunfermline. Settled at Noel, where his descendants still reside.

ANDREW WESLEY. Unknown.

---

* This list was drawn up about forty years ago by the late William McKenzie, Loch Broom.

CHARLES FRASER. A Highlander, though shipping at Glasgow. Lived at Cornwallis; afterward married and settled at Fishers Grant, where he bought out a soldier of the 82nd. Had one son and two daughters, whose descendants are on West River and elsewhere.

JOHN STEWART. Unknown.

FROM INVERNESSHIRE.

WILLIAM MCKAY AND FAMILY. Afterward Squire McKay; settled on the East River, where the Mines now are. Died 2nd March, 1828, aged 97, when his death was thus noticed in the *Colonial Patriot* :—" For a great many years he was a leading man among his countrymen. His house was always open and his table welcome to travellers and neighbours. The proverbial hospitality of Highlanders was never more fully exemplified than by Squire McKay, and in these early times his liberality must have prevented or alleviated the wants of many of his fellow men. He went to bed in his usual health, and was found dead about half an hour after." Had in the "Hector" four children: 1. *Donald*, who was the first settler on Frasers Mountain. His son, William McKay, the surveyor, was the author of a map of Nova Scotia, published in London, which has supplied information for all the map makers since. 2. *Alexander*, who afterward owned the land where New Glasgow is now situated. 3. *James*, who settled opposite the Loading Ground on the East River. 4. A daughter, *Sarah*, married to William Fraser, surveyor. He had two sons born in this country, John usually known as Collier, and William, who inherited his father's property, where the Halifax Company's works now are, but who afterwards moved to McLennans Brook, and a daughter married to John McKay.

RODERICK MCKAY AND FAMILY. He and three brothers, all of whom came to Pictou, were natives of Beauly, in Invernesshire. He took up land at the East River, where his grandson, J. C. McKay, now lives, being one of the first five who settled on the East River, the others being William McKay, Colin McKay, Donald Cameron, and his brother, Donald McKay. He was a blacksmith by trade, and through the influence of some of his wife's friends, afterward obtained a situation as head of the blacksmiths' work in the dockyard at Halifax. He and his wife travelled thither through the woods on foot, each carrying a child. Under his direction was made and placed the chain, which, during the war, was stretched across the north-west arm, to prevent the entrance of hostile vessels. He was a man of middle height, but thick-set and strongly built, distinguished for activity, determination and fertility in resources. His character may be seen from incidents already given, but one that took place before leaving Scotland, was deemed by his countrymen still more worthy of admiration. The gaugers had seized some whiskey which did not belong to them. Indignant at such an invasion of the rights of property, he interposed, and, perhaps, using some needful violence to these myrmidons of Saxon oppression, rescued it from their unworthy hands. So little, however, was his prowess appreciated by the Sassenach bodies, he was for this lodged in a jail in Inverness. His free-born spirit chafed under such restraints, particularly in a cause so good, and he was soon contriving schemes to secure his liberation. Having ingratiated himself with the jailer, he sent him one day to procure a quantity of ale and also of whiskey, in order duly to cement their friendship. The jailer on his return, advancing into the cell with both hands full, Roderick stepped behind him and out at the door. Closing it after him, he locked it and carried off the key, which some say he brought to America with him. The first of these feats would have given him an honorable place in the hearts of his countrymen, but the latter, added to it, was sufficient to make him their idol for ever. We suppose that some in this effeminate age will scarcely regard such affairs as creditable. But similar qualities, exercised for much worse purposes, have rendered Rob Roy the admiration of all lady readers of Sir Walter Scott, and his grave to be visited with veneration, even by Royalty.

When in Halifax he gained notoriety by another feat. An officer was paying some attention to a female inmate of his house, of which he did not approve. Roderick meeting them together near the Citadel Hill, upbraided him for his conduct, when the latter drew his sword, and before the former was aware, struck him a cruel blow on the head, cutting him so severely that he felt some of the effects as long as he lived. Telling the officer that he would meet him in an hour, he got his head dressed, and prepared within the time, stood before him with a good ash stick. The officer drew his sword, and a combat ensued, but Roderick was not only an adept in all Highland games, but like many Highlanders of that time, was acquainted with the sword exercise, and though his stick bore the marks of the officer's sword cuts, he soon disarmed him, and repaid him heartily for his former cowardly attack. He afterwards returned to his farm on the East River, where he died. One daughter, with him in the "Hector," was afterward married to Dr. McGregor. The rest of his family were born in this country. One daughter, mother of the late J. D. B. Fraser, Esq., our late custos, Robert McKay, Esq., his son.

COLIN McKAY AND FAMILY. Served in the Fraser Highlanders at the taking of Louisburg and Quebec. Settled on the East River below the Mines. Few of his descendants there. "McKay brothers," of Liverpool, G. B., his grandsons.

HUGH FRASER AND FAMILY. The following is a copy of a certificate in the possession of a grandson : " These do certify that the Bearer Hugh Fraser, Weaver, a married man was born of honest Parents in this Parish of Kiltarlity, where he has resided from his Infancy and behaved soberly and honestly free of any public scandal. So that now when he with his Wife and Family are to remove from our bounds we are at freedom to declare that we know no reason why they may not be received into any Christian Society or Congregation where Providence may order their Lot. Given at Kiltarlity this 29th day of June, 1773 years and attested in name of the Kirk Session by

" MALCOM NICOLSON, Minister."

Settled on McLellans Brook. Had three children in the "Hector"— 1, *Domald*, known as Donald Miller ; 2, *Mary*, married to Cameron, Merigomish, and 3, *Jane*, married to John Fraser, Merigomish. One son, *John*, long known as John Squire, having been taken sick with small-pox, did not arrive for some years after. Rev. Wm. Fraser, Bondhead, Ont., a grandson.

DONALD CAMERON AND FAMILY. The only Roman Catholic among the passengers, had also served at the taking of Quebec, settled on the East River where the Mines now are, on the farm afterward owned by Dr. McGregor. He was drowned in the river. His family removed to Antigonish county.

DONALD MCDONALD AND FAMILY. Settled on the Middle River, where his descendants are numerous. Had two children in the " Hector," the eldest, *Marion*, aged 10, afterward married to Alex. Fraser Elder (Middle River), the second, *Nancy*, who died the winter after arrival, and an orphan niece, Mary Forbes, afterward married to Wm. McLeod, who settled at McLellans Brook.

COLIN DOUGLASS AND FAMILY. Lost two children on the passage. His eldest daughter survived, afterward married to Peter Fraser, McLellans Mountain. Settled at Middle River, one son long known as Deacon Douglass.

HUGH FRASER AND FAMILY. Settled at West River about twelve miles from town, afterward an elder. His descendants numerous.

ALEXANDER FRASER AND FAMILY. Settled at Middle River, where and at various other places his descendants are numerous. He is said to have been connected with Lord Lovat, and the family were largely involved in the rising of forty-five. Had three brothers fighting for Prince Charlie at Culloden, of whom two were killed; was too young to serve himself, but followed them, and saw at least part of the scenes of that day. Afterward married Marion Campbell, youngest daughter of the Laird of Skriegh, in Inverness, who had raised a troop to fight for Prince Charlie, and at Culloden was wounded. After the battle he was set up against a wall by the English soldiers to be fired at. Missing him, one of them said, " You poor devil, you're ordained to live for some mischief," and struck him on the face with his musket, knocking out his eye. His wife found him among the dead and wounded, and though with the loss, it is said, of a leg, an arm, and an eye, he survived for some years.

Fraser was in comfortable circumstances, when an instance of Saxon oppression led him to seek for freedom in America. His horses and cart were seized by guagers, with some whiskey that they were carrying. What Highlander's soul will not boil, even at the hearing of such an outrage. The seizers took their plunder to Inverness, where they had it cared for at an inn, and then proceeded to enjoy themselves drinking. When they were comfortably disposed of for the night, the stable lad, who was a relation of Fraser's, took the horses and cart out, and driving across the country, restored them to the proper owner, who lost no time in taking them to some other part of the country, where he disposed of them as well as he could, and, determined to stay no longer in a country, where he was subjected to such treatment, was the first to engage a passage in the " Hector."

He settled on the Middle River, where Samuel Fraser now resides. He had five children in the " Hector." 1. *Alexander*, who occupied his father's farm, and was afterward an Elder. 2. *Simon*, particularly referred to in the history. 3. *Catherine*, married first to Alexander Ross, afterward to John Fraser (Squire.) 4. *Isabella*, married to David McLean, Esq., West Branch, East River. 5. *Hugh*, settled at Middle River Point, and the last survivor but one of the " Hector " passengers.

After arrival he had two sons, *David*, the first child born to the Highlanders after arrival, and *William*, the first child born on the Middle River.

JAMES GRANT AND FAMILY. Lived for some time in Kings County; first settled at Grahams Pond, Carriboo; afterward moved to Upper Settlement, East River; father of Alexander Grant, miller, and Robert Grant, Elder; grandfather of Dr. W. R. Grant, Professor of Anatomy in Pennsylvania Medical College. One daughter married to John Sutherland, Sutherlands River; another to McNeil, who removed into Antigonish County, and another to Fraser. The eldest son remained in the old country, and it is believed afterward emigrated to the United States. The supposition of the family is that he was the ancestor of President Grant.

DONALD MUNROE, went to Halifax, where he married, and had one son, Henry. Afterward settled at West Branch, East River, where he died, and was the first buried there. His descendants numerous in that neighborhood.

DONALD MC————. Name illegible, and history unknown.

FROM LOCH BROOM.

JOHN ROSS, agent. History unknown.

ALEX. CAMERON AND FAMILY. Was nearly eighteen years of age at the time of the rising in 1745. There was some badge, which the Highlanders were allowed to wear on their bonnets, on arriving at that age, as a sign of manhood. His brothers followed the Prince, but he being only seventeen years of age, was required to remain at home, where he was employed in herding. But drawn by the crowd who followed the Highland army to Culloden, he left his charge to accompany them. When he returned, his master being very angry, " went for him" to chastise him. He ran and his master pursued. The latter finding him too nimble, stooped down to pick up a stone to throw at him, and in doing so wounded himself with his dirk in the leg, so that he was obliged to remain for some time in hiding, lest he should be taken as having been at Culloden, by the soldiers who were scouring the country, killing any wounded stragglers from the field. Cameron settled at Loch Broom, to which he gave the name of his native parish. He died on the 15th August, 1831, when he must have been at least 103 years of age. He had two children in the " Hector"—1, *Alexander*, long an elder in the church, and 2, *Christiana*, afterward married to Alex. McKay, New Glasgow, and several children born after arrival.

ALEX. ROSS AND FAMILY. He and his wife advanced in life at arrival. Parents of the next.

ALEX. ROSS AND FAMILY. Settled at Middle River, at what has since been known as Olivers farm. Died when only 35 years of age, the youngest of any of the band. Believe the following his children : 1, *Donald*, who occupied his father's farm, but afterward moved to Ohio ; 2, *Alexander*, who settled at Middle River Point ; 3, a daughter married to Archibald Chisholm, East River, and another married to Blair, East River.

COLIN McKENZIE AND FAMILY. Settled on East River, about a mile above New Glasgow on the farm immediately above John McLellans. Said to have lived to 104. Had one child on board, *Duncan*, who died in 1871, in his 100th year, the last survivor of the band.

JOHN MUNROE AND FAMILY. History unknown.

KENNETH McRITCHIE AND FAMILY. Probably the same whose name appears in early lists as Kenneth McClutcheon, but know not what became of him.

WILLIAM McKENZIE, an intelligent man, who had enjoyed a better education than the rest, and who had been engaged as schoolmaster for the party, as they expected to settle together. He settled at Loch Broom, where some of his descendants still are.

JOHN McGREGOR. History unknown.

JOHN McLELLAN. Settled above New Glasgow, at the mouth of McLellans Brook, and gave his name to that stream and McLellans Mount. Properly the name however was McLennan, the two being quite distinct in Gaelic.

WILLIAM McLELLAN. Relative of the last, settled at West River, where some of his descendants still are.

ALEXANDER McLEAN. Settled at East River, above Irishtown. One son settled on McLennans Mountain, where his descendants still are.

ALEXANDER FALCONER. Settled near Hopewell.

DONALD McKAY, afterward the Elder, brother of Roderick. Settled at the East River, just above the Mines. His house on the same site as that now occupied by his grandson, Duncan. Another brother, Hugh, came afterward, but died without a family.

ARCHIBALD CHISHOLM. Believed to be the same person who settled at East River, after having served in the 84th Regiment.

CHARLES MATHESON. History unknown.

ROBERT SIM. After residing for some time in Pictou, removed to New Brunswick. Never married.

ALEXANDER McKENZIE. History unknown.

THOMAS FRASER. History unknown.

FROM SUTHERLANDSHIRE.

KENNETH FRASER AND FAMILY. First settled at Londonderry, but afterward moved to Pictou, where he settled on Middle River, above the bridge at Squire McLeod's. His descendants numerous on Green Hill, Mill Brook, Rogers Hill, &c.

WILLIAM FRASER AND FAMILY. History unknown.

JAMES MURRAY AND FAMILY. Removed to Londonderry, where his descendants still are.

WALTER MURRAY AND FAMILY. Settled in Merigomish, where a number of his descendants still are.

DAVID URQUHART AND FAMILY. Settled at Londonderry. One daughter, the late Mrs. Thomas Davidson, afterward resided in Pictou.

JAMES McLEOD AND FAMILY. Settled at Middle River on the farm which has descended to his relative, George McLeod, Esq, he having no children of his own.

HUGH McLEOD AND FAMILY. His wife died as the vessel arrived. Had three daughters on board, one of whom married in Cornwallis; another afterward Mrs. Donald Ross, and the third afterward Mrs. Shiels. Settled on West River. Married the widow of Alexander McLeod, by whom he had one son, David, long a highly esteemed Elder there.

ALEXANDER McLEOD AND FAMILY. Was drowned in the Shubenacadie. Had three sons on board; one died in the harbour after the vessel's arrival; another died unmarried; the third, the late Donald McLeod, settled at West River, on a farm still occupied by his descendants.

JOHN McKAY AND FAMILY. Settled at Shubenacadie.

PHILIP McLEOD AND FAMILY. Uncertain.

DONALD McKENZIE AND FAMILY. I believe settled at Shubenacadie.

ALEX. McKENZIE AND FAMILY. History unknown.

JOHN SUTHERLAND AND FAMILY. History unknown.

WILLIAM MATHESON AND FAMILY. First settled at Londonderry, but after Dr. McGregor came to Pictou, removed and settled at Rogers Hill, where John T. Matheson now resides. The eldest son, John, afterward the elder at Rogers Hill, was three years of age when they landed. His second, born after arrival, the late William Matheson, Esq.

DONALD GRANT. History unknown.

DONALD GRAHAM. History unknown.

JOHN McKAY, piper. History unknown.

WILLIAM McKAY. Went to work with McCabe, one of the old settlers, and thence got the name of McCabe, by which his descendants are still distinguished. Was drowned in the East River by falling from a canoe.

JOHN SUTHERLAND. Removed to Windsor, where he married. Returning settled at the mouth of Sutherlands river, which derived its name from him.

ANGUS McKENZIE. Then only sixteen years of age. Removed to Windsor, where he married. Returning to Pictou some years after, he settled first at the Beaches on the farm afterward owned by the Lowdens, and afterward on Green Hill, where some of his descendants still are.

D.

LIST OF DUMFRIES SETTLERS, WITH PLACES OF SETTLEMENT.

ON WEST RIVER.

CHARLES BLAIKIE. Settled on east side of the river, opposite Durham, on the farm now belonging to David Matheson.

DAVID STEWART. Settled farther up on same side, where his descendants still reside.

ANTHONY MCLELLAN. Settled on west side of the river, just at Durham.

WILLIAM CLARK. Settled above him on same side of the river, where his descendants still reside.

JOSEPH RICHARD. On same side of the river, below the Ten Mile House, where his descendants still are.

JOHN MCLEAN. Settled where his son, the late John McLean, Elder, lived; was one of the first Elders ordained by Dr. McGregor; the Rev. John McLean, of Richibucto, his grandson, and John S. McLean, of Halifax, his great grandson.

WILLIAM SMITH. Father of late Anthony Smith, Esq.; settled near the Ten Mile House.

ON MIDDLE RIVER.

ROBERT MARSHALL. Afterward the Elder. His house stood close by the bridge crossing McCullochs Brook, close by the road leading to Middle River.

JOHN CROCKETT. Settled where his grandson, W. P. Crockett, now lives.

ROBERT BRYDONE. Settled farther up the river.

JOHN SMITH. Had come out to Prince Edward Island as agent for some of the proprietors, earlier than the other settlers. He settled on the property since owned by Thomas Horn; was drowned, it is said, with a daughter, Mrs. McCulloch, and her child, which they were taking to Pictou to have baptized, and another woman.

ON EAST RIVER.

THOMAS TURNBULL. Settled on McLennans Brook.

ANTHONY CULTON. Settled above the Mines.

Besides these, we have already mentioned Wellwood Waugh, with whom came a half-brother, William Campbell, then a young man, who settled on the farm above him, a little below the town.

All these have left numerous descendants in various places.

E.

"A ROLL OF THE INHABITANTS OF PICTOU OR TINMOUTH CAPABLE TO BEAR ARMS.

" James Grant, William Campbell, Robert Jones,* Wm. McCracken, George McConnell, John Patterson, sen., James Patterson, David Patterson, John Patterson, jr., John Rogers, sen., James Rogers, John Rogers, jr., David Rogers, James McCabe, John McCabe, Anthony McLellan, James McLellan, Ed. McLean, Joseph Ritchie, William Clark, John McLean, Wm. Smith, David Stuart, John McKenzie, Hugh Fraser, Wm. McLellan, James McDonald, Charles Blaikie, John Blaikie, James Watson, Alex. Cameron, Colin Douglass, Don. McDonald, Robt. Breading (Bryden), John Breading, Alex. Ross, sr., Alex. Ross, jr., James McCulloch, Robt. Marshall, John Marshall, John Crockett, John Crockett, jr., Alex. Fraser, Alex. Fraser, jr., Simon Fraser, Colin McKay, Rod. McKay, jr., James McKay, Donald McKay, Donald McKay, jr., Donald Cameron, Anthony Culton, John Culton, Colin McKenzie, Alex. McLean, John Sutherland, Thos. Turnbull, John McLellan, Wm. McLeod, Hugh Fraser, sr., James Fraser, Esaias Horton, Stoatly Horton, Morton (Walter) Murray; George Morrison, Barnabas McGee.

" The above is a true list, given under my hand at Halifax, 12 February, 1783.   ROBT. PATTERSON, Captain."

The above begins at Carriboo, and passes up the harbour, and round the three rivers to Merigomish.

F.

LIST OF GRANTEES OF THE 82ND REGIMENT.

COL. ALEX. ROBERTSON. Obtained the big island of Merigomish as his share, hence sometimes known as Robertson's Island. Never lived on it himself, but some relatives of the name settled upon it. Employed an agent, who built a large house on it, which he called Struan House. At his death, his property in this county descended to his nephew, Oliphant, of Gask.

CAPT. JOHN FRASER. Lived at Frasers Point, appointed a magistrate October 15th, 1784. His wife and two sons followed him from Scotland, one of the latter, John, being afterward known as Collector Fraser, the other, Simon,

---

* This was a Welshman, who had served both at the capture of Fort Beau Sejour and Quebec, and lived about a mile below the town. He died unmarried.

·called also Major, and sometimes Colonel Fraser, afterward employed in bringing out passengers.

ALEX. MCDONALD. Unknown.

COLIN MCDONALD. I believe the same that known as Colo McDonald, who lived on the Big Island, near what is still known as Coles Brook.

DONNET FENUCANE. His land located to the west of Frasers Point, but his history unknown.

These three received each 500 acres.

JOHN MCNEIL. Received 300 acres, but history unknown.

NON-COMMISSIONED OFFICERS RECEIVING EACH 200 ACRES.

CHARLES ARBUCKLES. A native of Falkirk, moved afterward to the Ponds. Married to a daughter of B. McGee. His descendants numerous.

DAVID BALLANTYNE. Removed to Cape George, where his descendants are numerous.

GEORGE BROWN. Settled on Frasers Mountain.

JOHN BROWNFIELD. A native of Derry, in Ireland, and a Presbyterian, died near French River, where his descendants still are.

JAMES CARMICHAEL. A native of Perthshire. His descendants well known.

ROBERT DUNN. A native of Glasgow, settled on the property now owned and occupied by his sons.

JOHN FRASER. A Highlander from Inverness. One of 18 who survived out of a detachment of 111 men, employed in the Southern States during the war, the rest having been cut off by fever. He lived at Fishers Grant, where he was one of the first elders of Pictou congregation. Afterward removed to French River, where his descendants still are.

DUFFEY GILLIES. Believe the same as James Gillies, who lived where R. S. Copeland now resides, afterward removed to Big Island, where his descendants still are.

JAMES PEACOCK. Lived near Chance Harbour, but do not know what became of him.

JOHN ROBSON. From being able to bleed, and his skill otherwise, he was usually known as Dr. Robson. His descendants still there.

CHARLES ROBINSON, properly Robertson Was a son of the proprietor of the estate of Lude, at the foot of the Grampians. Was a student attending college when he enlisted. One daughter, married to Robert Patterson (Black.)

JOHN SCOTT. Sold out to John Fraser, 1785.

ROBERT SMITH. His lot where the Merigomish church now stands. His descendants still there.

DAVID SIMPSON. Had been a student at college, but he and some others having indulged in " a spree," some eighteen of them found themselves in the morning with the King's shilling in their pockets. Their professors endeavored to obtain their discharge, but without success. From his education, he obtained some office in the regiment. His lot, on which his descendants still live, the farthest up in Merigomish, in the grant. He was afterward employed as a schoolmaster in several parts of the county.

ROBERT STEWART. Usually known as Smashem, from this being a favorite expression in describing battle scenes. He acted as agent for Col. Robertson, and lived on the Big Island, at a point which has since received the name of Smashems Head.

Robert Miller, Gerrard Cullen, heirs of John Eves, John Fowler, John Foot, Thomas Loggan, Archibald Long, John Morton, Alexander McKinnon, John McNeil, Jun., George Oswald, James Robertson, Alexander Stewart, James Struthers, William McVie, William West, Archibald Wilson. History unknown.

RECEIVING 150 ACRES EACH.

JOHN BAILLIE. A native of Sutherlandshire; afterward took up land at the mouth of Baillies Brook, which received its name from him.

ARCHIBALD CAMERON. History unknown.

PRIVATES RECEIVING 100 ACRES EACH.

ANDREW ANDERSON. A native of East Lothian, and the first settler on Andersons Mountain. Died 3rd August, 1845.

JOHN BRADAW (properly Brady). Sold out.

DAVID BOGGEY. Died at Fishers Grant, leaving no family.

DUNCAN CHISHOLM. Removed to Baillies Brook, where his descendants still are.

WILLIAM CAMPIN or CAMPDEN. Sold out and removed to Truro.

JOHN COLLY. Suppose the same who afterward settled on Middle River, where his descendants still are. A native of Elgin.

JAMES DANSEY or DEMPSEY. An Irishman; settled at French River Bridge. His descendants still there.

BRITISH FREEDOM. Strange as this name is, there is in the Registrar's office in Pictou a deed from him of his lot, under this inspiring name. Hence I presume that he moved away.

HARDIN FERDINAND. A very stout, well-made Irishman, who afterward enlisted in Governor Wentworth's Regiment.

THOMAS FLEMMING or FLEEMAN. Sold out and removed away.

ROBERT FERRET, properly GERRARD. An Irishman; afterward removed to Rogers Hill.

ALEXANDER GORDON. "Died at Fishers Grant on the 18th inst, after an illness of eight days, which he bore with resignation to the divine will, for which he has always been exemplary, Mr. Alexander Gordon, aged 80 years, leaving a circle of relatives and friends to mourn their loss. He was of the old 82nd Regiment, and one of the earliest settlers in the district of Pictou."— *Bee*, August 31st, 1836.

JOHN IVES. A native of Nottingham, England, but married in the North of Ireland. Died in Halifax, and his children, the eldest, the late George Ives, Elder, then 12 years of age, came to take possession of their lot at Fishers Grant. His descendants well known.

WILLIAM KIRK. Afterwards removed to St. Marys, where his descendants are numerous. A grandson in the Dominion Legislature.

ANDREW MUIRHEAD. A Lowland Scotchman; first settled at the Ponds. His descendants at Little Harbour and other places.

HUGH MCCARTHY. A tailor. Sold out and removed to Truro.

JOHN McDOUGALL. Blacksmith in the Regiment. Lived at Fishers Grant. The ferrymen, Donald and William, his sons.

ANGUS MCQUEEN. A native of the Isle of Skye; settled at Little Harbour; a number of his descendants still in that neighbourhood. DONALD McDONALD (Lochaber), DONALD MCDONALD (Bann), ANGUS MCDONALD. These four the first settlers in Little Harbour.

CHARLES MCKINNON. From the Isle of Barra. Moved to Baillies Brook where his descendants still are.

JOHN MCNEIL, DONALD MCNEIL, MURDOCH MCNEIL, MATTHEW MCNEIL, JOHN MCNEIL, Jun. Isle of Barra men, most of whom removed to Antigonish County.

JOHN AND JAMES MCPHERSON. Settled at Fishers Grant; John dead in 1785. James at his death described as a native of Badenoch. Their descendants still there.

WILLIAM ROBINSON. A Scotchman. His descendants settled there.

WILLIAM SHARP. Died at the Beaches.

WILLIAM SYMPTOM. Married to Ives' widow, lived at lower part of Fishers Grant.

JOHN SMALL. Afterward the Elder, belonged to the Grenadier company. One of the 18 saved from the wreck of the Transport. For some time in an American prison, but with fifteen others made his escape; and passing through the American lines reached a British man-of-war. But afterward drowned near his own house, at a part of the harbour of which it was said that he knew every foot as well as his own farm.

JAMES TRUESTATE, properly Truesdale. Sold out and removed to Truro.

James Arthur, Wm. Adams, William Bilboa, William Branon, Michael Branon, George Brown, Charles Brown, John Brown, Archibald Cameron, Robt. Clawson, Finlay Campbell, Donald Campbell, sr., Donald Campbell, jr., Alex. Campbell, Matthew Campbell, John Chisholm, Archibald Cochrane, Thomas Connelly, Robt. Dewar, John Dickson, Alex. Dickson, Dennis Dirkham, Lawrence Donnachie, Charles Dunce, Francis Gobbiel, Angus McDonald 2nd, Angus McDonald 3rd, Peter McDonald, John McDonald, Roderick McDonald, William Gowe, Peter Gowe, Richard Griffin, James Gibes, Robert Gardner, Patrick Hayne, or Kane, Archibald Henderson, Wm. Hodges, John Holmes, Patrick Hunt, John Ives, Wm. Jack, Alex. King, Alex. Kennedy, John Little, John Lunn, Wm. Lamplash, Peter Lamplash, Thomas Matheson, John Munro, Hugh Miller, John Muir, John Morton, Samuel McBawe, Neil McCallum, John McCladdy, Bryan McDermaid, Archibald McGavy, John McGillivray, Alex. McKenzie, Robert McKenzie, Alexander McLean, Ewan McLean, Alexander McLean, Samuel McLean, John McLeod, Kenneth McLeod, John Patterson, William Riddle, Robert Reid, Thomas Ryan, George Robertson, Alex. Shaw, Charles Stewart, David Skervine, John Sovereign, Patrick Skey, James Struthers, John Scott, John Stevenson, Matthew Talbot, Thomas Townsend, Robt. Thyne, Wm. Wood, Thomas Wood, John Wright, Robt. Warren. Unknown.

We find also in the County Records, the names of the following as soldiers of the 82nd, whose names are not in the grant, viz. : John McIlvain, Robert Irvin, Wm. McKay, John McGarvie, Wm. Hogan, George Osborne, Thomas Crowe, as selling out their lots. There were others also who occupied theirs, such as Owen McEwan, or McKowan, a native of the County of Down, whose descendants are numerous.

G.

LIST OF GRANTEES OF 84TH REGIMENT ON EAST BRANCH.

ON THE EAST SIDE OF THE RIVER.

DONALD CAMERON. 150 acres. With his brothers, Finlay and Samuel, afterward mentioned, were natives of the parish of Urquhart. Served 8 years and 4 months. His son, Duncan, long the elder, was a drummer boy in the regiment, having served two years, and being fifteen years of age at his discharge.

ALEX. CAMERON. 100.

ROBT. CLARK. 100.

FINLAY CAMERON. 400. Enlisted in Canada with the view of joining his friends in Nova Scotia. Returned thither to bring his family at the peace. Was drowned shortly after his arrival, along with John Chisholm at the Narrows.

SAMUEL CAMERON, jr. 100 acres.

JAMES FRASER (Big). 350 acres. A native of Strathglass. Settled where his grandson Donald, lives, a little below St. Pauls.

PETER GRANT. The first elder in this settlement.

JAMES McDONALD. Long the Elder; said to have been the strongest man in the Regiment. Removed to the London district of Ontario. Hon. James McDonald his grandson.

HUGH McDONALD. 100.

ON THE WEST SIDE.

JAMES FRASER, 2nd. Usually known as Culloden; 100 acres; farthest-up settler on that side. His descendants there still. Rev. James W. Fraser descended from him.

DUNCAN McDONALD. 100 acres.

JOHN McDONALD. 250 acres; brother of James.

SAMUEL CAMERON. 300 acres; brother of Donald and Finlay, already mentioned.

JOHN CHISHOLM, Sen. 300. A Roman Catholic from Strathglass; drowned with F. Cameron, as mentioned; father of Mrs. John McKenzie, Sen., West River.

JOHN CHISHOLM, Jun. 200 acres. Son of the last.

JOHN McDONALD, 2nd. 250 acres.

H.

LIST OF GRANTEES AT WEST BRANCH AND OTHER PLACES ON THE EAST RIVER, 18TH DECEMBER, 1797.

WILLIAM FRASER. 350 acres. From Inverness, land situate at Big Brook, now owned by his grandchildren.

JOHN McKAY. 300.

JOHN ROBERTSON. 450. At Churchville.

WM. ROBERTSON. 200. Son of the last, also near Churchville.

JOHN FRASER. 300. From Inverness, Springville, now occupied by Holmes and others.

THOS. FRASER. 200. From Inverness. An elder and noted for piety. His lot was at the head of the West Branch.

THOS. McKENZIE. 100. Settled near Fish Pools.

DAVID McLEAN. 500. A sergeant in the army, or as some say a petty officer in the navy. Was captured by the Spaniards, and afterward exchanged as a prisoner. He was a better scholar than usually found among the settlers, was a surveyor, a magistrate, an elder in the church, and a leading man in that section of the county.

ALEX. CAMERON. 300.

HECTOR McLEAN. 400. From Inverness. Land still occupied by his descendants.

JOHN FORBES. 400. From Inverness. Land on East Branch River.

ALEX. McLEAN. 500. Brother of Hector. Land opposite Stellarton, part of it still occupied by his descendants.

THOS. FRASER, JR. 100.

JAS. McLELLAN. 500. From Inverness. Land above the Fish Pools, on the opposite side or the river, occupied by his descendants.

DONALD CHISHOLM. 350. From Strathglass, originally a Catholic, but became a Presbyterian. St. Columbas church built on part of his farm.

ROBT. DUNBAR. 450 ⎫ Three brothers from Inverness. All we believe
ALEX. DUNBAR. 200 ⎬ in the 84th. Their land still occupied by their
WM. DUNBAR. 300 ⎭ descendants.

JAMES CAMERON. 300. Of the 84th. Land still occupied by his descendants.

JOHN McDOUGALL. 250. In the Registrar's book, " J. M. Douglass."

JOHN CHISHOLM. 300.

DONALD CHISHOLM, jun. 400. From Inverness. Land occupied by his grandsons.

ROBERT CLARK. 150. Of the 84th, but moved away. Land now occupied by Mr. Thomas Fraser.

DONALD SHAW. 300. From Inverness. Land occupied by his grandsons.

ALEXANDER McINTOSH. 500. From Inverness. His land now partly occupied by Hopewell Village.

JOHN McLELLAN. 100. From Inverness. Land occupied by D. H. McLean and James Fraser.

The most of those marked as from Inverness were from the parish of Urquhart, in that county, and served in the 84th. In the record of the grant, dated 1st April, 1793, there are the additional names of Colin Robinson, William Robinson, William McKenzie, William Robertson, heirs of John Forbes, Hugh Dunoon, and Thomas Fraser, but Hector McLean's is omitted.

I.

LIST OF HIGHLAND EMIGRANTS BY HALIFAX IN 1784.

THOMAS FRASER. A native of Kirkhill Parish. Settled nearly opposite New Glasgow, where he was known as Deacon Thomas, and his descendants are still so distinguished.

WILLIAM FRASER (Ogg). Settled just above him.

—— FRASER. Usually known as "basin."

ALEXANDER McKAY. A brother of Roderick and Donald, already mentioned as passengers in the Hector. Had served in the Fraser Highlanders at the capture of Louisburg and Quebec. Near the latter place received a ball in his leg, which he carried till his death. Was a very powerful man. Lived to be 97 years of age, and almost to the last, a reference to the campaign at Quebec would stir up his blood. Settled near Fish Pools, where his son lately resided.

THOMAS McKENZIE. Settled near Fish Pools, where Thomas Grant now resides.

ALEXANDER FRASER OR McANDREW. From the parish of Kilmorack. Settled at McLellans Brook, but did not live long.

SIMON FRASER. Also an Elder. Settled on McLellans Brook, where his son, William, filled the same office for fifty years.

JOHN ROBERTSON. A brother-in-law of Roderick and Donald McKay. The first settler at Churchville. His first clearing was made where John Robertson, miller, now resides.

There was also a family of Frasers, who came, we believe, with this band. The father died, the widow married William Dunbar, and the only son moved into Antigonish County.

J.

LIST OF MINISTERS FROM PICTOU COUNTY.

PRESBYTERIAN.

Bayne, E. S. .... { Murray Harbor, P.E.I,
Blaikie, Alex., D. D...Boston.
Cameron, John.........Elmsdale, N. S.
" John.........Dunoon. Scotla'd.
" D. W.........Ontario.
" J. G...........Souris, P. E. I.

PRESBYTERIAN—(continued.)

Cameron, Alex.........Riverside, N.S.
" Robert J......Scotland.
" Alex. H......Heckston, Ont.
Campbell, John.........Sherbrooke, N.S.
" John.........Scotland.
" D. K.........United States.
Cumming, Thomas......Stellarton, N.S.

PRESBYTERIAN.—*Continued.*

Cumming, Robert.......Glenelg, N. S.
Dunbar, Hugh........... { New London, P.E.I.
Falconer, Alex...........Trinidad.
Forrest, John...........Halifax, N. S.
Fraser, Wm..............Bondhead, Ont.
" Simon............United States.
" J. W............Rogers Hill, N.S.
Geddie, John, D. D..... { Aneiteum, New Hebrides.
Gordon, D. M............Ottawa.
" John............Paisley, Ont.
Grant, George M.........Halifax, N. S.
" Charles M......Glasgow.
" Kenneth, J....... { Missionary to Trinidad
" Ed..............Stewiacke, N. S.
" Wm..............Earltown, N. S.
Gunn, Adam.............Kennetcook, N.S.
" S. C............St. Peters, P.E.I.
Livingston, John........Dundee, Ont.
Matheson, J. W.......... { Missionary to New Hebrides.
Melville. Peter........Kincardine N B.
Millar, E. D............Shelburne, N. S.
Morton, John........... { Missionary to Trinidad.
Murdoch, John L........Windsor, N. S.
Murray, Isaac, D. D.....Cavendish, P E I.
" James A........London, Ont.
" James D. ......N. South Wales.
" John D . . ...Buctouche, N.B.
" John...........Sydney, C. B.
" ———..........British Columbia
McBean, Alex........ . Halifax, N. S.
McCulloch, Wm., D. D..Truro, N. S.
McDonald, F. R........Scotland.
" D............Nottawasaga, On.
McGillivray, Angus......Springville, N.S.
" J. D........Newport, N. S.
" Daniel......Brockville.
McGregor, P. G., D. D. .Halifax, N. S.
" Simon........British Columbia
McKenzie, Alex.........Goderich, Ont.
" Kenneth.....Baddeck, C. B.
" J. W........ { Missionary to New Hebrides.
McKay, H. B............River John, N.S.
" J. McG........Economy, N. S.
" Alex.........Eldon, Ont
" A. W.........Streetsville, Ont.
" Kenneth.......Richmond, N. B.
McLean, John ....Richibucto, N.B.
" Alex.........Belfast, P.E.I.

PRESBYTERIAN.—*Continued.*

" James..........Londonderry, NS
" John D........Miss'y to Japan.
McLeod, John M......... { Charlottetown, P.E.I.
McMillan, Wm..........East River.
" John..........Truro.
McQuarrie, Hector......Wingham, Ont.
" A. N..........Quebec.
McRae, Donald.........St. Jo!n, N. B.
Patterson, R. S.........Bedeque, P.E.I.
" George, D.D..Green Hill, N. S.
" Isaac ....United States.
Richard, James.........Westport, N. S.
Robertson, Hugh A...... { Missionary to New Hebrides.
Roddick, George ....West River, N.S.
Ross, Hugh ....Tatamagouche.
" James, D. D......Halifax.
" Ebenezor ....Londonderry, NS
" Walter ....Beckwith, Ont.
" Donald ....Lancaster, Ont.
" William........... { Prince William N. B.
Simpson, Isaac.......... { Musquodoboit, N. S.
Sutherland, Alex.......Longwood, Ont.
" Wm........Strathburn, Ont.
" George ....N. South Wales.
" J.A.F......St. Croix, N. S.
" John M.....Pugwash.

CONGREGATIONALISTS.

Halliday, ———........United States.
McLeod, ———.........United States.

BAPTIST.

Beattie, Francis.........New Brunswick.
Clark, John......... ...United States.
Campbell, ———......... " "
Gunn, ———............. " "
Shaw, ———.. ....P. E. Island.

WESLEYAN METHODIST.

Burns, J..............———

EPISCOPAL.

Wilkins, L. M..........———

ROMAN CATHOLIC.

McGregor, D. M........Merigomish.
McKinnon, Donald.....Cape Breton.

Some of these were not born in Pictou, though brought up here; but there might be added others, as Revs. A. G. Forbes and J. D. Forbes, born in this County, though brought up elsewhere.

K.

LIST OF MEMBERS OF THE LEGISLATURE FROM PICTOU COUNTY.

MEMBERS OF NOVA SCOTIA HOUSE OF ASSEMBLY.

1840 to 1843. *County:* John Holmes and Henry Blackadar.
              Township: Henry Hatton.
1843 to 1847. *County:* John Holmes and George R. Young.
              Township: George Smith, elected in 1843, but unseated in 1844, when Henry Blackadar elected for remainder of term.
1847 to 1851. *County:* George R. Young and Andrew Robertson.
              Township: Henry Blackadar.
1851 to 1855. *County:* John Holmes and Robert Murray.
              Township: Martin I. Wilkins.
1855 to 1859. *County:* George McKenzie and A. C. McDonald.
              Township: Martin I. Wilkins.
1859 to 1863. *Eastern District:* George McKenzie and A. C. McDonald.
              Western do: A. C. McDonald and R. P. Grant.
1863 to 1867. *Eastern District:* James McDonald and James Fraser.
              Western do: Alexander McKay and Donald Fraser.

MEMBERS OF LEGISLATIVE COUNCIL.

Hons. George Smith, David Crichton, and John Holmes.

UNDER CONFEDERATION.

MEMBERS OF HOUSE OF COMMONS

1867 to 1872. James W. Carmichael.
1872 to 1873. Robert Doull and James McDonald.
1873.   James W. Carmichael and John A. Dawson.

SENATORS.

1867 to 1876. Hon. John Holmes.
1877.   Hon. R. P. Grant.

HOUSE OF ASSEMBLY.

1867 to 1871. Dr. George Murray, R. S. Copeland, and M. I. Wilkins.
1871.   Simon H. Holmes, Alexander McKay, and Hugh J. Cameron.

LEGISLATIVE COUNCIL.

1867.   Hon. James Fraser.

INDEX.

	PAGE.
Abenakis	34
Academy Pictou	322 et seq.
Acadia Coal Company	413 et seq.
Agriculture	436
Agricultural, first Society	296
Albion Mines	412
Alline, Rev. H	113
American War, first	98
" Settlers	99
Archibald, S. G. W	221
Artillery Company	261
Baillie, John	460
Baillies Brook	460
Ballast Pier	312
Baptists	446
Barneys River	108 283
Beaches	15
Bears	80
"Bee"	384
Bennet, Rev. James	113
Bible Society, first,	271
" Reorganized	312
Blanchard Settlement	231
Blanchard, Jotham	353 368 et seq.
Blue Mountain	284
Boundaries of County	9
Burke, Rev. Ed	263
Cameron, Alexander	454
" Donald	90
Campbell, John	413
" Donald, murder by	290
Cape John	237
Carmichael, James	274
Carriboo	11 239
Carr, Adam	398
Census, first, 67, later	442
Cheese Factories	430

	PAGE.
Churches, first	147
Churchville	123
Clearing Land	223
Coal	198 399 423
Coast	14
Cochrane Grant	153 240
Cock, Rev. Daniel	110 112
"Colonial Standard"	385
Copelands	158
Council of XII	332 371
Court, Inferior	215
" Supreme	216
Court House	216
Crerar, Peter	406
Crichton, David	363
Crown Brick Coal & Pottery Co.	420
Dalhousie, Earl of	253 331
Dalhousie Mountain	275
Davidson, James	69
" Thomas	256
Dawson, John	160 256
" James	309
" Robert	309
Denys, M., description by	24
DesBarres, Col. J. F. W	53 128
Dumfries Immigrants	94 160 231 276 457
Dunoon, Hugh	159 226
Earltown	277
East Branch East River	121 163 231 460
"Eastern Chronicle"	385
Eden, Garden of	284
Eighty-second Regiment	115 456
" Grant	115
Eighty-fourth Regiment	119 460
Eight Mile Brook	123

	PAGE.
Elders, first	133
Election of 1799	193
" 1830	380
Elliott, Rev. Charles	445
England, Church of	444
Emigration, how conducted	241
"Enterprise," brig	300
Fairbanks, C. R.	333
Falls	17
Fanners, first	231
"Favourite," voyage of the	234
Fisher, John's Grant	50
Fisheries	305
Fitzpatrick, James	238
Flogging	204
Foord pit	410
Four Mile Brook	235
Fraser, Alexander	188 454
" David	262
" Rev. Donald A..	317 343 362
" James D. B	412
" John	460
" " (Collector)	217
" Simon	178
" William (surveyor)	136
Free Stone	441
Free Church disruption	429
Free Port, Pictou made	365
French	24 38 42 43
Friends' Society of	449
Frost, year of	295
Gairloch settled	241
General Mining Association	400
Geology of county	18
Goderich, Lord, despatch from	353
Grant, Peter	157
Grants, first	49
Haliburton, T. C	338 351
Harris family	73 75
" Dr. John	73
" Matthew	74 100
Harris Thomas, Sr	74
" Thomas, Jr.	57 216
Hatton, Henry	309

	PAGE.
"Hector"	79 et seq.
" passengers	81 et seq. 450
Hines	303
Holmes, Hon. John	428
"Hope," brig	56 et seq.
Houquard, H	233
Howe, Hon. Joseph	376
Indians (See Micmacs)	
Inglis, Bishop	337
Intercolonial Coal Company	415
Iron ores	425
Iron Foundries	441
Jail erected	199
Kennedy, Wm	65
Knoydart	160
Lakes	17
Langills	132
Legislatures, members of	467
Library, subscription	312
Lighthouses	14 397
L. & S. Society	396
Lowden, Capt	102 168
Lowrey, Capt	309
Lulan	190
Lyon, Rev. James	76
McCabe, James	60 75
" John	178
McCara, A	238
McCulloch, Dr. Thomas	266 321 et seq.
" Michael	330
McDonald, Rev. James	443
" " Alex	443
" Dougald murdered	218
McFadyan, murder by	302
McGee, Bar.	65 108
McGregor, Dr. James	134 et seq. 392
McIntosh, murder by	218
McKay, Alex., sen.	467
" " jun	286 290
" John (Collier)	398
" Robert	309
" Roderick	85, 103, 451
" William	451

McKenzie, Captain George.....	431	"Observer," Pictou.............	384
" Rev. Kenneth J..	319	Olding, N. P...................	124
	343 355 362		
McKinlay, Rev John	330 361	Pagan, John...................	70
McLean, Hector................	256	" Robert	159
McLeod, Rev Norman..........	318		
" " Hugh	320	Patlass	189
McLennans Brook..............	228	Patrick, Rev. William..........	271
" Mount..............	228	Patterson, Abraham	304
McNutt, Col...................	48	" Governor	153
McPhail, John.................	302	" James and David....	73
		" John (deacon).. 111	154
Mail Carrying	228 397	" John (second).......	72
"Malignant"...................	106	" John (third)........	306
Manufactures	440	" Robert (squire) 56 70	99
Marshy Hope..................	284		110
Matheson, William.............	256	" Robert (black)......	307
Merigomish.................12	108	" Walter....·........	307
Mice, in Prince Edward Island ..	94	Peace, effects of...............	292
" year of	293	Petitpas, M....................	41
Micmacs.. 26 et seq. 42 58 91 106		Philadelphia Company..... 52 et seq.	
	183 186 188 192	Pictou Island........... 13 18	281
Mill Brook....................	225	Piedmont	283
Mills, James...................	237	Point Betty Island.............	27
Milne, William................	368	Poor, Account for.............	167
Militia	261	Pounds	201
Ministers	467	Presbytery of Pictou, first......	171
Mitchell, Rev. John............	269	Press Gangs...................	259
Mohawks......................	32	Privateers, American	105
Montbiliards	126	Prothonotaries.................	217
Moose Hunting................	177	Railroad, A Mines.............	406
Morrison, George	109	" Halifax and Pictou..	430
Mortimer, Edward........250 et seq.		River John.................12	130
" William.............	305	Roads......61 202 205 210 264	392
Mountains	17	Robertson, Col. Alex........115	460
Mount Thom settled...........	227	Rogers, John..................	75
Murray, Walter...............	109	Roman Catholic Immigration...	163
		" Church........	443
Name of Pictou................	21	Ross, Rev. Duncan.............	171
Natural History	19	" John	70
New Annan...................	279	"Royal William".............	394
New Glasgow..................	274	Rum Drinking...	247
New Lairg	236	Sabbath School, first...........	69
Nova Scotia Coal Company.....	418	" Society for.....	311
Oat Mill, first........-......,	296	Salmon Regulations.........201	206
" second........ ......:.	229	Saltsprings	273

Savage, Captain	27	Toney, Captain	42
St. Marys	285	Toney River	43 237
Sawmill, first	65	Tonge, W. Cottnam	193
Seaview Cemetery	39 241	Town, first laid off	61
Sessions, Court of	201	Town, present, commenced	1 4
Sheriffs	218	Town Gut bridge	212
Ship Building	169 430	Township boundaries	10
Shipping	431		
Simpson, David	461	Union of Presbyterians	315
Six Mile Brook	235		
Slavery	107 148	Vale Colliery	422
Smith, Hon. George	303		
Stage Coach	391	Wallace, Hon. M	195
Stellar Coal	413	Walmesley	116
"Stepsures" Letters	257 360	Waugh, W	94
Steam Navigation	394	Wentworth, Sir John	51 124 161
Stewart, Alex. (Post)	227		165
Stocks	200, 207	West Branch East River	122 461
Sutherland, John	124	" River John	238
		West India Trade	306
Tatamagouche	128	Westville	419
Tattrie, George	127	Williams, Richard	152
Temperance Societies	386	Witherspoon, Dr	79
Timber	92 98 244 366 436	Windsor College	324

POSTSCRIPT.

After the greater part of the foregoing work was printed, we received the following statement, as made by the late James Patterson :—

"When the "Hector" arrived here in 1773, there were but nine huts in the whole of the present county, viz., one at Howlets (a little above Norway House), just new, occupied by Robert Patterson ; one at John Brown's farm, occupied by Dr. Harris ; one at James Rae's, occupied by John Rogers ; one at Saw Mill, occupied by Kennedy and Blaisdell ; one at Fraser's farm (now Evans'), by James McCabe ; one at Dunoon's farm, on the hill, by Mr. Earl; one at David Lowden's farm, by Mr. Watson ; one at West River, at Davidson's farm, by Mr. Aitkin; one at West River, at Beck's farm, by Barney McGee."

We are satisfied that this is imperfect, as Matthew Harris was certainly here in 1773; also Horton, and probably some others. Watson lived, and soon after died, at West River, where Robert Stewart now lives. It would seem also that the place where we mentioned John Rogers as having located himself on Rogers Hill, was not where he first lived. We may have made a mistake at page 102, in describing the seizure of Captain Lowden as having taken place in the house of Mr. Waugh. The gentleman who gave us the history, the late Robert McKay, Esq., and who received it from Captain Lowden himself, said that it took place in the house of the man who lived on Dunoon's farm. Waugh, we knew, lived there, but Earl, who was a keen American sympathiser, may have been still there, and it may have taken place in his house.

BY THE SAME AUTHOR.

I.
Price $1.25.

Memoir of the Rev. James McGregor, D.D.,

First Missionary to Pictou, N.S.

II.
Price 75 cents.

Remains of the Rev. James McGregor, D.D.,

Containing Essays, Letters, &c.

III.
[Price, $1.25.

Memoirs of the Rev. S. F. Johnston, the Rev. J. W. Matheson, and Mrs. Mary J. Matheson,

Missionaries on Tanna, New Hebrides.

With Selections from their Diaries and Correspondence.

IV.
Price $1.

The Doctrine of the Trinity underlying the Revelation of Redemption.

JAMES McLEAN,

Bookseller, Stationer,

AND

NEWS AGENT.

Special attention paid to keeping on hand a full stock of Commercial and Fancy

STATIONERY,
LEDGERS,
BLANK BOOKS, &c., &c.,

PHOTOGRAPH,
AUTHOGRAPH, } ALBUMS,
AND SCRAP

POCKET BOOKS, PURSES, &c., &c.

Fairchild's Celebrated GOLD PENS & Pencil Cases.

SCHOOL BOOKS

and SCHOOL MATERIAL of every description constantly kept on hand in large quantities.

LAW & SHIPPING BLANKS, CHARTS, &c. &c.

NEW AND STANDARD PUBLICATIONS

on hand in great variety.

BOOKS, &c.,
IMPORTED TO ORDER.

Weekly parcels of Books, &c., received from the United States ; fortnightly one from Britain.

TWO DOORS EAST FROM THE POST OFFICE,

WATER STREET, - - - PICTOU, N.S.

ISAAC A. GRANT,

IMPORTER AND DEALER IN

DRY GOODS,

Opposite the Market,

WATER STREET, - PICTOU, N.S.

Alvays in Stock, the newest Styles in

DRESS GOODS,
    SHAWLS,
        MANTLES,
            GLOVES,
HOSIERY,
    SMALL WARES,
        LADIES' MOURNING GOODS,
CARPETS and HOUSE FURNISHING GOODS.

GENTLEMEN'S OUTFITTINGS.

Also, in connection with the above,

First Class Fashionable Tailoring Department,

Under the superintendence of skilled workmen.

GENTLEMEN'S GARMENTS

MADE TO ORDER AT SHORT NOTICE,

From one of the

LARGEST AND BEST SELECTIONS OF CLOTHS

IN THE PROVINCE.

FRED. MACLENNAN,
MANUFACTURER AND IMPORTER OF

Every Description of Sheet Iron and Tin Wares.

ALWAYS ON HAND,
Or Imported to Order,

Stoves, Ranges, and Furnaces.

A SPECIALTY MADE
OF LATEST IMPROVED

Cooking, Parlour, Office, Hall, and Ship
STOVES.

Now Importing or Manufacturing:
DAIRY OUTFITS,
KITCHEN FURNISHINGS,
OIL CANS, COAL HODS,
CONFECTIONERS' MOULDS,

TIN-WARE AND STOVE PIPE,
Of best Native and Foreign Material.

TOILET WARE
In Bath Tubs, Foot Tubs, Pails, &c.

WOODEN & SHIPS WARE

Highest Price paid for Wool and Wool Skins.

PEDDLERS SUPPLIED ON REASONABLE TERMS.

FRED. MACLENNAN,
Sign of the Stove, - - PICTOU, N.S.

R. McGREGOR & SONS,
NEW GLASGOW, N.S.

WHOLESALE DEALERS
IN
FLOUR, TEA, SUGAR,
Molasses,
SALT, OILS, &c., &c.

G. W. UNDERWOOD,
IMPORTER AND DEALER IN
British, Canadian, and American
DRY GOODS,
ALSO,
FAMILY GROCERIES,
FLOUR, &c.

☞ COUNTRY PRODUCE BOUGHT AND SOLD. ☜

PROVOST STREET,
NEW GLASGOW, N.S.

"The Daily Telegraph,"

ST. JOHN. N.B.

Is Published EVERY DAY (except Sunday), and delivered by Carriers in the City, or sent by Mail, at

FIFTY CENTS A MONTH, OR SIX DOLLARS A YEAR,

PAYABLE IN ADVANCE.

"The Weekly Telegraph,"

RECENTLY INCREASED IN SIZE,

And now REDUCED IN PRICE and Improved in Form—it now being and eight-paged Paper, is published at

ONE DOLLAR A YEAR.

WILLIAM ELDER,
Editor and Publisher.

PRINCIPAL DAWSON'S WORKS.

THE DAWN OF LIFE: being the History of the oldest known fossil remains, "Eozoon Canadense."—Illustrated with Plates and Woodcuts. $2.00.

HANDBOOK OF ZOOLOGY, with examples from Canadian Species, recent and fossil.—275 Illustrations. $1.25.

ACADIAN GEOLOGY; the Geological Structure, Organic Remains, and Mineral Resources of Nova Scotia, New Brunswick, and Prince Edward Island; with Geological Maps and numerous Illustrations. $3.50.

THE AIR BREATHERS OF THE COAL PERIOD; a description of the remains of Land Animals found in the Coal Formations of Nova Scotia. $1.00.

REPORT ON THE GEOLOGY AND MINERAL RESOURCES of Prince Edward Island. 75c.

DAWSON BROTHERS,
Booksellers and Publishers, MONTREAL.

www.ingramcontent.com/pod-product-compliance
Lightning Source LLC
Chambersburg PA
CBHW051851300426
44117CB00006B/354